URBAN
WILDLIFE
MANAGEMENT

URBAN
WILDLIFE
MANAGEMENT

Clark E. Adams
Kieran J. Lindsey
Sara J. Ash

 Taylor & Francis

Taylor & Francis Group

Boca Raton   London   New York

A CRC title, part of the Taylor & Francis imprint, a member of the
Taylor & Francis Group, the academic division of T&F Informa plc.

Published in 2006 by
CRC Press
Taylor & Francis Group
6000 Broken Sound Parkway NW, Suite 300
Boca Raton, FL 33487-2742

International Standard Book Number-10: 0-8493-9645-X (Hardcover)
International Standard Book Number-13: 978-0-8493-9645-8 (Hardcover)
Library of Congress Card Number 2005051481

On the cover this fox squirrel *(Sciurus niger)* was phototgraphed eating a candy bar on the Texas A&M University campus (College Station, TX). The squirrel was radio-collared as part of an ongoing study by the University's Department of Wildlife and Fisheries Sciences to understand the diet, population dynamics, movements and behavior of urban squirrels. The fox squirrel is a common urban resident, but we chose this photo for the cover because it was one of the first subjects of radio telemetry by Clark E. Adams in the early 1970s. (Photo taken by Robert McCleery.)

Library of Congress Cataloging-in-Publication Data

Adams, Clark E. (Clark Edward), 1942-
    Urban wildlife management / Clark E. Adams, Keran J. Lindsey, Sara J. Ash.
       p. cm.
    Includes bibliographical references and index.
    ISBN 0-8493-9645-X
    1. Urban wildlife management. 2. Urban ecology (Biology). I. Lindsey, Kieran J. (Kieran Jane).
    II. Ash, Sara J. III. Title.
QH541.5.C6A33   2005
639.9'09173'2—dc22   2005051481

Taylor & Francis Group
is the Academic Division of T&F Informa plc.

**Visit the Taylor & Francis Web site at
http://www.taylorandfrancis.com**

**and the CRC Press Web site at
http://www.crcpress.com**

We dedicate this book to all those individuals, organizations, and agencies on the front lines addressing urban wildlife management problems. They represent the unsung heroes of wildlife management who receive little recognition or peer acceptance for their attempts to confront a growing wildlife management phenomenon. They are the futurists, involved in the cutting-edge aspects of human/wildlife interactions in urban environments. In other words, we dedicate this book to all those who realize that urban wildlife management goes far beyond controlling raccoons in garbage cans.

Preface

We began the adventure of writing this book because we were unable to find a single publication that addressed all the pertinent issues related to urban wildlife management; rather, information was scattered throughout various books, journal articles, conference proceedings, government documents, websites, data sets, and within the anecdotal tales of our colleagues. Now, with the end of the journey in sight, we confess to having been somewhat naïve about the breadth and depth of information sources available on this subject. None of us expected to spend two years preparing this manuscript. Invariably, though, in attempting to track down a single piece of information we would stumble across a dozen others, each of which led to still others — all of which made our seemingly simple task of synthesizing the available information into a conceptual, rather than in-depth, presentation a lengthy and formidable one.

Nevertheless, unearthing, accumulating, and organizing the wealth of information has turned out to be an extremely enjoyable and intellectually stimulating task. The path has led us to well over 500 professional manuscripts and countless articles and stories in the popular media. We examined secondary data sets that contained a wealth of information relevant to the story we wanted to tell in this book. We convinced some of our colleagues to share their expertise in urban wildlife research and management in several sections of this book. Even some of our students contributed to the cause. We continue to discover new information, but at some point one must stop reading and start writing!

We believe the challenges and opportunities related to urban wildlife are beginning to take front stage in the wildlife profession. As a result, this book is a much-needed tool for teaching and learning. In twelve chapters, we examine a range of issues that explain human interactions with wildlife in urbanized environments. We begin with a discussion of the past, present, and future directions of wildlife management in the U.S. — what we have come to see as the changing landscape of wildlife management. Selected lessons in ecology relevant to understanding the presence or absence of wildlife species in urban communities include ecosystem structure and function, population dynamics, and the water cycle, with particular reference to the impacts of urbanization. Urban habitats and hazards are discussed in terms of the unique features of green and gray spaces, urban streams, and urban soils. Issues of particular importance in urban wildlife management — the human dimension, stakeholders, and legal considerations — are explained. The ecology and management of selected species are discussed in terms of conflicts between predators (e.g., coyotes and bears) and humans, management of endangered species, species that roost in large numbers, feral species, and overabundant animals including resident Canada geese and urban deer.

We're pleased to say this book represents the first published synthesis of the multifaceted issues surrounding the human and wildlife interactions in urban communities. It tells "the rest of the story" by addressing those issues that have been left out of manuscripts that focus primarily on how to alleviate the problems associated with nuisance urban wildlife (e.g., M. Conover. 2002. *Resolving Human–Wildlife*

Conflicts: The Science of Wildlife Damage Management; J. Hadidian et al. 1997. *Wild Neighbors: The Humane Approach to Living with Wildlife*). The book provides a basic framework of information that will give the reader an understanding of factors that promote or prevent the presence of wildlife in urban communities.

We believe our readers will include wildlife management professionals, students in the wildlife sciences, and at least some members of the general public. We hope it will serve as a guide for urban wildlife management courses in colleges and universities. Additionally, we think it can serve as a compendium of information for urban planners and land managers in urban areas.

The development of this book was a team effort, and we have many people to thank for their contributions. Debra Cowman helped us assimilate the relevant literature on environmental toxicants on wildlife population dynamics. John Davis wrote the "biotic communities" section of Chapter 3 and Perspective Essay 3.1, the impacts of stream channelization and urbanization on soil structure and function in Chapter 7, Perspective Essay 8.1, and the "local ordinances" section in Chapter 10. John also provided many of the photos used in this text. Thanks to Rob Denkhaus and Suzanne Tuttle for their case study on feral hogs in Chapter 11. Fran Gelwick reviewed and edited the urban streams section in Chapter 7. Roel Lopez wrote the key deer case study in Chapter 11. Ardath Lawson wrote the section on cemeteries in Chapter 5 and Perspective Essay 5.1. Robert Meyers conducted an analysis of the Wildlife Services Management Information Systems data set that led to the national and regional overview of the species of most concern and economic impacts of animal damage in Chapter 2. The Quality Deer Management Association (Bogart, GA) was the source of information concerning white-tailed deer densities in each state as provided in Figure 12.4. Tamala Schields and Kara Stanfield helped track down a lot of the literature sources used in this book. Linda Tschirhart wrote the first draft of Chapter 10 and provided several photos. We are grateful to each of these individuals for their talents and contributions.

Clark E. Adams, Kieran Lindsey, and Sara Ash

Authors

Clark E. Adams is a professor in the Department of Wildlife and Fisheries Sciences at Texas A&M University in College Station, Texas. He has a B.S. in biology and education from Concordia Teachers College, Seward, Nebraska, an M.S. in biology and education from the University of Oregon, and a Ph.D. in zoology from the University of Nebraska–Lincoln. Clark chaired the Conservation Education Committee for The Wildlife Society (TWS), edited the newsletter for the Human Dimensions of Wildlife Study Group, is now a member of the Urban Wildlife Management Working Group, and has chaired many committees for the Texas Chapter and TWS. He is a certified wildlife biologist with TWS. He is past president of the Texas Chapter of TWS and current president of the TWS Southwest Section. Since 1981 he has directed the Human Dimensions in Wildlife Management research laboratory. He and his students have conducted and published many national, regional, and statewide studies on the public's activities, attitudes, expectations and knowledge concerning wildlife. He developed the degree option in urban wildlife and fisheries management for the Department of Wildlife and Fisheries Sciences and developed and teaches the senior-level urban wildlife management course.

Kieran J. Lindsey is a principal in The Wildlife Information Group, a business dedicated to human–wildlife interface resources and research based near Albuquerque, New Mexico. She has a B.S. and an M.S. in wildlife and fisheries sciences from Texas A&M University in College Station, Texas, where she is currently pursuing a Ph.D. in the same field. Kieran worked on the front lines of urban wildlife management as the director of a nonprofit urban wildlife shelter and education center in Houston, Texas. During this time her newspaper column, "The Urban Jungle," was a regular and popular feature in *The Houston Chronicle*. In 1989, Kieran received a regional Emmy Award from the National Academy of Television Arts and Sciences as coproducer of the feature segment "Rehab at the Mall." After moving to New Mexico, she became the executive producer, senior writer, and host of *Wild Things Radio,* a weekly educational program that aired on KUNM-FM. She has taught students of all ages, in both formal and informal education programs, about living with and enjoying wildlife. Kieran is a member of The Wildlife Society and its Urban Wildlife Management Working Group.

Sara J. Ash is an assistant professor and chair in the Department of Biology at Cumberland College in Williamsburg, Kentucky, where she teaches courses in ecology and conservation biology and introductory biology. She has a B.S. in biology from Cumberland College and an M.S. and Ph.D. in wildlife and fisheries sciences from Texas A&M University in College Station, Texas. She is a member of The Wildlife Society's Kentucky Chapter and Urban Wildlife Management Working Group. Sara sees urban wildlife management as both a complex challenge and rich opportunity. She believes that urban and suburban habitats can be used as outdoor classrooms to teach ecological concepts and principles to both the general public and college students. Additionally, she believes that scientists can elucidate some of the puzzles of ecology by studying urban ecosystems.

Contents

INTRODUCTION

A New Wildlife Management Paradigm

In the future, we're all going to be urban biologists.

Timothy Quinn, Chief Habitat Scientist, Washington Department of Fish and Wildlife

---

CONTENTS

I.1 A SNAPSHOT OF THE URBAN WILDLIFE MANAGEMENT LANDSCAPE

Americans may like to think of themselves as a primarily rural nation, but nearly 80% of those who dwell in the lower 48 states live in areas classified by the U.S. Census Bureau as *urban:* "a large central place and adjacent densely settled census blocks that together have a total population of at least 50,000" (U.S. Census Bureau 2001). These areas can also be thought of as places where most of the land is devoted to buildings, roads, concrete, grassy lawns, and other elements of human use. A substantial potion of the land is covered with impervious surfaces, in the form of various kinds of buildings and pavement, and constructed "green" spaces.

Urbanization and the encroachment of humanity into former wild habitats will continue into the foreseeable future. This population shift from rural to urban has changed, and will continue to change, the landscape and the agenda concerning wildlife management. In fact, within the next 10 to 20 years, we believe management of urban wildlife will become the dominant focus of wildlife professionals.

As Americans have become more urbanized, their curiosity about wildlife has increased, as has their interest in attracting wildlife in order to improve quality of life. If you doubt this is the case, consider the fact that Americans involved in wildlife watching now outnumber those involved in traditional consumptive-use activities such as hunting and fishing (U.S. Census Bureau 2001). Interest is so high, in fact, that wildlife watchers are able to support their own advertising-free bimonthly magazine, *Birds 'n Blooms* (Reiman Publications, Greendale, Wisconsin). Even the home-building industry has taken note, promoting close proximity of greenspace and wildlife to prospective customers has become a common marketing strategy for developers.

Concurrently, there is a growing concern about human–wildlife encounters, especially those perceived to endanger the health and safety of humans and their companion animals. Outbreaks of zoonotic diseases such as West Nile virus, hanta virus, and chronic wasting disease make headlines in both local and national newspapers. The number of nuisance wildlife complaints continues to rise, as does the number of private wildlife control businesses. Clearly, life in the urban wilds is not a return to Eden. Which begs the question—who's tending this garden?

I.2 THE NEED FOR A COMPREHENSIVE TREATMENT OF URBAN WILDLIFE MANAGEMENT

Change often must begin at the grassroots level. The National Institute for Urban Wildlife was the first formal organization of individuals who recognized and wanted to address urban wildlife management issues. In order to start a dialogue, the Institute hosted three national symposia on urban wildlife (Chevy Chase, Maryland; Cedar Rapids, Iowa; and Bellevue, Washington), and two proceedings were published (Adams and Leedy 1987, 1991). The fourth symposium was held in Tucson, Arizona, in 1999 (Shaw et al. 2004). Subsequently, The Arbor Day Foundation, in cooperation with the Urban Wildlife Working Group of TWS, took on the task of organizing and hosting biannual national urban wildlife management conferences (Lied Conference Center, Nebraska City). A variety of topics are presented at these conferences, including managing wildlife in urban environments, human–wildlife conflicts, public education on urban wildlife, stakeholder recognition, and a host of others. No records or proceedings have resulted from these meetings thus far.

However, during the 2003 conference, one attendee who represented a city government asked if there was any publication that summarized the issues and complexities of urban wildlife management under one cover. At that time, there was no such document. This question increased our interest in producing a book that addressed what we perceived to be a basic need by those involved in urban wildlife issues.

Ideally, a comprehensive book on urban wildlife management should have been available at least 20 years ago. Lowell Adam's seminal work, *Urban Wildlife Habitats: A Landscape Perspective* (1994), which focused on urban and suburban wildlife habitat, was a first and important step in the right direction. Other authors have addressed individual aspects of urban wildlife management, such as urban ecology and sustainability (Whiston-Spirn 1985; Platt et al. 1994); human dimensions (Decker et al. 2001), including human–wildlife conflicts (Hadidian et al. 1997; Conover 2002); urban wildlife law (Rees 2003); urban planning (Tyldesley 1994); and even urban species identification (Landry 1994; Shipp 2000). An all-inclusive book on the important issues that comprise urban wildlife management, however, had yet to be written. Our intention is to remedy this omission.

This is not a book on how to address specific urban wildlife conflict issues — other authors (Hadidian et al. 1997; Conover 2002) have addressed these issues admirably. Rather than providing a prescription for short-term, reactive methods that address only symptoms, the information included here will provide professionals in wildlife management and related fields with the understanding required to set and achieve long-term, proactive management goals and objectives that focus on the root cause of urban wildlife management challenges.

This book came to be written because one of its authors had been teaching an urban wildlife management class at Texas A&M University for several years without an appropriate textbook. The textbook is commonly considered the fundamental curriculum guide in academic instruction. Rarely are college-level courses taught without a textbook; if a textbook is not available, much needed classes, including urban wildlife management, often will not be taught. Perhaps this is one reason urban wildlife management tends to be a missing component in university curricula used to train wildlife biologists. We hope that situation will change now that university instructors and their students will have access to an instructional tool—one that provides an introductory exposure to the information required to understand urban wildlife issues.

There are two key questions a teacher has to answer when preparing a class: (1) "What am I going to teach?" and (2) "How am I going to teach it?" Applying these decisions to an undergraduate university class on urban wildlife management prior to the writing of this book was a formidable task. There is a growing body of literature in scientific journals and the popular media (both print and broadcast) about urban wildlife. The public, and to some degree even wildlife professionals, are unaware of the information on urban wildlife presented in the scientific literature, while the primary focus of popular media is entertainment rather than education. A curriculum for training urban wildlife biologists emerged as we examined the full range of urban wildlife issues in the context of human history and society, natural history, ecology, politics, law, and economics. We knew what we wanted to teach, but the information we wanted to provide students, as well as those currently involved in various wildlife management professions, was strewn about in journal articles, conference proceedings, government documents, websites, other books, secondary data sets, and the personal experiences of colleagues. This book gathers the essential information together under one cover, providing a synthesis document for academic, community, and professional development. Case studies are included to provide further illustration of the concepts. Our literary approach was to tell a story that has not yet been told, based on a review of hundreds of references, but by no means all of the pertinent literature. The information we provide in each chapter is meant to be an overview of the subjects discussed, not an exhaustive treatment. Many of our chapters have been, or could be, the subject of an entire book. We hope our readers will take the opportunity to expand their understanding of the concepts introduced here.

We expect this text to find an audience outside of the university classroom as well. There are many individuals working for the government, nonprofit organizations, and for-profit businesses whose responsibility or job description is to address some aspect of urban wildlife management. Developers, for example, may fail to take into account the surrounding wildlife community, which can lead to some of the management challenges addressed in this book. An understanding of the cause-and-effect outcomes could significantly change the "business as usual" development process, leading to a planning process that allows for increased interaction between humans and wildlife while avoiding potential conflicts. Other fields that may benefit from a greater understanding of urban wildlife management issues include public health, parks and recreation, sanitation, tourism, transportation, and animal control. This book provides the fundamental information that these individuals need to know in order to design a holistic urban wildlife management plan that is more proactive than reactive.

Our goals in developing this book were to (1) compile a body of information that stimulates the reader's curiosity about the urban world in which most of them live, (2) expand the reader's knowledge about how differently natural and urban ecosystems work, (3) challenge the reader to examine the role their personal actions may have in causing at least some of the urban wildlife issues covered in this book, and (4) give the reader an opportunity to apply their new knowledge and understanding through personal actions that promote sustainable approaches to urban wildlife management.

I.3 UNDERSTANDING AND MEETING THE FUTURE CHALLENGES OF WILDLIFE MANAGEMENT

Unfortunately, the wildlife profession is ill-prepared for this shift in public interest and need. In 1999 the Urban Wildlife Working Group of The Wildlife Society conducted a national survey of state wildlife management agencies and land-grant universities that offered a degree in wildlife science (Adams 2003). The survey was designed to determine how well the above-mentioned entities were prepared to address urban wildlife management issues. The results were disturbing in many respects, but the general conclusion was that the infrastructure for urban wildlife management is missing in state departments of natural resources and land-grant universities.

The role of the wildlife manager has changed in the twenty-first century. So has the role of those responsible for providing accredited personnel to serve the public—but many wildlife management faculty members have yet to notice the change in public interest and demand that is bearing down on them like a runaway train (Figure I.1). Often they are either completely oblivious regarding urban wildlife management or grossly misinformed as to what the field entails. A good example of the latter was observed by one of the authors of this book during a meeting at a large university with one of the premier departments of wildlife and fisheries in the U.S. The topic

Figure I.1   The change in public interest and demand is bearing down on the wildlife management profession much like a runaway train. (L. Causey)

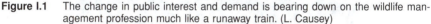

of discussion was whether or not to add an urban wildlife management option to the department's undergraduate curricula. One faculty member, a nationally recognized figure in ecology and conservation biology, commented that he could see no reason for developing a curriculum about raccoons in garbage cans and, consequently, voted against its inclusion. University faculties must begin to recognize the need to develop new instructional paradigms that meet the challenges of urban wildlife management or they may find they have become largely irrelevant.

Students, too, must become aware of this paradigm change. When we visit with students interested in wildlife professions, we often see a reflection of ourselves at their age and level of educational development. Many are drawn to working with wildlife because they enjoy nature and are looking for professional pursuits that do not require "direct interaction" with people. They picture themselves collecting and analyzing data in the middle of a remote forest, far removed from the aggravation of the human race. They want to "fight the good fight," gathering information that could someday be an integral part in saving a species or ecosystem from extinction (Figure I.2).

In our formative years and early in our careers we were drawn to the study of wild things for the sake of knowledge and also because, in our opinion, they were so much more interesting than people. Since those early days we have come to understand that we can make a much more significant impact on the world if we direct more attention on understanding people and their relationships with wildlife and educating them about the wildlife that surrounds them (Figure I.3).

Figure I.2   The traditional image of a wildlife biologist is that of an individual collecting and
analyzing data in the middle of a remote forest, possibly even saving a species
from extinction. (John and Karen Hollingsworth/USFWS)

We recognize that the complex nature of urban wildlife issues prevents "quick
fix" resolutions. Therefore, an effective urban wildlife biologist needs academic
training and preprofessional experience in both the traditional curricula now used
to train wildlife biologists and nontraditional curricula. The naturalist approach to
understanding urban wildlife issues needs to be resurrected in academia. Too much
time is spent memorizing the names of stuffed and preserved animals from museum
collections, and too little time is spent on understanding why the animal lives where
it does and its relationship to both the habitat and other species found there! Related
to this is the need for preprofessional students to have more time in the field
experiencing the situations they will encounter in the real world.

In addition to basic core courses in zoology, botany, taxonomy, genetics, and
chemistry, potential urban wildlife biologists need exposure to courses in ecology,
conservation and management of wildlife, urban forestry, urban and land use planning, environmental education, public speaking, and conflict resolution. A thorough

Figure I.3   As the general population becomes more urbanized, wildlife biologists are being asked to have greater interaction with people, particularly in the area of public education. (John M. Davis/Texas Parks and Wildlife Department)

understanding of wildlife laws and the legal ramifications of urban wildlife management at the community, city, county, state, and federal levels also is crucial. This type of academic preparation will help students understand how urban communities function and how wildlife management issues arise. In fact, extensive knowledge and experience will be needed in all the areas mentioned above if potential job candidates are to have any hope of rising to the challenges they face after graduation (see Sidebar I.1—Job Description of an Urban Wildlife Biologist).

We are convinced that the future scope and purpose of wildlife managers will be shifting from rural to urban issues. This shift will occur even though we currently have at least two generations of deer, turkey, and quail managers in the field or waiting in the wings. As we point out in Chapter 2, practicing urban wildlife biologists are still relatively rare, and few new professionals are being trained to meet the demand of an ever-increasing urban population. Students with an interest in urban wildlife as a future profession will likely be confronted with some negativity by more traditional wildlife biologists. As pointed out by Witter et al. (1981:424) several decades ago, "Plying the wildlife trade in the world of high-rises and suburbia is one with which more wildlifers are not overly comfortable. Conventional wisdom holds that wildlife management is most productive when conducted far removed from metropolitan environments and concentrations of people."

While others may regard your professional pursuits with a huge measure of naiveté and disdain, we encourage you to take heart. The national organization for U.S. wildlife professionals, The Wildlife Society (TWS), has begun to pay some attention to the need to address the urban wildlife issues presented in this book by forming an Urban Wildlife Working Group. In addition, one entire issue of the peer-reviewed *Wildlife Society Bulletin* (WSB, 1997:25) was devoted to urban wildlife management issues. This is usually the first step in the professional recognition of a subdiscipline within the larger context of wildlife management. These are encouraging signs because the TWS membership (by and large academicians) usually follow the lead of the parent organization in the policy and decision-making process regarding curriculum development and change.

It is highly unlikely that state departments of natural resources and land-grant universities will embrace urban wildlife management in the immediate future. In large part, this is because they have fixed agendas, no budget for additional management or research obligations, no personnel who are trained in urban wildlife research or management, and little recognition of urban wildlife management as their responsibility.

Where then does the future urban wildlife biologist find a job? The lack of attention to urban wildlife issues by government agencies has spawned a grassroots movement of sorts in which responsibility for wildlife management is shifting away from state and federal agencies, with privatization becoming a growing trend. We believe that future job opportunities will emerge in the private sector with environmental consulting firms and private wildlife control businesses, with city and county government departments, including urban planning and recreation and parks, at the federal level in agencies such as APHIS Wildlife Services, and with nonprofit organizations, including urban nature centers, wildlife rehabilitation organizations, and environmental education centers. In addition, we have found that students trained in a wildlife and fisheries curriculum make excellent high school biology teachers (Adams and Greene 1990).

For now, college students will need to tailor their academic training around a holistic management paradigm. Traditional management practice focuses on either increasing (in the case of game species) or decreasing (in the case of "nuisance" species) animal populations. The focus of urban wildlife management, however, is in large part to provide urban residents with the knowledge and skills required to avoid problems, properly assess human–wildlife encounters, and enjoy the wilderness that exists in their own backyards.

An additional challenge will be the art of managing wildlife populations in urban ecosystems in a manner that allows both wildlife and humans to coexist in a sustainable context. If we can figure out how to manage people and wildlife in the city in a sustainable, proactive way, we may actually slow the rate of urban sprawl and indirectly save some habitats and species. Additionally, if urban residents

Figure I.4   Aldo Leopold, the father of wildlife conservation in America, recognized that people are part of the wildlife management equation. (Aldo Leopold Foundation)

recognize the potential for positive interactions with wild things in the city, it may help to ease the pressure on the last remaining true wild areas we have left.

In the late 1930s Aldo Leopold recognized that meaningful and significant wildlife management requires management of *people* (Figure I.4). This fact is not often acknowledged within the more traditional wildlife management fields, but there is no way to escape it when dealing with urban wildlife. As we demonstrate in this book, urban areas are home to entire communities of wildlife. This affords future wildlife professionals the opportunity to better understand the human–wildlife relationship and make significant contributions to all areas of wildlife conservation.

SIDEBAR I.1: JOB DESCRIPTION FOR AN URBAN WILDLIFE BIOLOGIST

An urban wildlife biologist is an individual who works primarily in metropolitan (nonrural) environments, focusing on nondomestic vertebrate and invertebrate species and interactions between humans and wildlife (Adams 2003). During the course of a normal day, an urban biologist may be called on to handle anything from development of a plant or wildlife species conservation plan to assisting with conflict resolution between a developer and the local city planning board. Job titles and descriptions vary depending on the emphasis an employer places on specific tasks. Texas Parks and Wildlife Department, for example, states that the duties of its urban biologists include "providing opportunities for urban residents to reconnect with the natural systems, presenting educational programs for adults and students on a variety of habitat/wildlife issues, serving as technical advisors on multi-agency conservation planning initiatives, and assisting landowners with habitat restoration or enhancements" (http://www.tpwd.state.tx.us/expltx/eft/careers/urban.htm, March 21, 2005). The North Carolina Wildlife Resources Commission, on the other hand, stresses "forging and working with partnerships of government and private entities to achieve conservation objectives" as well as wildlife damage management and public outreach in an announcement for a contract position (NCWRC, Urban Biologist Contract Position # 04-FD-JM1). The following job listing from the Missouri Department of Conservation will provide additional insight into the scope of knowledge and diversity of skills needed by those interested in this profession.

Community Conservationists, Missouri Department of Conservation

Salary: $33,024–$58,764

Qualifications:

Graduation from an accredited college or university with a Bachelor's Degree in Fisheries, Forestry or Wildlife Management, Environmental Planning, Biological Sciences, or applicable field of study and at least three (3) years of progressively responsible professional experience; or an equivalent combination of education and experience. **Experience in community development and natural resource planning is highly desirable.**

Duties and Responsibilities:

- Promotes the conservation of fish, forest, and wildlife in the urban environment through interaction with government officials, planning and zoning boards, land use committees, development committees, urban development industries, park boards, recreation planning committees, developers, public conservation groups, and others;
- Educates citizens about social, environmental and economic impacts of urban sprawl and the benefit of alternatives;
- Encourages the creation and support of local nonprofit environmental groups; capitalizes on existing opportunities to deliver natural resource conservation messages by attending conferences, festivals, symposia, and workshops as presenters and attendees;
- Provides information, education, materials, research time, and coordination with citizen(s) that are willing to advocate for proper natural resource protection/management

by local government; develops and assists with the development of educational materials, plans, grant applications, etc.;

- Provides technical assistance to local governments for natural resource management/protection and takes advantage of opportunities for involvement in community planning, assisting with plat review, planning and zoning code revisions, ordinance revisions, greenway establishment, green space conservation, aquatic resource management, storm water master plans and long range land use master plans;
- Educates local governments of the negative impacts of sprawl, especially economic impacts through presentations to city councils, planning and zoning codes and ordinances;
- Utilizes cost share docket, other Missouri Department of Conservation (MDC) funds and outside financial resources to promote the establishment and/or maintenance of natural areas;
- Provides technical assistance on land such as parks, natural areas, riparian corridors, storm water facilities, nature trails, institutional campuses, etc.;
- Builds strong working relationships with local governments;
- Advocates for and provides technical assistance on the use of conservation development techniques that maintain or restore healthy fish, forest, wildlife or natural community resources;
- Shows professionals how conservation development can be an economically sound alternative;
- Assists with easements, deed restrictions, and other methods of insuring long-term environmental protection;
- Provides training to inter-agency personnel concerning urban natural resource management; completes reports;
- Performs other duties as required.

REFERENCES

Adams, C. E. (2003). The infrastructure for conducting urban wildlife management is missing, *Transactions of the North American Wildlife and Natural Resource Conference,* 68: 252–268.

Adams, C. E. and Greene, J. (1990). Perestroika in high school biology education, *American Biology Teacher,* 52:408–412.

Adams, L. W. (1994). *Urban Wildlife Habitats: A Landscape Perspective,* University of Minnesota Press, Minneapolis, MN.

Adams, L. W. and Leedy, D. L. eds. (1987). *Integrating Man and Nature in the Metropolitan Environment,* National Institute for Urban Wildlife, Columbia, MD.

Adams, L. W. and Leedy, D. L. eds. (1991). *Wildlife Conservation in Metropolitan Environments,* National Institute for Urban Wildlife, Columbia, MD.

Conover, M. (2002). *Resolving Human–Wildlife Conflicts: The Science of Wildlife Damage Management,* Lewis Publishers, Boca Raton, FL.

Decker, D. J., Brown, T. L., and Siemer, W. F. (2001). *Human Dimensions of Wildlife Management in North America,* The Wildlife Society, Bethesda, MD.

Hadidian, J., Hodge, G. R., and Grandy, J. W. (1997). *Wild Neighbors: The Humane Approach to Living with Wildlife,* Fulcrum Publishing, Golden, CO.

Landry, S. B. (1994). *Peterson First Guides: Urban Wildlife,* Houghton Mifflin Company, New York.

Platt, R. H., Rowntree, R. A., and Muick, P. C. (1994). *The Ecological City: Preserving and Restoring Urban Biodiversity,* University of Massachusetts Press, Amherst.

Rees, P. (2003). *Urban Environments and Wildlife Law,* Blackwell Publishers, Oxford, UK.

Shaw, W. W., Harris, L. K., and Vandruff, L. eds. (2004). *Proceedings of the 4th International Symposium on Urban Wildlife Conservation.* May 1–5, 1999. Tucson, AZ.

Shipp, D. (2000). *Urban Wildlife Spotter's Guide,* Usborne Publishing Ltd., London, UK.

Tyldesley, D. (1994). *Planning for Wildlife in Towns and Cities,* English Nature, Peterborough, UK.

U.S. Census Bureau. (2001). 2001 National Survey of Fishing, Hunting and Wildlife-Associated Recreation, United States Census Bureau, United States Government Printing Office, Washington, DC.

Whiston-Spirn, A. (1985). *The Granite Garden: Urban Nature and Human Design,* Basic Books, New York.

Witter, D. J., Tylka, D. L., and Werner, J. E. (1981). Values of urban wildlife in Missouri, *Transactions of the North American Wildlife and Natural Resources Conference,* 46:424–431.

SECTION I

Urban Landscapes

CHAPTER 1

Wildlife Management: Past and Present

The forest stretched no living man knew how far.

Willa Cather, 1931

CONTENTS

1.1 A BRIEF HISTORY OF WILDLIFE MANAGEMENT IN NORTH AMERICA

For hundreds of generations, conditions for wildlife populations in North America were nearly ideal, although, as with any healthy ecosystem, constantly in flux. Indigenous peoples used wildlife at sustainable levels for food, clothing, and shelter, and the animals played a large role in their culture and spiritual life (Decker et al. 2001). This situation began to change with the arrival of European immigrants approximately 400 years ago. These new Americans carried with them a culture of human domination over nature, at least in theory; most Europeans did not have the legal right to exercise domination over wildlife. Those privileged few who owned land also owned the wildlife living on that land.

Immigrants to North America found a seemingly infinite supply of natural resources. The abundance of wildlife during the 1600s and early 1700s must have been staggering to those who left behind lands that had been overhunted for centuries. What's more, there were no legal restraints on the exploitation of these resources. In North America, wildlife was a *commons*, a resource owned by all. Although even the earliest settlers depended on domestic livestock and cultivated crops, wild game provided variety and an essential food source when crops failed (Root and De Rochemont 1994).

By the mid 1800s wildlife populations in the East were suffering under the combined effects of subsistence use, market hunting, and habitat loss. The migration of Americans westward during this time created a wave of similar pressures on wildlife. White-tailed deer, elk, black bear, most species of waterfowl, and wild turkey were extirpated in many areas of the East and Midwest (Decker et al. 2001). The passenger pigeon, a species that had once been so abundant that when flocks flew overhead the sky grew dark, was effectively extinct in the wild by 1900.

1.2 RISE OF THE AMERICAN CONSERVATION MOVEMENT

As the extent of damage to wildlife became more apparent, concerned citizens spent much of the nineteenth century working to develop a meaningful and effective system for protecting this disappearing resource. In 1844, a group of about 80 militant conservationists formed the New York Sportsmen's Club, the sole purpose of which was "the protection and preservation of game."

The club had three primary targets: sale of game for market, spring shooting of game birds, and lax game laws (Figure 1.1). A majority of the club's members were attorneys; they developed an effective strategy of suing poachers, dealers, and hotel proprietors for the sale or possession of game killed out of season. They tracked down violators by following tips from informants and using private detectives. The group was so effective in its crusade against game-law violators that their approach was adopted by newly formed sportsmen's clubs in Massachusetts, Rhode Island, and Ontario. Prior to 1870, however, most efforts were local rather than regional or national in scope (Trefethen 1975).

The American conservation movement began in earnest during the late 1800s with the emergence of two views of nature and the country's wildlife heritage. Sustainable-use advocates, often termed "progressives," included within their ranks Gifford Pinchot and Theodore Roosevelt, while John Muir and other "romantics" represented the preservationist standpoint (Lutts 1990) (Figure 1.2). Originators of the conservation movement at this time often embraced both the sustainable-use and preservationist worldviews. It was due to the two major influences of this movement that the U.S. Congress created what would become the National Forest System (sustainable-use) and the first National Parks (preservation).

Figure 1.1   Market hunting of waterfowl and other wildlife species was one impetus behind the enactment of the first wildlife laws. (USFWS.)

(a)   (b)   (c)

Figure 1.2   Conservation pioneers (a) John Muir (Edward Hughes, 1902 [public domain]), (b) Theodore Roosevelt (USFWS), and (c) Gifford Pinchot (Underwood and Underwood, 1921 [public domain]).

During this period, state agencies were created to manage wildlife for current and future citizens to use and enjoy. Laws to restrict harvest and protect habitat were passed by legislators across the country, but implementation of these laws proved more difficult. Reliable funding for enforcement was limited or nonexistent for at least another 50 years.

In 1937 Congress took an unprecedented step to address the funding problems associated with wildlife management by passing the Federal Aid in Wildlife Restoration Act, also known as Pittman-Robertson for the legislators who sponsored it. Pittman-Robertson funnels an 11% federal user fee on hunting rifles, shotguns, and ammunition to the U.S. Fish and Wildlife Service. A 1970 law added a tax on handguns and archery equipment for wildlife management.

In 1950 Congress passed the Federal Aid in Sport Fish Restoration Act to address similar issues regarding fisheries management. Dingell-Johnson, as it is commonly known, collects a 10% manufacturers' user fee on fishing equipment and tackle. The 1984 Amendment to Dingle-Johnson expanded the tax to include new motorboat fuel taxes and duties on imported tackle and boats.

User fees established by the Restoration Acts, and collected by the U.S. Treasury from manufacturers, go into trust funds administered by the Department of the Interior. A maximum of 8% of the funds may be retained by the Service for administration, and the rest is allocated to states based on a formula that considers the state's land or water area and the number of licensed hunters or anglers. These are cost-reimbursement programs; the states cover the full amount of an approved project and then apply for reimbursement through Federal Aid for up to 75% of the project expenses. States must provide at least 25% of the project costs from non-federal sources. Appropriate state agencies are the only entities eligible to receive Federal Aid grand funds.

The purpose of establishing the Restoration Acts was to provide funding for restoration, rehabilitation, and improvement of wildlife habitat and fisheries, wildlife and fisheries management research, and information distribution. Pittman-Robertson was amended in 1970 to include funding for hunter training programs and development, operation, and maintenance of public target ranges. Dingell-Johnson funds purchase of land for boating, fishing and fish production, research and inventory projects, and public education about fish and their habitats. Both of these Acts served the needs of wildlife management agencies and consumptive users well for decades.

1.3 CHANGING WILDLIFE VALUES

The "typical" American has changed dramatically over the past 50 years. In contrast to the country's rural, agricultural heritage, 80% of Americans now live in areas classified as urban by the U.S. Census Bureau. Life in cities and suburbs has changed Americans' attitudes and expectations concerning wildlife. Many, if not most, are several generations removed from a culture of living close to the land. They are more likely to value wildlife similarly to the way they value companion animals and people (Mankin et al. 1999) than as a consumptive-use resource. Americans are now more likely to be involved in wildlife-related recreation such as observing, feeding,

Figure 1.3 Participation in traditional consumptive-use activities (top) has decreased over time while interest in nonconsumptive wildlife recreation (bottom) is on the rise. (Bill Harper [top], USFWS [bottom])

and photography (Figure 1.3) than traditional activities (Figure 1.2) such as hunting and fishing (U.S. Department of the Interior 2001).

Due in large part to a historic focus on consumptive issues, governmental agencies are not well positioned to address the concerns of an increasingly urbanized population. State wildlife agencies have a legislative mandate to manage all wildlife within their borders as a public resource, but in all but a few states funding for nontraditional wildlife management issues is extremely limited. As a result, a trend is developing toward increased privatization of wildlife management, particularly in urban and suburban settings. Nongovernmental organizations, such as private wildlife control businesses, conservation groups, humane societies, and wildlife rehabilitators have stepped in to address public demand unmet by governmental agencies. This paradigm shift away from a system of managing wildlife as a commons for the good of the resource and toward private, profit-driven systems is not, by and large, the result of specific policy decisions. Rather, it has evolved as a grassroots response while agencies struggle to adapt to changing public expectations and funding limitations.

1.4 A NEW KIND OF WILDLIFE

Many Americans readily acknowledge that animals such as pigeons, starlings, house sparrows, rats, mice, and tree squirrels are "urban animals," but far fewer are aware that a wide diversity of other wild species live in human cities, towns, and suburbs. For example, a study in Tucson described urban and suburban populations of coyotes (*Canis latrans*), burrowing owls (*Athene cunicularia*), Harris hawks (*Parabuteo unicinctus*), Cooper's hawks (*Accipiter cooperii*), javelina (*Tayassu tajacu*), and deer (*Odocoileus* spp.), just to name a few (Shaw et al. 2003).

1.4.1 Categorizing Wildlife

Early in the development of management agencies, wildlife species were categorized into three groups based on their utility to people: game (including "furbearers" of interest to trappers), nongame, and nuisance. This approach reflected the interests of both agencies and consumptive-use stakeholders. Game species were managed to produce the greatest possible harvest, and nuisance species were targeted for lethal control. Nongame species received little attention.

In the early 1970s, growing public interest in the environment, concern over vanishing species, and legislation enacted as a result of these trends expanded the wildlife lexicon. The term *threatened or endangered* (T/E) allowed wildlife professionals to classify species based on their abundance or scarcity. In spite of this change, wildlife categories remained sharply drawn; for example, once a game species is listed as T/E, management strategies shift and hunting is prohibited, so for all practical purposes it is no longer a game species. If, however, the species recovers and is delisted, it returns to its original classification as game.

With the publication of a book entitled *Urban Wildlife Habitats: A Landscape Perspective* (Adams 1994), the term "urban wildlife" was formally introduced. Rather than classifying wildlife based on their usefulness to humans or their population status, urban wildlife species are categorized as such based on whether populations can be found living in and around human settlements. Developing a definition of urban wildlife has been difficult, in large part because it attempts to categorize nondomesticated species in a way that is radically different from traditional classification methods. Some examples of definitions used by other authors include the following:

- All native, nondomestic, wild animals found in or around urban areas (Schaefer 2004)
- Any wild creature that lives in an urban environment or an urban-rural interface, including birds, reptiles, amphibians, mammals, fish, insects, and worms (U.S. Department of Agriculture Forest Service 2001)
- Any nondomestic animals that live in cities, suburbs, or other urban areas (Maurizi and Friedner 1997)
- Wild vertebrates local to the region that may occur in the urban area (Fletcher 1994)

For the purposes of this book, *urban wildlife includes all nondomestic vertebrate species, e.g., fin, feather, and fur, with populations in areas classified as urban.* Wild invertebrates will receive only cursory attention in this publication because they are usually considered a separate study area by academia. Management of these species is a specialty in its own right, considered to be the responsibility of entomologists (insects), arachnologists (spiders), and malacologists (mollusks) rather than wildlife biologists.

Categorizing wildlife as urban blurs the sharp distinctions established by traditional categories. For example, deer can be classified as both a game and as an urban species. To some degree, traditional categories are less applicable in urban ecosystems (Case Study 1.1). While both game and nongame species can be found in urban habitats, game species are rarely hunted in commercial or residential areas. Conversely, nongame species, such as grackles (*Quiscalus* spp.), prairie dogs (*Cynomys* spp.), and bats (*Chiroptera* spp.), often are the target of lethal control measures, individually or as communities. Species listed as threatened or endangered may experience localized abundance within urban habitats, thanks to greater abundance of food and water or lack of predation pressures.

1.5 THE UNIQUE ECOLOGY OF URBAN WILDLIFE

Within a species, individuals living in urban habitats may exhibit distinct behavioral differences compared to animals living in exurban environments. Activity patterns may change; for example, Canada geese (*Branta canadensis*) have historically migrated up to 3000 miles to nest, but some individuals and flocks have abandoned migration altogether and have become year-round urban residents (Hope 2000; see also Chapter 12). Reproductive behavior may change in urban environments. This phenomenon can be observed in Georgia, where 73% of the state's 1,270 known breeding pairs of least terns (*Sterna antillarum*) nest on gravel rooftops compared with only 1% using traditional beach nesting grounds (Youth 1999).

Many urban species become habituated to human presence and lose at least some of their natural wariness. A survey of New Mexico urban and semirural residents found that bobcat (*Felis rufus*) sightings were more frequent in areas of high-density housing than in "traditional" habitat, and that 70% of sightings were less than 25 meters from a house (Harrison 1998). Species living in urban areas often tolerate higher population densities than individuals living in rural habitats. Raccoon density has been found to be higher in urbanized landscapes, possibly due to increased survival and reproduction rates and greater site fidelity in urban habitats (Prange et al. 2003).

Feeding strategies also may change in urban habitats when animals make use of new sources of food. Ring-billed gulls (*Larus delawarensis*) in Ohio, for example, have become more dependent on landfills than on fish from historic Great Lakes feeding areas (Belant et al. 1998; see also Chapter 6).

1.6 SPECIAL CHALLENGES FOR WILDLIFE MANAGEMENT WITHIN URBAN SETTINGS

Wildlife management does not have an extensive history within metropolitan areas because, prior to the mid-twentieth century, America was a predominantly rural society. As a result, most wildlife management practices were developed with rural landscapes in mind. Urban and suburban environments present a host of special challenges for wildlife professionals (Decker et al. 2001). A thorough discussion of the effects of urbanization on the wildlife management profession is presented in Chapter 2.

1.6.1 Urban Ecosystems

Urbanization is the process of transforming wild lands to better meet the needs of humans. Usually this consists of clearing much of the native vegetation and replacing it with exotic species. Some, if not all, of the topsoil layer is removed, and much of the remaining soil is either compacted, "waterproofed" by creating an impervious barrier to precipitation in the form of concrete or asphalt, or both. In the process, the land's habitat potential is significantly altered.

An "urban ecosystem" can be thought of as a system influencing, and being influenced by, human attitudes, human behaviors, regulatory policies, and a sense of resource control throughout areas where humans live, work, and recreate at moderate to densely populated social scales (U.S. Department of Agriculture 1995). Certain species are naturally well-suited to the altered environment, some are able to adapt to these changes, and others decrease in number or disappear through a combination of mortality and emigration. An extensive examination of urban ecosystems and population dynamics can be found in Chapters 3 and 4.

1.6.2 Urban Habitats

Urban environments do not consist of one type of habitat. Examples of unique habitats within urban areas include parks, cemeteries, vacant lots, streams and lakes, residential yards, schools grounds, corporate campuses, golf courses, airports, bridges, parking structures, and landfills. In addition, the extent of habitat diversity occurs within a much smaller area than would normally be found in rural landscapes. Chapters 5 through 7 provide a large- and small-scale analysis of selected urban habitats.

1.6.3 Sociopolitical Factors

Most Americans like wildlife. They believe that hearing birds sing and watching squirrel acrobatics adds to their quality of life. But this love affair with wildlife quickly can turn sour when animals "cross the line" beyond what is perceived as acceptable versus unacceptable behavior (Schmidt 1997). The line is different for each person. Human reactions to wildlife include a broad spectrum of emotions and reactions based on previous exposure to both formal and informal education programs and personal experience (Kellert 1980).

Natural resource agencies have a vested interest in shaping public understanding of wildlife management and conservation (Mankin et al. 1999), and they have had success identifying and communicating with specific, traditional clienteles (Hesselton 1991; Kania and Conover 1991; Jolma 1994). However, the number of stakeholders who want explicit consideration in management has grown over the last several decades (Decker and Enck 1996).

While traditional stakeholders tend to have similar expectations for wildlife management, several regional studies have indicated that urban residents often have diverse and conflicting goals, such as the desire to reduce human–wildlife conflict while enhancing wildlife viewing opportunities (Conover 1997). The wildlife profession has had difficulty communicating effectively with the general public (Decker et al. 1987; Gray 1993). In fact, Rutberg (2001) found that public trust in these agencies is declining, often because they are perceived as overseeing, regulating, and promoting traditional uses of wildlife, such as hunting and trapping.

Urban wildlife professionals must constantly consider both biological and sociopolitical factors when developing management strategies. Identifying stakeholders and attempting to understand their perspectives, expectations, and demands is a daunting task. Chapters 8 and 9 delve into the application of human dimensions in urban wildlife management and the use of a stakeholder approach to management, respectively.

Throughout most of the wildlife profession's history, management activities took place in rural settings, primarily on public lands and large tracts of private agricultural and forest land. Farmers, ranchers, and game managers have much in common; game management grew out of an agricultural mindset of making land produce harvestable crops, including wildlife. Managers have long worked to motivate private landowners in rural areas to enhance their properties for wildlife (Decker et al. 2001). In contrast, management of urban wildlife requires working with many private owners holding small parcels of land. Consider, for a moment, the logistics of gaining access to land in private subdivisions for management purposes — this would involve the immensely complicated task of contacting possibly hundreds of property owners (Figure 1.4).

Figure 1.4   Urban wildlife habitat may consist of a vast array of small plots of land classified as private property, easements, rights-of-way, and parks, and each with a different owner or regulatory jurisdiction. (John M. Davis/Texas Parks and Wildlife Department)

Public lands in metropolitan areas tend to consist of easements, rights-of-way, and parks, all of which fall under the jurisdiction of a multitude of municipal and county governments and private entities.

Wildlife professionals must learn to navigate a maze of legal considerations in urban and suburban areas. The sociopolitical landscape is cluttered with laws, regulations, ordinances, and policies, and varying levels of enforcement. Within city limits there exists a minimum of four layers of jurisdiction: federal, state, county, and municipal. Often there will be overlapping areas of responsibility within these jurisdictions. The issue of laws and jurisdiction, and their effect on urban wildlife management, are discussed in greater detail in Chapter 10.

1.7 MANAGEMENT CONSIDERATIONS

Ecological and sociological factors combine to create urban wildlife management challenges. On the wildlife side of the equation, a species' ecology and behavior can be used to predict generalities such as the presence and abundance of resident populations, while the specific circumstances often are tied to the geographic areas. On the human side of the equation, issues such as culture, economics, and politics can predict how humans will respond to a species' presence and abundance.

Chapter 11 examines four categories of management scenarios that commonly play out in the urban/suburban ecosystem: conflicts between predators and humans, management of endangered species, species that roost in large colonies, and overabundant game animals. Chapter 12 takes an even closer look at the management implications of two prevalent urban species: resident Canada geese and urban deer.

CASE STUDY 1.1: DOWNTOWN DEER

Missouri has more than 800,000 deer in the state, and some of the state's highest deer densities are recorded in urban and suburban areas. High numbers of deer and high numbers of people in the same area results in conflicts. In metro areas, cars are probably the most efficient predators of urban deer. Out of 130 deer carrying transmitters as part of a study of urban deer in the St. Louis area, for example, at least 15 have been killed by vehicles in the last 6 months.

Deer aren't attracted to the cities the way that people are. Instead, their numbers tend to increase where food is plentiful and predators are few. Where survivability is high, deer numbers can increase rapidly. Our urban areas provide abundant food and protection from hunters and other predators.

Missourians currently are overwhelmingly in favor of protecting deer. They like to see them, to hunt them. But in many other communities, especially east of the Mississippi, deer have gone beyond the cultural carrying capacity of their environment (Figure 1.5).

People often propose birth control chemicals to keep deer from propagating. We have the histories of contraception and sterilization efforts tried in other communities to limit deer, and the results are not promising. Others propose moving the deer out

Figure 1.5   In many communities deer have gone beyond the cultural carrying capacity of their environment. (T. Gunderson)

into the country, where they would be more welcome. There are places in Missouri that could support more deer, but deer, even in an urban environment, are not easy to catch, and they don't take to confinement well.

Most communities have eventually accepted the fact that the only way to control deer numbers is to regularly eliminate some of the deer. Some hire contractors or sharpshooters to knock the population down. Certainly, hunting is not the answer in all urban areas. Discharging firearms in some cities is likely to bring squad cars, and some cities even have ordinances against the discharge of archery equipment, even though shooting accidents in archery hunting are rare.

Managed hunts, where hunters are allowed into certain areas on designated days, using specific weapons, concentrate carefully monitored hunting activity into a short time period, with minimal disruption to other uses of the areas. Controlling deer populations in all of our metropolitan areas may not be feasible through hunting alone, however.

"It's a dilemma," Lonnie Hansen, the state's deer biologist, said. "We'd rather do it through hunting, but some areas are not conducive to hunting, and we'll probably have to go in with trapping, sharpshooters, or pest control companies authorized to do deer control."

The Conservation Department is able to provide recommendations and technical assistance to communities and individuals suffering from deer problems, but many people balk at management of any kind. "It's difficult to do proactive deer management in urban areas," Hansen said. "Many people will accept deer management only after they have been directly and negatively affected by deer overpopulation. We're reduced to after-the-fact management — after the damage is done and people have come to dislike them."

October 1997
Excerpted from Missouri Conservationist *magazine*
Written by Tom Cwynar, Editor

REFERENCES

Adams, L. W. (1994). *Urban Wildlife Habitats: A Landscape Perspective,* University of Minnesota Press, Minneapolis, MN.

Belant, J. L., Ickes, S. K., and Seamans, T. W. (1998). Importance of landfills to urban-nesting herring and ring-billed gulls, *Landscape and Urban Planning,* 43:11–19.

Conover, M. R. (1997). Wildlife management by metropolitan residents in the United States: practices, perceptions, costs, and values, *Wildlife Society Bulletin,* 25:306–311.

Decker, D. J., Brown, T. L., Driver, B. L., and Brawn, P. J. (1987). Theoretical developments in assessing social values of wildlife: toward a comprehensive understanding of wildlife recreation involvement, in *Valuing Wildlife: Economic and Social Perspectives,* Decker, D. J. and Goff, G. R., Eds., Westview, Boulder, CO, pp. 76–95.

Decker, D. J., Brown, T. L., and Siemer, W. F. (2001). *Human Dimensions of Wildlife Management in North America,* The Wildlife Society, Bethesda, MD.

Decker, D. J. and Enck, J. W. (1996). Human dimensions of wildlife management: knowledge for agency survival in the 21st century, *Human Dimensions of Wildlife,* 1:60–71.

Fletcher, D. (1994). Urban wildlife: an overview, Urban Animal Management Conference Proceedings, Canberra, Australia.

Gray, G. (1993). *Wildlife and People: The Human Dimensions of Wildlife Ecology,* University of Illinois, Urbana.

Harrison, R. L. (1998). Bobcats in residential areas: distribution and homeowner attitudes, *Southwestern Naturalist,* 43:469–475.

Hesselton, W. T. (1991). How governmental wildlife agencies should respond to local governments that pass anti-hunting legislation, *Wildlife Society Bulletin,* 19:222–223.

Hope, J. (2000). The geese that came in from the wild, *Audubon,* 102:122–126.

Jolma, D. J. (1994). *Attitudes Toward the Outdoors: An Annotated Bibliography of U.S. Survey and Poll Research Concerning the Environment, Wildlife, and Recreation,* McFarland, Jefferson, NC.

Kania, G. S. and Conover, M. R. (1991). Another opinion on how governmental agencies should respond to local ordinances that limit the right to hunt, *Wildlife Society Bulletin,* 19:222–223.

Kellert, S. R. (1980). American's attitudes and knowledge of animals, *Transactions of the North American Wildlife and Natural Resources Conference,* 45:111–124.

Lutts, R. H. (1990). *The Nature Fakers: Wildlife, Science, and Sentiment,* Fulcrum, Golden, Colorado, CO.

Mankin, P. C., Warner, R. E., and Anderson, W. L. (1999). Wildlife and the Illinois public: a benchmark study of attitudes and perceptions, *Wildlife Society Bulletin,* 27:465–472.

Maurizi, R. and Friedner, T. (1997). The effect of city living on urban wildlife, ASCI. http://www.uvm.edu/~rmaurizi/oldStuff/urban.html.

Prange, S., Gehrt, S. D., and Wiggers, E. P. (2003). Demographic factors contributing to high raccoon densities in urban landscapes, *Journal of Wildlife Management,* 67:324–333.

Root, W. and De Rochemont, R. (1994). *Eating in America: A History*, Ecco Press, New York.

Rutberg, A. T. (2001). Why state agencies should not advocate hunting or trapping, *Human Dimensions of Wildlife*, 6:33–37.

Schaefer, J. (2004). Florida's definition of urban wildlife, http://www.wec.ufl.edu/faculty/SchaeferJ/3401/November24.html.

Schmidt, R. (1997). Drawing the line for wildlife, *Wildlife Control Technology*, 4:6–7.

Shaw, W. W., Harris, L., Livingston, M., Charpentier, J.-P., and Wissler, C. (2003). Wildlife habitats in urban environments, Effects of Urbanization Symposium: Wildlife Habitats in Urban Environments, University of Arizona, School of Renewable Natural Resources, Tucson.

Trefethen, J. B. (1975). *An American Crusade for Wildlife,* Boone and Crockett Club, Alexandria, VA.

U.S. Census Bureau, Department of the Interior, Fish and Wildlife Service (2001). National survey of fishing, hunting and wildlife-associated recreation, United States Government Printing Office, Washington, DC.

U.S. Department of Agriculture (1995). Urban community ecosystems: a national action plan, United States Government Printing Office, Washington, DC.

U.S. Department of Agriculture, Forest Service, Southern Region, Southern Research Station, and the Southern Group of State Foresters (2001). *The Urban Forestry Manual,* Southern Center for Urban Forestry Research and Information, USDA Forest Service, Athens, GA.

U.S. Department of the Interior, Fish and Wildlife Service and U.S. Department of Commerce, Bureau of the Census (2001). *National Survey of Fishing, Hunting, and Wildlife-Associated Recreation*, United States Government Printing Office, Washington, DC.

Youth, H. (1999). These rare birds flock to a shopping mall in search of home furnishings, *National Wildlife,* 37(4):18–19.

CHAPTER 2

The Changing Landscape of Wildlife Management

There are two spiritual dangers in not owning a farm. One is the danger of supposing that breakfast comes from a grocery store, and the other that heat comes from the furnace.

Aldo Leopold 1949

CONTENTS

This chapter is a revision and extension of a paper presented by Adams (2003) at the North American Wildlife and Natural Resources Conference. Permission to use presentation information in this chapter was given by the Wildlife Management Institute.

2.1 DEMOGRAPHIC FACTORS THAT SET THE STAGE FOR URBAN WILDLIFE MANAGEMENT

The convergence of three societal events at the end of World War II set the stage for urban wildlife management. First, automobiles and homes became more affordable and available, and returning soldiers had the means to purchase both. Second, Cold War mania provided a stimulus for the passage of the Highway Revenue Act of 1956, which created the Highway Trust Fund. This legislation enabled the development of the network of superhighways through, around, and out of the cities. The original political justification for extensive highway construction was to provide city populations a rapid escape mechanism in the event of nuclear attacks. Lastly, America began to shift from a largely agrarian to a primarily urban society around 1945, as people moved away from family farms to the city for work (Figure 2.1). Highways, affordable automobiles, really cheap gas (about 10 cents/gallon), and an increasingly urban population created the opportunity for mass human migration from self-contained communities in cities and small towns to suburban developments several miles from the city core.

These three events led to a phenomenon called "urban sprawl," a process whereby the perimeters of the city are extended outward into the countryside, one development after the next, with little plan as to where the expansion is going and no notion as to where it will stop (Wright 2004). In 1990 there were 274 metropolitan areas in the U.S., covering 20% of the country's land area. While populations in and around some of these urban centers grew by only 3 to 5% in the past decade, the area of land developed grew by more than 50% during that same time (Heimlich and Anderson 2001). This phenomenon set the stage for a human–wildlife interface that

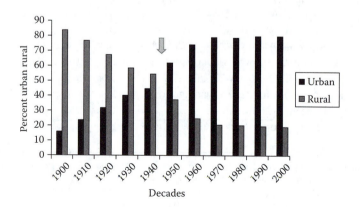

Figure 2.1 Change from rural to urban residence in the U.S., 1900 to 2000.

Figure 2.2 Urban sprawl. (John M. Davis/Texas Parks and Wildlife Department)

had not existed before and reinforced a separation between people and wildlife (Figure 2.2). But the trend toward urbanization is not limited to the U.S., or even North America. According to a study done by the United Nations (Department of Economic and Social Affairs — Population Division) in 2003, 74% of people living in developed countries could be classified as urban. The same study found that in less developed regions 42% of the population qualified as urban in 2003, and this number is expected to reach 57% by 2030 (http://www.un.org/esa/population/publications/wup2003/2003Highlights.pdf).

2.2 THE SEPARATION OF PEOPLE AND NATURE

One might conclude that if people moved to the city fringe they would take advantage of the opportunity to reconnect with nature and wild things. However, the opposite reaction seems to occur; people bring the structure of the city ecosystem to the country, simplifying and destabilizing their surroundings. This is accomplished by changing the natural landscape, removing and replacing natural vegetation and endemic wildlife with exotic and domestic species.

A detailed description of the structural, biotic, and socioeconomic features of urbanization was developed by McDonnell and Pickett (1990:1233). Structural features consist of dwellings, factories, office buildings, warehouses, roads, pipelines, power lines, railroads, channelized waterways, reservoirs, sewage disposal facilities, dumps, gardens, parks, cemeteries, and airports. Crops, ornamentals, domestic pets, pests, and disease organisms are the dominant biotic features of urbanization. Socioeconomic features include changes in human values, wealth, lifestyles, resource use, and waste.

Urban dwellers are usually unaware that, by changing the structural and functional components of natural ecosystems, they are creating alternative habitats for a wide array of both invertebrate and vertebrate animals that are quick to take advantage of the habitats provided. This commonly results in increased human–wildlife interactions (e.g., coyotes in city parks). A variety of these wildlife encounters are discussed in detail throughout this book.

Leopold's prophetic quote at the beginning of this chapter is a succinct evaluation of how an urbanized society runs the risk of using the most simplistic explanations for complex phenomena. Decades and generations of human isolation from the natural world through urbanization have produced a society that lacks a connection to the natural world and their relationship to that world. For example, few people in contemporary society appreciate what their last breakfast cost wildlife in terms of food production, transportation, packaging, and distribution. Leopold points out a second disconnect between society and nature in terms of their understanding of the costs to the environment and wildlife just to provide the energy required for home heating, e.g., surface coal mining.

In order to more fully understand the separation between people and wildlife that occurs with urban sprawl, one needs to compare the interactions between people and wildlife in a rural community with what most readers experience on a daily basis in their urban communities. The first perspective essay in this book was written by one of the authors, who recalls human–wildlife interactions in a small agrarian community in Iowa in the early 1950s (Perspective Essay 2.1). Of particular importance in this essay is the description of people's attitudes, activities, knowledge, and expectations concerning wildlife in their day-to-day lives at that time. Furthermore, it is instructive to reflect on the past from time to time in order to recall where we have been and where we might be heading concerning our associations with wild things.

It is important to consider some comparisons because it is unlikely that urban societies will ever return to the condition described in Perspective Essay 2.1. Furthermore, it seems imperative that urban society understands its impacts on species diversity, interrelationships with wild things and natural habitats, natural cycles (water and biogeochemical), and how the production, provision, and utilization of energy resources affects wildlife and their habitats. Urban society should also realize that unsustainable use of natural resources endangers human society as well as wildlife. These comparisons and impacts will be addressed in subsequent chapters.

2.3 THE NEED FOR WILDLIFE MANAGEMENT IN URBAN AREAS

There is strong evidence justifying the need for wildlife management in urban areas. This need is based on decades of population shifts from rural to urban areas (Figure 2.1), a lack of wildlife management paradigms focused on urban people and wildlife, changing animal damage control issues, a growing body of literature (professional and popular) about wildlife in urban areas, career opportunities in the private sector, and generations of human isolation from their natural environments. Furthermore, from 1985 to 2001, the frequency of the U.S. population 16 years of age and older that reported taking time to observe (35 to 73%) and/or feed (46 to 84%) wildlife has nearly doubled (U.S. Department of the Interior 1985, 1991, 1996, and 2001).

The expansion of urban areas into formerly natural environments has caused an increase in human–wildlife encounters, resulting in a variety of human emotions, explanations, and reactions, mostly conceived in an intellectual and experiential vacuum. Some state agencies are now advertising available positions in various aspects of urban wildlife management (e.g., wildlife damage biologist, urban wildlife biologist, and urban outreach supervisor), but most wildlife biologists employed by state and federal agencies and the universities have been caught off guard. The traditional wildlife management curricula produce wildlife biologists who focus their expertise on game, nongame, or threatened and endangered species in nonurban habitats.

One study found that the degree to which contemporary wildlife curricula, nationally, are used to train urban wildlife biologists was, at best, a token effort (Adams, L.W. et al. 1987). The criteria that defined the training and tasks expected of an urban wildlife biologist emphasized the human dimensions of wildlife management (Tylka et al. 1987). The results of a more recent national study on the degree to which land-grant universities and state agencies are addressing urban wildlife management is presented in the next section of this chapter.

The number of peer-reviewed manuscripts on urban wildlife management research in the two major journals of The Wildlife Society (TWS), i.e., *The Journal of Wildlife Management* (*JWM*) and *The Wildlife Society Bulletin* (*WSB*), has grown steadily since 1997. One entire issue, 58 articles, of the *WSB* (1997, Volume 25, Number 2) was dedicated to research on white-tailed deer overabundance in urban communities. Newspaper articles about wildlife in urban areas, such as coyotes in city parks, cougars in people's backyards, and urban deer herds, are becoming more frequent. The complex nature of conducting field research in urban settings was explained by VanDruff et al. (1996). Again, the emphasis was on the considerations of habitat structure in urban compared to rural environments and the problems resulting from frequent contacts with humans during the research process.

Finally, typical urban residents are unable to identify common wildlife species, do not know why particular species of wildlife occur in their backyards or how to deal with a problem species, and lack an understanding of interrelationships between people and wildlife. Typical misconceptions about urban wildlife include:

- All snakes are "poisonous."
- Solitary fawns (*Odocolius virginianus*) have been abandoned and need human care (Figure 2.3).
- Missing pet cats and puppies have strayed off or have been stolen.
- Feeding wildlife is a necessary activity (see Case Study 2.1).
- Wild predators do not exist in urban habitats.

Figure 2.3   Deer fawn lying in the grass waiting for it mother to return. (Wildlife Services)

2.4 THE NEED FOR PUBLIC EDUCATION PROGRAMS ABOUT URBAN WILDLIFE

Several indicators justify the need for educational programs that inform urbanites about the wildlife around them. One indicator is found in an examination of what formal public school curricula teach students about how the natural world works. An analysis of urban high school students' knowledge of wildlife conducted by C. E. Adams et al. (1987) provided guidance in educational program development. Observations of how people react to wildlife in their backyard provided further insight. Furthermore, an inventory of the types of programs by state, federal, and private agencies and how they were delivered was considered.

One might expect the biology class to be the most logical place in the public school curricula to learn something about urban environments, plants, and animals. Adams and Greene (1990) suggested an alternative biology curriculum built around a conceptual framework that provided both teachers and students with a socially relevant approach to biology education. Many of the changes they suggested would convert the traditional content of biology courses into material that promotes student awareness, knowledge, and actions concerning societal (urbanite implied) impacts on biological resources. Their suggested revisions would have made the biology curriculum, in part, a course in survival training and lifelong learning and would have provided a connection between people and wild things. In the same issue of the *American Biology Teacher*, Adams (1990) provided examples of classroom activities involving urban wildlife (e.g., tracking radio-collared fox squirrels around their urban habitat).

Human reactions to wildlife in their backyard represent a broad spectrum of emotions and perceptions based on previous information (formal and folklore), experience, the kind of animal involved, and what it is doing. It is difficult to categorize this spectrum of emotions and reactions into a static taxonomy that ignores the dynamics of human–wildlife encounters in urban environments (Kellert 1980). Clearly, a person's emotional response and reactions will be different when they encounter a snake or a swallow, a seemingly helpless animal or one that appears to be aggressive, a juvenile or an adult.

Examples such as these provide some guidance in the types of educational programs needed. One of the most basic lessons urbanites need regarding the wildlife around them is how to determine whether the animal is behaving normally. However, humans may apply an anthropogenic interpretation of unusual behavior (e.g., "the animal wants to be my friend" or "it's in trouble and knows I want to help"). Cases in point include feeding wildlife, rescuing "orphaned" baby deer, and encouraging the nearly exponential proliferation of some charismatic species (e.g., resident Canada geese and urban white-tailed deer — see Chapter 12). Schmidt (1997) suggested that human reactions to wild animals can be based on their perceptions of whether the animal has "crossed the line," that is, whether the human–wildlife encounter might result in the real or perceived potential for damage.

Since the mid 1980s, there has been a proliferation of educational programs about urban wildlife and habitats by state and federal agencies and private organizations. Nearly all of the programs were designed for informal delivery systems methods of information transfer that can occur in any setting to individuals for whom participation is a personal choice (Adams and Eudy 1990). Some programs were designed to address the educational needs of teachers and their students, including Project WILD, Aquatic WILD, Project WET, and Project Learning Tree. These programs are offered as activity-based learning systems to supplement the state-mandated public school curricula. Community-based programs on urban wildlife and their habitats include Master Naturalist, Master Gardner, and Watchable Wildlife. The latter program led to an annual event called Eagle Days in several communities throughout Missouri, Crane Mania in Grand Island, Nebraska, and the Gulf Coast Birding Trail in Texas. Nearly all of the programs available at the time this book was written relied on volunteer facilitators and participants to conduct training workshops. The long-term success of these programs is based on a steady recruitment and retention of qualified volunteers who have high achievement and altruistic values, identify with program goals, and have a high interest in and concern for the environment (Greene and Adams 1992).

The findings of C. E. Adams et al. (1987) discouraged any assumptions of an enlightened public concerning wildlife, particularly urban high school students. The students could not correctly identify many common urban wildlife species (i.e., opossum, *Didelphis marsupialis* vs. rat, *Rattus norvegicus*), the relative numbers of selected mammals within Harris County, Texas (i.e., raccoons, *Procyon lotor,* were considered rare while cougars, *Felis concolor,* were thought to be abundant), the eating habits of 8 of 16 mammals and the effect of human presence on the relative abundance of 12 of 16 mammals. The students' lack of knowledge might be attributed to a lack of contact with wildlife, given their lifetime residency in an urbanized environment and related lifestyle. One needs to consider how urban residents learn about wildlife in order to understand some of the study results.

The majority of urban residents learn (fact and folklore) about wildlife through television programs filmed in exotic locations around the globe, under circumstance quite different from those described in Perspective Essay 2.1. Why wouldn't young people consider cougars abundant in urban settings when they are so common in the popular media? Why wouldn't they consider these animals to be harmless when they see, on television, a former Secretary of the Interior sitting right beside an unrestrained adult cougar? Other examples of how urbanites learn about wildlife in the utilitarian worlds of advertising, the fashion industry, and shopping malls have been provided by Price (1999). Adams, C.E. et al. (1987:85) suggested that urban wildlife education programs focus on the basic principles of cycles, interrelationships, and diversity exemplified with human–wildlife interactions and include wildlife-related activities using species common to the urban environment.

2.5 OUTCOMES OF HUMAN–WILDLIFE ENCOUNTERS

Stories of human–wildlife encounters in urban areas are becoming increasingly a newsworthy event in both print and electronic media. Particular attention has been given to human encounters with predatory mammals (e.g., coyotes, cougars, and bears). For example, the occurrence of a coyote in Central Park New York, cougars in the backyards of El Paso, Texas, residents, and coyote, cougar, and bear attacks on humans in several locations in the U.S. have been newsworthy events. The particular importance of these observations was the realization that urban residents lack the tools required to make informed decisions about how to interpret and react to wildlife encounters.

The best people to interview concerning how people respond to the particular wildlife species they encounter, their knowledge and attitudes about wildlife based on the questions they ask, and their frustration in their inability to find information quickly are wildlife rehabilitators and animal damage control (ADC) specialists. These individuals, in most states, are the ones the public most often turns to for answers to their questions about the wildlife around them.

2.6 URBANITES NEED TO RECONNECT WITH THE NATURAL WORLD

The predominant method used by urbanites to reconnect with the natural world around them is to construct environments, by installing feeders and by wildscaping their yards (Damude et al. 1999). Texas Wildscapes are habitats that provide the essential ingredients for a variety of wildlife — food, water, shelter, and space. This is done by planting and maintaining native vegetation, installing birdbaths and ponds, and creating structure. Feeders can supplement native vegetation, but should never replace it. The goal is to provide places for birds, small mammals, and other wildlife to feed and drink, escape from predators, and raise their young.

Nearly all of the constructed environments lead to a paradox in urban wildlife management; wildscapes are designed to invite wildlife, and once the habitats are provided the animals will come. But when too many animals respond to the invitation they are seen as a nuisance and become the target for management efforts designed to reduce their numbers (see Chapter 12). In spite of this problem, wildscaping programs tend to be viewed as unequivocally positive. Some alternative considerations concerning the value of constructed habitats for urban wildlife are discussed in Chapter 8.

Another dilemma is caused by supplemental feeding of wildlife and the provision of alternative housing. Urbanites spend billions of dollars annually to purchase food and accoutrements to facilitate the feeding process (U. S. Department of the Interior 2001). The majority of feeders are designed to attract seed-eating birds, but other species are attracted to the seed, including squirrels, mice, rats, opossums, raccoons, and even bears. Additionally, some species are attracted to the animals feeding on the seed, such as snakes, domestic cats, skunks, owls and hawks (see discussion of food chains in Chapter 3). It is reasonable to expect that providing animals with a continuous, reliable, and high-calorie food source will lead to higher than normal birth and survival rates. High recruitment rates lead to population increases that exceed the carrying capacity of a natural environment and other concerns discussed in Chapter 4.

An additional way for urbanites to reconnect with the natural world around them is to utilize the benefits provided in urban open areas (e.g., parks, vacant lots, cemeteries, greenways, abandoned railroad tracks, lakes, streams, and rivers). The structure of urban habitats already provides plenty of food, water, and shelter for many wildlife species. In fact, peregrine falcons have been brought back from virtual extinction because of the resources provided in urban environments (see the peregrine story in Chapter 6). Many urban open spaces (e.g., parks and cemeteries) provide excellent birding opportunities during the height of migration. As discussed in Chapter 5, such places offer urban birders the opportunity to see a great variety of birds without ever leaving the city limits.

2.7 URBAN WILDLIFE SPECIES ARE INCREASING, SOMETIMES TO NUISANCE LEVELS

Many wildlife species of animals have adjusted so well to the urban life style that they have "crossed the line" (Schmidt 1997) and are now perceived by some segments of the population as nuisance animals. Warranted or not, animal damage control (ADC) experts often are called in to address the problems associated with these species. ADC professionals may be employed by USDA-APHIS Wildlife Services (WS), local governments, or by private companies that are members of the National Wildlife Control Organizations (NWCOs).

Some community differences in the common nuisance species should be expected, given the difference in types of resources (food, water, shelter) provided by urban communities, patterns of urban sprawl into natural habitats, age of community, and patterns of land fragmentation during development. The types of species that provoke large numbers of calls to ADC agencies and organizations will vary regionally (e.g., alligators and nutria in the Southeast, bear and cougar in the West, polar bears in extreme Northwest, and cattle egrets in the South Central U.S.). An investigation of the common nuisance species within the urban areas of each state would also provide some interesting comparisons. In addition, what these animals do that qualifies them as a nuisance species provides some extremely interesting examples of how these species have adapted to the urban environments.

2.8 SOME INSIGHTS INTO THE MAGNITUDE OF URBAN WILDLIFE PROBLEMS

As part of our research in producing this book, we requested and obtained data from a national database maintained by USDA-APHIS Wildlife Services (WS). These records (Management Information System, MIS) are reports completed by WS personnel. The MIS database contains records of the number of public inquiries (requests for assistance) that came from urban and rural residents, identification of nuisance species, and the economic impact (losses) reported to WS by the public resulting from damage caused by all species in four damage categories including agriculture, human health and safety, property, and natural resources. The MIS database provided longitudinal (fiscal years 1994 to 2003) and regional (Figure 2.2) scales of analysis of the magnitude of urban wildlife issues across the continental U.S.

The MIS data analysis yielded 350,661 projects conducted by WS in response to requests for assistance by the public with urban wildlife damage problems. In addition, MIS data provided nearly 1.2 million records of damage occurrence, which were used to evaluate urban damage losses. Each record represented one occurrence of a wildlife damage conflict in the form of actual damage or a perceived threat that was observed by Wildlife Services or reported to them by the public. These data are a good index for determining which wildlife species are causing, or are perceived as causing, damage to resources considered important to the public, but may not represent population sizes of those damaging species. However, empirical observations by WS employees suggest that damage levels in specific habitats or environments appear to correlate as directly proportional to relative population sizes within these habitats for the species or species groups causing damage.

It should be noted that the resource subcategory "human health and safety" is always underreported for loss values in the WS MIS database. Difficulty in deriving actual damage losses for this resource resulted in a value of zero being reported frequently. In addition, even when some values can be derived, all primary and collateral damage costs can rarely be assessed. This is especially true where human lives have been lost (USDA, APHIS, Wildlife Services 2004). Economic damage data from the MIS database was collected from October 1, 1993 to September 30, 2003.

The MIS database was queried for the urban species of most concern by various regions in the U.S. The regions were those used in the 2001 National Survey of Fishing, Hunting, and Wildlife-Associated Recreation (Figure 2.4). There were some disclaimers using the regional breakout in Figure 2.4 as one scale for data analysis. For example, there were regional differences in geographic size, the number of WS outlets, and the level of WS personnel reporting activity. However, the size of the MIS data source may offset these problems.

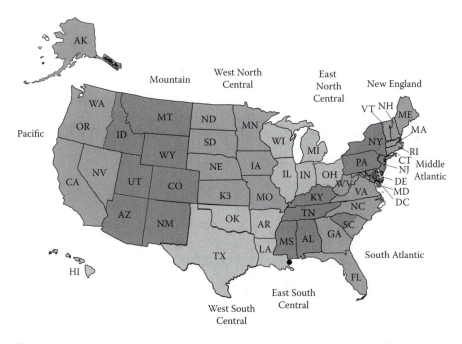

Figure 2.4   Regions used in the analysis of urban species of most concern. (U.S. Department of the Interior)

The national and regional reviews of the urban species of most concern were not provided in the form of scientific names; rather, they are presented as common names or species groups (e.g., deer, blackbirds, geese). Such "lumping" is sometimes desirable because several species collectively cause identical damage in the same habitats to the same kinds of resources, and evaluation of the magnitude of damage to these resources is more accurately reflected when the impacts of these combined groups are considered. For instance, data reveal that, collectively, puddle ducks (dabbling ducks) cause significant damage to some urban resources, but when one species of these waterfowl is analyzed alone, it does not appear to cause damage of concern. Again, if only white-tailed deer are analyzed for damage impacts, it may appear that a national level of damage, or damage by deer in Western states, is not significant. But, when this species is lumped with mule deer, exotic deer, and all other deer species, a better picture is immediately apparent. In this evaluation scheme, only the top ten species or species groups were presented in both national and regional analyses, even though all species reported were used in the analysis (Table 2.1).

2.8.1 Urban Species of Most Concern: National Analysis 1994 to 2003

The national overview of top ten urban species of most concern provided a baseline to compare on the regional level. Listed in order of greatest to least magnitude of damage caused by each species on a national scale were raccoons, coyotes, skunks, beaver, deer, geese, squirrels, opossums, foxes, and blackbirds (Table 2.1).

Table 2.1   National and Regional Records of Ten Species Most Recorded As Causing Damage to Urban Environments by Wildlife Services during FY 1994–2003

Region	Top Ten Species / Number of Records									
Nationwide	Raccoons 73,084	Coyotes 56,081	Skunks 54,198	Beaver 45,958	Deer 36,943	Geese 34,808	Squirrels 28,975	Opossums 21,871	Foxes 20,635	Blackbirds 19,228
Pacific	Skunks 26,476	Raccoon 22,871	Coyote 18,964	Opossum 8,972	Coots 5,614	Foxes 4,801	Beaver 4,479	Geese 3,804	Bobcat 3,670	Cougar 3,508
Mountain	Skunks 3,202	Woodpeckers 2,919	Raccoons 2,537	Pigeons 2,244	Gophers 1,994	Coyotes 1,429	Squirrels 1,128	Beaver 1,061	Mourning Doves 1,036	Foxes 935
West North Central	Skunks 2,144	Beaver 1,861	Geese 1,404	Hawks/ Owls 835	Raccoons 770	Blackbirds 588	Coyotes 526	Pigeons 514	Mink 497	Squirrels 462
West South Central	Coyotes 29,227	Raccoons 15,654	Skunks 8,635	Beaver 8,545	Squirrels 8,241	Opossums 7,012	Black Rats 6,299	Bobcats 4,759	Armadillos 4,357	Pigeons 3,755
East South Central	Blackbirds 2,707	Geese 1,808	Pigeons 1,680	Beaver 1,518	Hawks/Owls 1,271	Raccoons 905	Bats 616	Vultures 520	Skunks 520	Ducks 502
East North Central	Deer 29,149	Raccoons 13,276	Hawks/owls 10,322	Black Bears 10,089	Beaver 7,808	Geese 6,847	Skunks 6,632	Squirrels 5,537	Woodchucks 5,349	Crows 4,315
New England	Beaver 8,491	Raccoons 8,289	Skunks 4,673	Foxes 3,491	Geese 3,212	Bats 3,193	Squirrels 2,889	Black Bears 2,766	Woodchucks 1,827	Deer 1,811
Middle Atlantic	Geese 9,393	Crows 2,675	Blackbirds 1,572	Gulls 1,037	Sparrows 582	Raccoons 404	Blue jays 395	Vultures 393	Hawks/owls 378	Pigeons 330
South Atlantic	Beaver 12,190	Raccoons 8,378	Woodchuck 7,911	Geese 7,601	Squirrels 7,577	Blackbirds 6,889	Foxes 5,853	Deer 4,594	Snakes 4,013	Vultures 3,185

2.8.2 Urban Species of Most Concern: Regional Analysis 1994 to 2003

As might be expected, the species of special concern were different by region and did not always reflect the nationwide list. For example, birds rather than mammals were the species of special concern in the Middle Atlantic region, and no reptiles were reported in the top ten list except for the South Atlantic region. In the Pacific region, black bears were reported (3,225 records) nearly as often as cougars, but they were not included in the top ten list.

There were differences in number of public reports (records) for each region. For example, the level of public reports in the Pacific, West South Central, East North Central, and South Atlantic regions were nearly ten times greater than in the other regions. This difference might be attributed to the size of the urban population, and those factors listed above that might bias the MIS data set. Nevertheless, there is no other national synthesis that demonstrates which species are causing damage to resources in urban environments.

2.8.3 Economic Impact of Damage to Resources by Urban Wildlife: National Overview

There was a reported economic damage to resources by urban wildlife of $550.8 million from FY 1994 through 2003. Nearly 75% of the damage was to personal property ($397.3 million), followed by agricultural losses (e.g., gardens) ($88.7 million), human health and safety ($50.2 million), and natural resources (e.g., loss of native flora or fauna) ($14.1 million). The level of economic loss increased steadily during the 10-year period (Figure 2.5).

2.8.4 Economic Impact of Damage to Resources by Urban Wildlife: Regional Overview

The increasing level of reported economic loss may be correlated with a higher level of public reports of urban wildlife damage in years succeeding 1994 (e.g., public awareness of knowing who to report to), rate of urban sprawl into natural habitats

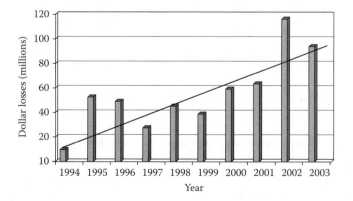

Figure 2.5 Trends in reported dollar losses caused by urban wildlife, FY 1994 to 2003.

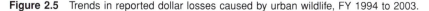

Table 2.2 Sum of Dollar Losses Due to Damage Caused by Various
          Species of Wildlife in Nine U.S. Regions (1994 to 2003)

Region	Sum of Dollar Losses
South Atlantic	119,096,379
New England	102,597,126
Pacific	92,143,657
West South Central	57,274,703
East North Central	55,931,311
Mountain	38,542,420
Middle Atlantic	33,888,330
East South Central	31,092,058
West North Central	20,208,391

with increased human–wildlife conflicts and consequent damage, and more rigorous pursuit of these data by WS personnel. Figure 2.5 does not include a FY 1995 outlier, which was a $1.2 million single-occurrence plane crash.

There was an increasing trend in dollar losses in every region from FY 1994 through 2003. At the end of the 10-year period, the South Atlantic, New England, and Pacific regions reported the highest dollar losses due to damage caused by various species of urban wildlife (Table 2.2).

There can be no doubt that there is a growing trend of urban wildlife management issues that will need professional attention. The degree to which the public can expect professional attention to these issues is addressed later in this chapter.

A nuisance species could be defined as any individual that causes negative impacts on human health or economics. For example, an urban coyote that preys on urban rodents will probably attract little attention from its human neighbors, especially if it stays out of sight. But small dogs and cats provide an easy meal for coyotes, and at this point the predator is likely to be viewed as a nuisance, or worse, by pet owners. Large flocks of birds may create large amounts of droppings that cover trees and buildings or pollute water impoundments. Raccoons, opossums, squirrels, skunks, armadillos, foxes, and many other species find den spaces for living and raising their young in houses and other buildings. In short, if there is a resource that can be exploited, urban wildlife will find and utilize it. More often than not their exploitive behavior is aided and abetted by humans who, directly or indirectly, invite wildlife into their backyards or homes.

2.9 WILDLIFE AS A DOMINANT FOCUS OF WILDLIFE PROFESSIONALS

There is great satisfaction in knowing or at least assuming that one's efforts are on the cutting edge or present a new paradigm for action. The issues addressed above represent a current wildlife management need that is sure to remain into the foreseeable future. If there was ever a time to conduct urban deer management, it is now. Nearly every urban community within the range of white-tailed deer populations has an overabundant deer problem. There is probably a greater need for wildlife biologists with state and federal agencies to apply their management skills to urban rather than rural deer herds. Interestingly, some developers of residential areas have marketed the concept of close proximity of deer to prospective customers. It may be an attractive advertising ploy, but the wildlife management and conflict resolution problems that ensue are formidable. Conversely, management of urban wildlife may focus on species restoration of threatened or endangered species (e.g., peregrine falcon, Chapter 6).

Urbanization and the encroachment of humanity into former wildlife habitats will continue. Every state wildlife agency has been given the legislative mandate to manage the state's wildlife resources, without any caveat as to whether the wildlife reside in the city or country. This mandate does not apply to species covered under federal laws or under federal jurisdiction (e.g., migratory species, threatened or endangered species, and marine mammals). Still, the degree to which these agencies are prepared to embrace the challenges of urban wildlife management is a matter for speculation. Few of the wildlife professionals within these agencies have been trained to manage wildlife in an urban setting. Few land-grant universities that offer at least a Bachelor of Science degree in wildlife sciences include even one course in urban wildlife management. Traditional wildlife management strategies applied in rural areas are not always appropriate in urban areas where people become a larger part of the equation (Van Druff 1996).

2.10 THE INFRASTRUCTURE FOR URBAN WILDLIFE MANAGEMENT IS MISSING

In 1999 the Urban Wildlife Working Group of The Wildlife Society conducted a national survey of state departments of natural resources and land-grant universities that offered at least a bachelors degree in wildlife science (management). The survey was conducted by faculty and students associated with the Human Dimensions in Wildlife Management Laboratory in the Department of Wildlife and Fisheries Sciences at Texas A&M University in College Station, Texas. The e-mail survey was designed to determine how well land-grant universities that offer at least a B.S. degree in wildlife sciences and 48 state wildlife agencies in the continental U.S. were prepared to address urban wildlife management issues.

Survey questions were designed to determine the status of urban wildlife management in each respondent's state related to the:

1. Relevant urban wildlife management issues,
2. Number of urban wildlife biologists,
3. Qualifications and tasks that differentiate urban from other wildlife biologists,
4. Number of urban wildlife biologists, degree of respondent (university or state agency) responsibility for urban wildlife management,
5. How urban wildlife management issues are addressed,
6. Number of universities and colleges that offer at least a B.S. degree in wildlife sciences and/or courses in urban wildlife management, and
7. Future need for urban wildlife biologists.

Response rates were 80% from universities ($n = 37$) and 90% from state wildlife agencies ($n = 46$), respectively. Survey results can be considered as an accurate portrayal of national trends.

Only one university respondent reported that there were no urban wildlife management issues relevant to his/her state. Of the remaining questions, there was little difference in the response frequencies by university and state agency respondents. For example, 85% admitted that urban wildlife was a growing management concern, and 90% agreed that urban human populations needed educational programs about the wildlife around them.

Two-thirds (62–67%) felt that there was a growing human curiosity about the wildlife in their urban habitats. Yet nearly two-thirds to over three-quarters (60–83%) of the respondents agreed that:

1. There was a growing concern about dangerous human–wildlife encounters in urban environments.
2. There is a need to show urban humans how to reconnect with the natural world around them.
3. Urban habitats provided plenty of food, water, and shelter for many wildlife species.
4. The number of many species of urban wildlife were increasing to nuisance levels.

The lower levels of agreement with the above issue statements were always by the university respondents. Finally, few (10–11%) of the respondents agreed that urban wildlife management will probably become the dominant future focus in their states.

Based on the definition of urban wildlife biologist provided in the Introduction, university respondents reported the employment of 7 urban wildlife biologists out of a total of 545 traditional wildlife biologists. Agency respondents reported 46 urban wildlife biologists out of a total of 5,409 traditional wildlife biologists.

Nearly half (44%) of the respondents said there were few to no qualifications that differentiated urban from other wildlife biologists. However, given a list of qualifications provided by Tylka et al. (1987) respondent groups had different levels of agreement. For example, 65% of the university compared to 29% of the agency respondents felt that an urban wildlife biologist should be able to evaluate effects of urbanization on habitat. Only 43% of the university and 32% of agency respondents said the ability to solve wildlife damage problems should be a qualification. Few of the university (27%) and agency (15%) respondents identified ability to evaluate public attitudes as a necessary qualification.

Thirty-six percent of the university and 29% of the agency respondents said there were few to no tasks that differentiated urban from other wildlife biologists. However, given a list of tasks provided by Tylka et al. (1987), respondent groups had different levels of agreement. For example, 35% of the university compared to 21% of the agency respondents felt that a task of an urban wildlife biologist should be animal damage control. Forty-one percent of the university and 27% of the agency respondents identified the establishment of urban wildlife habitats as a task unique to urban wildlife biologists. Sixty-two percent of the university and 31% of the agency respondents identified conducting research on urban wildlife management as a task appropriate for an urban wildlife biologist.

Respondents were asked to identify the level of responsibility (all, some, none) their university or agency had for urban wildlife management. None of the university and 54% of the agency respondents said all urban wildlife management was their responsibility. Sixty percent of the university and 46% of the agency respondents said that they had some responsibility for urban wildlife management. Forty and zero percent of the university and agency respondents, respectively, said urban wildlife management was none of their responsibility.

The degree to which there is an academic infrastructure in place to train urban wildlife biologists was determined by asking survey respondents how many colleges or universities in their state offered at least a bachelor of science (B.S.) degree in wildlife science, or courses in wildlife management or urban wildlife management. University respondents could identify only 67 universities or colleges that offered at least a B.S. in wildlife sciences compared to 78 identified by agency respondents. Both respondent groups said that there were 111 colleges and universities that offered courses in wildlife management, but only 6 or 7 that offered courses in urban wildlife management. These were national totals!

The estimated future need by 2004 for urban wildlife biologists employed in land-grant universities was 20 compared to 170 by agencies. However, less than 25% of the respondents expected their needs to be met in the next five years.

This study showed that the infrastructure for urban wildlife management is not well developed within academia or wildlife agencies and agrees with the findings of Adams, L.W. et al. (1987). The degree to which land-grant universities and state agencies were prepared to embrace the challenge of urban wildlife management was less than encouraging. Few of the wildlife professionals within these agencies have been trained to manage wildlife in an urban setting (i.e., qualify as urban wildlife biologists). Few universities or colleges offer even one course in urban wildlife management. Overall, study results pointed out that universities and agencies:

1. Are relying on conventional management philosophies and skills to address urban wildlife management issues,
2. Are applying token effort to enormous if not insurmountable urban wildlife management problems,
3. Do not recognize urban wildlife management problems as their responsibility,
4. Are somewhat oblivious to present and emerging urban wildlife management problems.

PERSPECTIVE ESSAY 2.1: HUMAN–WILDLIFE INTERACTIONS IN THE 1950s

I spent the first 14 years of my life, from 1942 to 1958, in Algona, Iowa, which at that time had a population of about 4,000 people. A half-hour walk would put anyone into the country to conduct any number of activities including hunting; fishing; catching insects; building forts, rafts, and dams; looking for rocks or Indian artifacts; collecting mushrooms, nuts, or berries; making tools, musical toys, or weapons from the surrounding vegetation; visiting your special place for solitude

Figure 2.6   Fishing in the 1950s. (USFWS)

and reflection; hiking; snow sliding and a host of other activities that cost nothing but your time.

We fished year-round in the Des Moines River nearly every weekend and after school. Fishing included so much more than catching fish! River fishing usually resulted in wet, hot, dirty, insect-bit, barefooted and sunburned anglers (Figure 2.6) and is a far cry from fishing programs currently popular in urban areas (Figure 2.7). The spring rains caused night crawlers (*Lumbricus terrestris*) to emerge after dark, and it was common to collect a gallon can full of worms to be used or sold as fish bait during the summer. Night crawler husbandry was a familiar practice. In addition, youngsters knew how to raise tadpoles to adult frogs, care and feed baby owls until they fledged, catch pocket gophers for ten cents bounty, and live capture thirteen-lined ground squirrels (*Citellus tridecemlineatus*) with string.

Figure 2.7   Fishing from a well stocked swimming pool is common in urban cities. (C. E. Adams)

Rabbits (*Sylvilagus floridanus* and *Lepus californicus*), squirrels (*Sciurus niger*), ring-necked pheasants (*Phasianus colchicus*), and waterfowl were common fare on our dinner table. During open season we hunted nearly every weekend and after school. In fact, the importance of hunting in our community was demonstrated each year during the opening day of pheasant season. Work stopped at noon. My mother made a special meal, after which all the men and boys headed for the cornfields to hunt pheasants. Hunting was so universally accepted as a recreational and/or subsistence pursuit that we often took our .22-caliber rifles or .410 shotguns to school and stored them in our lockers so we could head to our hunting areas as soon as the last bell rang.

We learned the behavior patterns of the animals in our community, including their escape routes, how they tried to avoid detection, defense mechanisms, where they slept, and what they ate. Trapping wild animals for the fur trade was a common fall and winter enterprise by both young and adults. This meant checking trap lines at 4 a.m., dispatching the captured animals, skinning them and stretching the pelt on the appropriate board for the species, storing them for future sale, and still making it to school or work on time. Trapping animals also led to lessons in animal anatomy; for example, knowing the location of the scent gland in skunks (*Mephitis mephitis, Mephitis putoris*) and mink *(Mustela vison)* was critical! Other things we learned included special physical adaptations, e.g., badger (*Taxidea taxus*); the differences between carnivores, herbivores, and omnivores; and predator/prey relationships. Pet ownership was as much a matter of function as companionship. Dogs were expected to perform specific duties beyond eating and sleeping.

Unlike contemporary urbanized societies, the human–wildlife interface that existed during this time in my life was based on a utilitarian human attitude of domination and sustained yields. The extrinsic value of wildlife was a far more important consideration than their intrinsic value. Our decisions concerning wildlife were not constrained by physical, psychological, societal, economic, political, or structural barriers. There were nearly daily interactions with wildlife, but people seemed more aware of the reasons behind these interactions and what actions, if any, were appropriate and necessary to promote or prevent the interaction.

Clark E. Adams

CASE STUDY 2.1: NEIGHBORHOOD MOOSE KILLED BY KINDNESS

An Alaska state biologist says the people who feed wild animals in their backyards are responsible for the death of a moose in early March, 2005. On March 3, police shot and killed a moose after a 6-year-old Anchorage boy was stomped on the head. Rick Sinnott says he's seen too many people feeding animals and not thinking about what could happen. At the same time, a wildlife conservation group says it's an example of how things need to change.

It happened in a Muldoon neighborhood where trash cans without lids can tempt a hungry moose. But officials say the moose that hung out in the neighborhood wasn't just dumpster diving, it had been hand-fed.

"We know that scores, if not hundreds, of people are feeding moose every day in Anchorage, and they're creating dangerous situations," Sinnott says. "But for us, we've got jobs other than just being the moose monitors. And we do our other jobs. If we can, we try to bust these people. But it's almost impossible."

Shannon Eldridge lives just a few homes away from where it happened. "Pretty scary, because I have two little kids that play outside. It's just a sad situation that the little boy got hurt. They're pretty dangerous now. There's a lot of them on the road. They're causing accidents, they're tromping on kids and people. They need to be taken care of a little bit, I think."

The group Defenders of Wildlife says they may have an answer to that. At this week's Board of Game meeting, they'll testify against a proposal to create a limited moose hunt in Anchorage. Instead they want to create an Anchorage moose committee to find other ways for city residents to coexist with moose.

Karen Deatherage says a similar idea for creating an Anchorage bear committee has been successful. "And it was really a successful program because we sat down at a table, we rolled up our sleeves, and we came up with long-term solutions to some of the conflicts in Anchorage with wildlife."

State officials agree that the Anchorage bear committee has helped residents better understand the issues raised when so many people live so close to bear habitat. There's plenty of agreement that the moose issue is just as serious.

March 3, 2005
Excerpted from an article at http://www.ktuu.com
by Jeffrey Hope

REFERENCES

Adams, C. E. (1990). Resource ecology activities for introductory high school biology, *American Biology Teacher,* 52:414–418.

Adams, C.E. (2003). The infrastructure for conduction urban wildlife management is missing, *Transactions of the North American Wildlife and Natural Resources Conference,* 68:252–265.

Adams, C. E. and Eudy, J. L. (1990). Trends and opportunities in natural resource education, *Transactions of the North American Wildlife and Natural Resource Conference,* 55:94–100.

Adams, C. E. and Greene, J. (1990). Perestroika in high school biology education, *American Biology Teacher,* 52:408–412.

Adams, C. E., Thomas, J. K., Lin, P.-C., and Weiser, B. (1987). Urban high school students' knowledge of wildlife, in *Integrating Man and Nature in the Metropolitan Environment,* Adams, L. W. and Leedy, D. L., Eds., National Institute for Urban Wildlife, Columbia, MD, pp. 83–86.

Adams, L. W., Leedy, D. L., and McComb, W. C. (1987). Urban wildlife research and education in North American colleges and universities, *Wildlife Society Bulletin,* 15:591–595.

Damude, N, Bender, K., Foss, D., and Gowen, J. (1999). *Texas Wildscapes: Gardening for Wildlife,* University of Texas Press, Austin, TX.

Greene, J. S. and Adams, C. E. (1992). An evaluation of volunteerism in selected conservation education programs, *Transactions of the North American Wildlife and Natural Resource Conference,* 57:175–184.

Heimlich, R. E., and Anderson, W. D. (2001). Development at the urban fringe and beyond, impacts on agriculture and urban land, U.S. Department of Agriculture, Washington, DC.

Kellert, S. R. (1980). Americans' attitudes and knowledge of animals, *Transactions of the North American Wildlife and Natural Resources Conference*, 45:111–124.

Leopold, A. (1949). *A Sand County Almanac*, Oxford University Press, Oxford.

McDonnell, M. J. and Pickett, S. T. A. (1990). Ecosystem structure and function along urban-rural gradients: an unexploited opportunity for ecology, *Ecology*, 71:1232–1237.

Price, J. (1999). *Flight Maps: Adventures in Nature in Modern America*, Basic Books, New York.

Schmidt, R. (1997). Drawing the line for wildlife, *Wildlife Control Technology*, December: 6–7.

Tylka, D. L., Shaefer, J. M., and Adams, L. W. (1987). Guidelines for implementing urban wildlife programs under state conservation agency administration, in *Integrating Man and Nature in the Metropolitan Environment*, Adams, L. W. and Leedy, D. L., Eds., National Institute for Urban Wildlife, Columbia, MD, pp. 199–205.

U. S. Department of the Interior, Fish and Wildlife Service and U. S. Department of Commerce, Bureau of the Census (1985, 1991, 1996, and 2001). National Survey of Fishing, Hunting, and Wildlife-Associated Recreation, United States Government Printing Office, Washington, DC.

VanDruff, L. W., Bolen, E. G., and San Julian, G. J. (1996). Management of urban wildlife, in *Research and Management Techniques for Wildlife and Habitats*, 5th ed., rev., Bookhout, T. A., Ed., The Wildlife Society, Bethesda, MD, pp. 507–530.

Wright, R. T. (2004). *Environmental Science: Toward a Sustainable Future*, 9th ed., Prentice-Hall, Upper Saddle River, NJ.

SECTION II

Urban Ecosystems

CHAPTER 3

Ecosystems: Principles, Structure, Function, and Services

The famous balance of nature is the most extraordinary of all cybernetic systems. Left to itself, it is always self-regulated.

Joseph Wood Krutch, *Saturday Review,* **June 8, 1963**

CONTENTS

3.1 ECOLOGICAL PRINCIPLES

Ecology is the study of relationships between the biotic and abiotic components of an ecosystem. Ecology is a complex and complicated subject, but for our purposes here we can distill it down to four fundamental ecological principles: diversity, interrelationships, cycles, energy (DICE).

3.1.1 Diversity

Diversity is defined as the variety within ecosystems (Wilson 1992:393 in Folke et al. 1996). Natural ecosystems tend to be complex, highly diverse, and highly stable. Urban ecosystems, on the other hand, are the products of human manipulation, which tends to simplify and destabilize natural ecosystems. Figure 3.1 and Figure 3.2 demonstrate habitat simplification in suburban and metropolitan environments, respectively. Perspective Essay 3.1 provides an example of how the typical urban lawn is living testament to habitat simplification and destabilization. The remainder of this chapter points out several ways how human manipulation of natural habitats destabilizes ecosystem structure and function.

    To better understand the value of diversity, think of an ecosystem as a Boeing 707 jet. The structural components of our 707 ecosystem include sheet metal, rivets, wires, and a host of other materials. Now, imagine you have entered this ecosystem (boarded the plane) and have claimed your territory (found a seat). You look out the window and notice someone is removing rivets (*Rivetus solidus*) from the wing of the plane. You call this to the attention of the flight attendant who assures you that there is nothing to worry about. There are thousands of rivets in the plane and the person is just harvesting a few of them. You continue your trip without any mishaps.

Figure 3.1   Typical suburban landscaping causing simplification of habitat. (C. E. Adams)

On your return trip you notice that the same person is removing even more rivets from the plane. Again you voice your concern to the flight attendant, who again assures that there are more than enough rivets left. But how many members of the rivet species can be removed before the entire structure collapses?

Perhaps the 707 is a diverse ecosystem with several kinds of fasteners (species) whose function is to hold the plane together. If so, the loss of many or even most rivets might not be disastrous. But what if rivets are the only species performing this function on the 707? What if rivets are a *keystone species*, a part without which the 707 ecosystem cannot exist? Apply this same analogy to the ecosystem of your choosing, or even the whole planet, and you'll have a good idea of why diversity is so important.

Figure 3.2   Typical metropolitan area with high rise buildings, concrete surfaces, and automobile traffic. (John M. Davis/Texas Parks and Wildlife Department)

Table 3.1 Interrelationships between Ecosystem Components

Interrelationships	Example
abiotic × abiotic	temperature and rainfall = climate
abiotic × biotic	water temperature predicts fish species
biotic × abiotic	human pollution of the environment
biotic × biotic	predator/prey relationships, life cycles

Some of the most compelling arguments for preserving biodiversity are those concerning the effects of eroding biodiversity on human health. Both scientists and public health advocates have asserted that the preservation of biodiversity is crucial for present and future human health (Ostfeld and Keesing 2000:723). On the other hand, human sources of diversity in the urban landscape include introduction of exotic species, modification of landforms and drainage networks, control or modification of natural disturbance agents, and the construction of massive and extensive infrastructures (Pickett et al. 1997 as cited by Zipperer et al. 2000).

3.1.2 Interrelationships

"Everythingisconnectedtoeverythingelse" is a run-together sentence used as an analogy for ecological interrelationships. Ecosystems are composed of biotic (living) and abiotic (nonliving) components. Table 3.1 provides several illustrations of how these components interact. Humans, in general, do not recognize how much our presence affects ecological relationships. Urban development, for example, increases the amount and type of pollutants released into the environment, changing the abiotic and biotic structure of aquatic and terrestrial habitats, and altering plant and animal relationships. Without question, ecosystems are complex systems. In order to really understand complex systems one needs to integrate population, community, and ecosystem processes across multiple scales (Ostfeld et al. 1998). This integrative approach examines the complex interrelationships that exist between oaks, white-footed mice, deer, ticks, and gypsy moths causing Lyme disease, which is discussed in Chapter 12.

3.1.3 Cycles

Cycles are the mechanisms that make a finite amount of substance infinitely available in natural ecosystems. Nitrogen, carbon, hydrogen, oxygen, phosphorus, and sulfur (N-CHOPS) are the fundamental elements required by all living organisms, but only so much of each element exists on the planet Earth. Later, we illustrate trophic levels and nutrient cycling in terms of the major actors, and explain the roles they play in the cycle, and the human impacts on the nutrient cycles. Each of these elements is recycled through producers, consumers, and decomposers making them available forever in just the right amounts to sustain plant and animal life. In contrast, urban systems were designed around the principle of one-way trips! Examples of one-way trips can be found in the huge landfills for urban garbage, where many products that could be recycled including metals, glass, plastics, and organic materials are lost forever.

A cycle of a different nature is the water cycle. The water cycle does not affect the quantity of water like the cycles discussed earlier. The water cycle exists to provide a continual supply of *purified* water by using a system of three filters consisting of air, soil, and plants. Urban development affects several aspects of the water cycle (e.g., infiltration and runoff) and the purification process (e.g., pollution of air and soil). The water cycle is discussed in greater depth later in Chapter 7.

3.1.4 Energy

Energy becomes available to every aquatic and terrestrial ecosystem by trapping sunlight energy through the process of photosynthesis. *Sunlight energy is free, everlasting, and nonpolluting.* However, urban developments do not take advantage of these characteristics of sunlight energy. Instead, the primary sources of energy for urban ecosystems are fossil fuels (e.g., coal and gas), which have adverse effects on terrestrial and aquatic ecosystems in both urban and rural settings.

3.2 ECOSYSTEM STRUCTURE

Ecosystems, no matter how unique, have certain basic characteristics in common. All ecosystems are **biotic** communities (living plants, animals, and microbes) inter-acting with one another and their environment. When we use the term "environment" we are really talking about **abiotic** or nonliving factors that act upon the biotic community. To put it even more simply, an **ecosystem** is the marriage between biotic communities and the abiotic conditions they live in (Wright 2004:30).

The abiotic factors — including temperature, water, chemicals, and soils — are what cause the differences we observe between various ecosystems. For example, in regions with a high average annual temperature and rainfall, we can expect to find the plant, animal, and microbe species typical of a tropical rainforest. The opposite end of the ecosystem spectrum would be a region with low average temperatures and rainfall, commonly referred to as tundra (Odum and Barrett 2004:432–457). Additionally, the primary difference between freshwater and marine ecosystems is the absence or presence, respectively, of certain chemicals — in this case, sodium chloride (NaCl) or salt.

Natural ecosystems are inherently stable, self-contained units that derive energy from the sun, recycle matter (e.g., nutrients), and support complete food webs. Urban ecosystems, on the other hand, are inherently unstable, derive little of their energy directly from the sun, remove rather than recycle matter, and have incomplete food webs (Adams 1994). "Realistically, urban-industrial environments are parasites on the biosphere in terms of life-support resources" (Odum and Barrett 2004:408).

In order to fully understand the challenges of urban wildlife management, we need first to explain the individual components of natural ecosystem structure and function and how they compare to those of urban ecosystems. Ecosystem structure refers to basic components and how they fit together. As was explained above, these components can be organized into two categories that affect the structure of every ecosystem — the abiotic components and the biotic community. Population dynamics is discussed in Chapter 4. The two chapters in this section are intended

Table 3.2   Examples of Abiotic Factors

Conditions	Resources
pH (acidity vs. alkalinity)	Chemical nutrients (N, P, CO_2)
Salinity	Light (intensity and wavelength)
Temperature	Oxygen
Turbidity (cloudiness of water)	Space to live, feed, reproduce
Wind	Water or moisture

to be an introduction for some readers, and a review for others, of the ABCs of ecosystems.

3.2.1 Abiotic Structure

The abiotic structure of an environment, sometimes referred to as its inorganic components, is made up of various physical and chemical factors (Table 3.2). It can be helpful to divide abiotic factors into two types: *conditions* and *resources*. **Conditions** represent the state of the habitat for a given abiotic factor. Examples include climate (temperature and moisture), pH (acid or alkaline), turbidity (cloudiness), and wind (speed and direction). **Resources** determine the habitat conditions. As a result, the resources that are available or absent will determine which plant and animal species exist in a particular area (Begon et al. 1996).

For example, abiotic components predict the types of plants that will grown in an area, and the plants, in turn, predict the types of animals that will exist there (Figure 3.3). Temperature and moisture are the abiotic factors that most often define ecosystems. As we discussed in the introduction to this chapter, it is the combination of these two components that determine the dominant vegetation type (e.g., trees, shrubs, grasses, or succulents), while temperature determines differences within related vegetation types (e.g., rainforest, temperate forest, tundra).

Interestingly, it is rare to see a reference to regional climatic differences between urban environments (e.g., desert-urban, temperate-urban, tropical-urban) discussed in

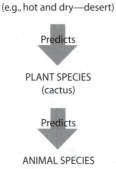

ABIOTICS
(e.g., hot and dry—desert)

Predicts

PLANT SPECIES
(cactus)

Predicts

ANIMAL SPECIES
(desert tortoise)

Figure 3.3   How ecosystems are formed.

Table 3.3   Effects of Cities on Urban Climate

Climate Element	Compared to Countryside
Temperature	1 to 6°F (3.3°C) higher
Relative humidity	6% less
Precipitation	5 to 15% more
Wind speed	20 to 30% less
Contaminants	5 to 25 times more
Clouds	5 to 10% more
Sunshine	5 to 15% less

Source: From Landsberg (1981) and Herold (1991), in Adams, L. W., University of Minnesota Press, Minneapolis, MN, 1994.

the literature, even though these variations certainly exist. Perhaps this is because the term "urban ecosystem" is relatively new. One need only visit the cities of Phoenix and Seattle to recognize that temperature and moisture variations are at work even in the city, resulting in at least some differences in the species living in each.

Often cities and even suburbs become microclimates, with weather that is significantly different from that of the surrounding area. Cities become heat islands — areas in which local temperatures are measurably higher than those in the nearby countryside. Variations in relative humidity, precipitation, and wind speed also are found within municipalities (Table 3.3). The increased temperatures are the result of human modification of the landscape, particularly an increase in thermal mass through the use of brick, concrete, and asphalt. Higher temperatures result in a longer growing season for plants. Additionally, city lakes and ponds may remain ice-free longer than in nearby rural bodies of water, which may influence the use of habitat by wildlife. Some studies have shown that cities have more cloud cover and more precipitation that surrounding areas (although most of this moisture is quickly lost due to increased runoff — see Chapter 7). Wind speed can be affected by the architectural features within human-created landscapes (Adams 1994).

3.2.2 Biotic Structure

Biotic structure refers to the way living organisms interact with one another. Biotic communities are composed of plants (everything from algae to trees), animals (mites to monkeys and anything in between), and microbes (bacteria, fungi, and protozoans). Ecosystem organisms interact with each other through feeding and nonfeeding relationships called trophic levels and food webs.

Members of the biotic community can be placed into three **trophic** (feeding) categories: (1) producers, (2) plant and animal consumers, and (3) decomposers. These three groups work together to produce food, move it through the food chain, and return the abiotic components to the environment.

Life on Earth depends on energy from the Sun. Energy is a one-way trip! **Producers**, also known as **autotrophs** (self-feeders), are able to transform the Sun's energy into organic matter. Green plants are the primary producers (Table 3.4). All

Table 3.4 Matching Organisms to Trophic Levels

Organism	Producer	Primary Consumer	Secondary Consumer
Plants	X		
Rabbits		X	
Foxes			X
Opossums		X	X
Snakes			X
Decomposers		X	X
Detritus feeders		X	X
Parasites	X	X	X

other organisms feed on organic matter produced by plants as their source of energy. Through a process called **photosynthesis**, plants capture the Sun's kinetic energy using specific wavelengths of light and chemically convert it into potential energy (e.g., sugar). A variety of molecules are used by plants during photosynthesis, but the most important one is the chlorophyll molecule, which has the capability to capture various wavelengths (e.g., blue, red, yellow, orange) of light energy. Interestingly, the reason chlorophyll is green in color is because plants do not absorb the green wavelength of light for photosynthesis. Rather, green light is reflected. All terrestrial ecosystems depend on green plants, and each ecosystem has specific producers that carry on the crucial work of photosynthesis. The potential energy stored in leaves, stalks, and stems becomes available to other organisms when they consume plants, or when they eat the animals that consume plants. All energy is ultimately lost from the ecosystem in the form of heat.

The **consumers** or **heterotrophs** (other-feeders) are a diverse group of organisms — including everything from bacteria and fungi to shrews and sharks (Table 3.4). Because this is such an extensive group, biologists have found it helpful to divide the consumers into subgroups based on their food source. The animals that feed directly and only on plants are called **primary consumers or herbivores** (herb-eaters). Mice, rabbits, and deer are examples of this group. The animals that feed on herbivores are referred to as **secondary consumers or carnivores** (meat-eaters). Coyotes, cougars, and cats are examples of this group. **Decomposers** (e.g., bacteria and fungi) recycle the essential nutrients of life by decomposing dead plant and animal materials and their wastes. These nutrients are returned to plants to use again in another round of photosynthesis. An alternative method of recycling nutrients is through a group of organisms called **detritus feeders**. The prime example of a detritus feeder is the earthworm (*Lumbricus terrestris,* see Perspective Essay 7.1 on Darwin's Earthworms). The primary diet of these organisms is dead plant or animal material. Both decomposers and detritus feeders occupy all trophic levels because their services are needed in each one. Decomposers and detritus feeders are not the most glamorous members of the biotic community, but without their ability to recycle nutrients the whole system would come to a grinding halt.

Herbivores and carnivores usually feed only at one trophic level, while other animals eat a little of everything (Table 3.4). The smorgasbord samplers are **omnivores** (all-eaters). Raccoons, opossums, and humans are generally recognized as omnivores. However, some carnivores such as coyotes, cats, and bears will abandon their carnivorous habits to feed on more abundant food sources such as garbage.

Parasites are a special group with feeding adaptations that allow them to feed on a "host" species without killing it (at least not immediately) or share what the host has eaten (Table 3.4). Parasites can feed on either plants or animals. Parasites can feed on (leeches and ticks) or in (tapeworms and roundworms) their host species. Some plant parasites (e.g., mistletoe) also carry on the process of photosynthesis which places them in a producer category.

3.2.3 Food Chains and Webs

Trophic levels and how they affect the interrelationships between the organisms in an ecosystem can be examined in greater detail by exploring the concepts of **food chains** and **food webs.** A food chain is a simplistic, straight-line model of the way organisms feed on one another. A typical grassland ecosystem food chain is illustrated below. Grass, as a producer, is the starting point of the chain, followed by various consumers:

$$\text{Food Chain} = \text{Grass} \rightarrow \text{Grasshopper} \rightarrow \text{Frog} \rightarrow \text{Snake} \rightarrow \text{Owl}$$

Of course, ecosystems are not this simple. Owls do not just eat snakes — they also prey on lower-order consumers such as rabbits and mice, as do snakes! Virtually all food chains are interconnected, forming complex webs of feeding relationships, also known as food webs. The food web is a convenient method of demonstrating the complexity of natural ecosystems.

Figure 3.4 illustrates a typical food web within a natural ecosystem. Mice eat grass seeds, and the mice are then consumed by owls. Of course, mice feed on other types of seeds as well as the occasional insect, just as owls may consume rabbits in addition to mice. Insects are even an important food for some smaller owl species (e.g., screech owls).

Urban food webs paint quite a different picture, as can be demonstrated by constructing a web of common urban producers and consumers, both native and introduced. For example, common representatives of the producer level in a suburban area might include flowers, trees, and shrubs used in landscaping; as well as sunflower seed mix, sugar water, edible garbage, and pet food. Urban herbivore representatives would be rabbits, mice, deer, geese, squirrels, hummingbirds, and several

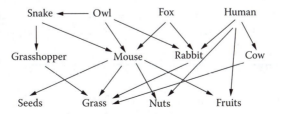

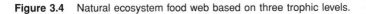

Figure 3.4   Natural ecosystem food web based on three trophic levels.

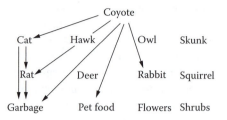

Figure 3.5   Urban ecosystem food web based on four trophic levels.

species of granivore (seed-eating) birds. Humans, raccoons, and opossums are all urban omnivores, while urban carnivores include domestic dogs and cats, coyotes, foxes, raptors (e.g., meat-eating birds such as hawks and owls), and in some parts of the U.S. and Canada, black bears, polar bears, and cougars.

Figure 3.5 illustrates a typical urban food chain — garbage fed on by rats that are fed on by cats which are fed on by coyotes (which comes as quite a shock to many urban residents when they find out their pets have been preyed upon by coyotes!). Note the food web of the coyote can also include the garbage and the rat, as well as rabbits and pet food. Cases have even been reported of rats and cats feeding side-by-side on garbage (Sullivan 2004).

3.3 SYMBIOTIC RELATIONSHIPS

Symbiotic (which translates as "living together") relationships are common in natural ecosystems. Not all of these relationships are beneficial to one species and fatal for the other — often species interact in ways that are mutually beneficial or, at the very least, have a neutral effect on one or both. There are four types of symbiotic relationships, based on whether the results are positive or negative: *mutualism, commensalism, parasitism,* and *competition.*

3.3.1 Mutualism

Mutualism is a form of symbiosis in which both species benefit from the interaction. A fine example of mutualism is the relationship between the pronuba moth (*Tegeticula* spp.) and the soaptree yucca plant (*Yucca elata*). In exchange for the pollination services that the moth performs as she deposits her eggs inside the flower, the yucca provides shelter and food for the moth larvae. Both species not only benefit from this arrangement, their survival is now dependent on the existence of one another.

3.3.2 Commensalism

Commensalism is the symbiotic relationship whereby one species benefits from the interaction, while the other is unaffected. This can be demonstrated using the example of the remora (*Remora* spp.) and larger fish. The remora is a long, slender fish that has a dorsal fin modified as a sucker-like attachment organ, with which it grasps the side of another fish or a turtle. Remoras use their host as transportation while at the same time taking advantage of food fragments left behind as the host eats. The host animal is not harmed by the remora's activities.

3.3.3 Parasitism

Parasitism is the symbiotic relationship in which one species benefits and the other is adversely affected. Technically, predation and disease are forms of parasitism, but for most people, the term parasite calls to mind creatures like ticks and fleas (external or *ecto*-parasites) or tapeworms (internal or *endo*-parasites). As long as the host animal is generally healthy, this type of parasite does not usually kill the host. However, if the host animal's health is compromised in some way (e.g., an injury or disease), or if the parasite is carrying a pathogen (e.g., ticks carrying Lyme disease [see Chapter 12] or mosquitoes carrying West Nile virus), death can result.

It may come as a surprise to some readers to learn that their companion animals can harbor several different types of parasitic diseases that use humans to complete part of their lifecycle. This is one reason veterinarians strongly recommend that puppies be tested and treated for parasitic nematodes such as roundworms (*Ascaris* spp.) and hookworms (*Ancylostoma* spp. and *Necator* spp.).

When both species in a symbiotic relationship are adversely affected, the interrelationship is called competition. Competition for food, habitat, and mates is part of the natural order of life, but it always results in an inequitable distribution of available resources among the competing species. However, "all out war" between species is reduced by specialization and adaptation to a particular *niche*. An animal's **ecological niche** refers to what the animal feeds on, where it feeds, when it feeds, where it finds shelter, where it nests, and how it responds to abiotic factors. Niches allow potential competitors to coexist within the same ecosystem. Both plants and animals are involved in competitive relationships.

3.4 BIOTIC COMMUNITIES

The first step to take when investigating a biotic community may be to catalog all the species present. By doing so, you are likely to notice that the individuals of any particular species in the ecosystem are part of a **population** — the interbreeding group that live within a given area. Natural ecosystems are wonderfully diverse, and it is their diversity that creates stability. For example, most herbivores consume many different kinds of plants, so if one particular plant species become temporarily or even permanently scarce, other food sources are still available. The same applies for omnivores and carnivores — because they are able to feed on many different species of plants and/or animals, and because there is a great variety of species available upon which to feed, fluctuations in populations are generally minor and, to some degree, predictable.

Once again, the situation is quite different in urban ecosystems. The species diversity, sometimes called **richness**, is much lower in urban areas than in natural ecosystems. This may come as a bit of a shock to urban and suburban residents who have observed large flocks of birds at feeders or roosting on buildings, or have marveled at the number of squirrels that can be seen in one small city park. While it is true that wildlife densities tend to be significantly higher in urban areas, the individual animals represent a relatively small number of different species. To put it more simply, there are lots of plants and animals living in the city, but there aren't a lot of different *kinds* of plants and animals living there.

3.4.1 Urban Flora

As an area undergoes urbanization, a pattern begins to emerge that categorizes urban plant communities into three broad groups: *natural remnants, derelict lands,* and *planted communities* (Adams 1994). This chapter introduces these plant communities. They are discussed in more detail in Chapter 5.

Natural remnants of vegetation persist even in the midst of heavy development. These remnants retain the characteristics of the native plant community that existed before urbanization. Remnants face many threats once they become surrounded by urbanization. As the patches become fragmented — smaller in size and further apart — pollination and dispersal of seeds becomes increasingly difficult for the plants contained within. As forest remnants are carved up, specimens once found inside the relatively protected interior are forced to deal with the harsher conditions found at the edge of habitat (e.g., increased wind, greater temperature fluctuations). Urbanization affects the local hydrology, which inevitably affects the plant communities in the remnant patches. Species that are able to tolerate the changes may persist; those that can't disappear. Introduced exotic species create yet another threat. Table 3.5 provides examples of both native plants and the exotics that may take their place as development occurs.

Derelict lands, sometimes referred to as vacant lots, are characterized by having been disturbed by construction at some point. Depending on how long it has been since the disturbance, the plant species that are likely to be found in these patches include invasive exotic species, including some listed in Table 3.5, and native pioneer species (e.g., ragweed). Like remnants, these small areas of habitat are fragmented and disconnected from one another. The biodiversity within these patches tends to be low, although those species represented may exist in large numbers.

Planted communities make up the majority of urban areas. They are largely artificial, making extensive use of high-maintenance species in a highly controlled manner. Many of the species found in planted communities come from Europe, Asia, and South America — boxwood, nandina, crepe myrtle, photinia, ornamental pear, Bermuda grass, and Asian jasmine.

3.4.2 Urban Fauna

Changes in soil, hydrology, and vegetation created by urbanization exert pressure on wildlife populations. Shifts in species composition are, at least to some extent, predictable. For example, urban ecosystems select for small- to medium-sized, highly adaptable predators. Large predators require large territories within which to hunt and, as mentioned earlier, urban habitat patches tend to be small and fragmented.

Table 3.5   Exotic Plants and the Native Species They Replace

Vegetation type	Exotic Species	Native Species
Forest systems	Chinese privet (*Ligustrum sinense*)	Green ash (*Fraxinus pennsylvanica*)
	Chinaberry (*Melis azedarach*)	Flowering dogwood (*Cornus florida*)
	Mimosa (*Albizia julibrissin*)	American elm (*Ulmus americana*)
	Siberian elm (*Ulmus pumila*)	Sweet gum (*Liquidambar styraciflua*)
	Chinese tallow (*Sapium sibiferum*)	Live oak (*Quercus virginiana*)
	Japanese honeysuckle (*Loincera japonica*)	Maple (*Acer* spp.)
Prairie Systems	Tall fescue (*Festuca arundinacea*)	Big bluestem (*Andropogon gerardii*)
	Johnsongrass (*Sorghum halepense*)	Indian grass (*Sorghastrum nutans*)
	Purple scabies (*Scabiosa atropurpurea*)	Switchgrass (*Panicum virgatum*)
	Sweetclover (*Melilotus officinalis* & *M. alba*)	Rose Gentian (*Sabatia campestis*)
	Japanese brome (*Bromus japonicus*)	Prairie coneflower (*Ratibida pinnata*)
Aquatic Systems	Giant salvinia (*Salvinia molesta*)	Pondweed (*Potamogeton* spp.)
	Water hyacinth (*Eichhornia crassipes*)	Water stargrass (*Heteranthera dubia*)
	Hydrilla (*Hydrilla verticellata*)	Arrowhead (*Sagittaria* spp.)
	Parrot's feather (*Myriophyllum aquaticum*)	Spatterdock (*Nuphar luteum*)
	Eurasian watermilfoil (*Myriophyllum spicatum*)	Pickerel weed (*Pontederia cordata*)

Source: Brockman, C. F., Golden Press, Racine, WI, 1986; Borman, S. and Korth, R., University of Wisconsin-Stevens Point Foundation, Stevens Point, 1998; Davis, J. M., in Haggerty, M. M. Ed., Texas Parks & Wildlife Department, College Station, 2003; and Cushman. R. C. and Jones, S. R., Houghton Mifflin, New York, 2004.

Raccoons and coyotes are perfect examples of predators that do well in urban ecosystems. These species actually increase in numbers living in close proximity to humans; they don't require large amounts of space and they find ample and diverse prey (rats, mice, rabbits, cats, and small dogs) within the city limits.

As a general rule, specialists decline in urban areas, while generalists thrive. Specialists are those species with very strict survival requirements regarding diet, habitat, and/or nesting sites. Urbanization, by its very nature, significantly alters the existing habitat, creating more edge habitat while decreasing available interior habitat. Edge-loving species are attracted to these areas; generalists, with their wider range of acceptable living conditions, adapt; and the interior specialists disappear. As the specialists leave, the decrease in interspecies competition for resources allows the generalists to not only survive, but to increase in numbers.

Humans play an important role in which species are able to make the transition from life in the country to city living. Garbage is an important food source for many urban generalists, particularly omnivores like opossums and raccoons. Providing

Table 3.6   Common Urban Wildlife Species

Invertebrates	Birds
Black widow spider (*Latrodectus spp.*)	Northern cardinal (*Cardinalis cardinalis*)
Garden spider (*Argiope spp.*)	Blue jay (*Cyanocitta cristata*)
Silverfish (*Lepisma saccharina*)	Mourning dove (*Zenaida macroura*)
American cockroach (*Periplaneta americana*)	Ruby-throated hummingbird (*Archilochus colubris*)
House cricket (*Acheta domestica*)	American robin (*Turdus migratorius*)
Praying mantis (*Mantis religiosa*)[i]	European starling (*Sturnus vulgarius*)[i]
Mosquito (*Aedes* spp. and *Culex* spp.)	Killdeer (*Charadrius vociferous*)
House fly (*Musca domestica*)	Chimney swift (*Chaetura pelagica*)
Cabbage butterfly (*Pieris rapae*)	Purple martin (*Progne subis*)
Tent caterpillar moth (*Malacosoma spp.*)	Pigeon (aka rock dove, *Columba livia*)[i]
Pharaoh ant (*Monomorium pharaonis*)	House sparrow (*Passer domesticus*)[i]
Paper wasp (*Polistes spp.*)	Mallard (*Anas platyrhynchos*)
	Canada goose (*Branta canadensis*)
Fishes	American coot (*Fulica americana*)
Common carp (*Cyprinus carpi*)[i]	Great blue heron (*Ardea herodias*)
Goldfish (*Carassius auratus*)[i]	American kestrel (*Falco sparverius*)
Catfish (*Ameiurus spp.*)[i]	Peregrine falcon (*Falco peregrinus*)
Green sunfish (*Lepomis cyanellus*)[i]	Screech owl (*Otus* spp.)
Bluegill (*Lepomis macrochirus*)	Vulture (*Cathartes aura* and *Coragyps atratus*)
Largemouth bass (*Micropterus salmoides*)[i]	
	Mammals
Amphibians	Opossum (*Didelphis virginiana*)
Spotted salamander (*Ambystoma maculatum*)	Eastern mole (*Scalopus aquaticus*)
Bullfrog (*Rana catesbiana*)[i]	Little brown bat (*Myotis lucifugus*)
Green frog (*Rana clamitans*)	Red bat (*Lasiurus borealis*)
Leopard frog (*Rana pipiens*)	Cottontail rabbit (*Sylvilagus spp.*)
Woodhouse's toad (*Bufo woodhouseii*)	Norway rat (*Ratus norvegicus*)[i]
	House mouse (*Mus musculus*)[i]
Reptiles	Tree squirrel (*Sciurus spp.*)
Snapping turtle (*Chelydra serpentine*)	Nutria (*Myocastor coypus*)[i]
Red-eared slider (*Trachemys scripta*)[i]	American Beaver (*Castor canadensis*)
Mediterranean gecko (*Hemidactylus turcicus*)[i]	White-tailed deer (*Odocoileus virginianus*)
Green anole (*Anolis carolinensis*)	Red fox (*Vulpes vulpes*)[i]
Fence lizard (*Sceloporus* spp.)	Gray fox (*Urocyon cinereoargenteus*)
Five-lined skink (*Eumeces inexpectatus*)	Raccoon (*Procyon lotor*)
Common garter snake (*Thamnophis sirtalis*)	Striped skunk (*Mephitus mephitus*)
Gopher snake (*Pituophis catenifer*)	Coyote (*Canis latrans*)
Rattlesnake (*Crotalus spp.*)	Mountain lion (*Felis concolor*)

[i] Introduced in some or all regions of the U.S.

Source: Landry, S. B., Houghton Mifflin Company, New York, 1994; Davis, J. M., in Haggerty, M. M. Ed., Texas Parks & Wildlife Department, College Station, 2003.

supplemental food for wildlife has become a popular hobby — and a lucrative business — in the U.S. (Hope 2000). The number of seed-eating (granivore) birds rises in urban environment due to the availability of bird feeders that provide an endless supply of easily attainable food.

Urbanization, and human actions, affects the behavior of wild species living in this type of system, too. Artificial food supplies (e.g., birdseed, garbage, pet food) reduce the need for individual animals to move around in search of something to eat. The result is smaller home ranges for urban animals compared to their rural cousins. The heat islands discussed earlier in this chapter cause a number of behavioral changes, from longer breeding seasons (Adams 1994) to the disappearance of traditional migration activity (Hope 2000). Other impacts of urbanization and human activities on the population dynamics of urban wildlife are discussed in the next chapter.

The wildlife species found in cities and towns across the U.S. are remarkably similar (Burger 1999), in spite of significant regional variations in climate and terrain. Table 3.6 lists a selection of animal species — native and introduced — commonly found living in and around human-created landscapes. Additionally, people may introduce exotic wildlife into the system, accidentally or intentionally, which can have profound effects on the introduced and native species (see Perspective Essay 3.2). The spread of invasive exotic species can be correlated with the spread of humans (Withers et al. 1998).

3.5 ECOSYSTEM FUNCTION

In order for ecosystems to function properly they need to include a critical number of community members (e.g., producers, consumers, and decomposers) as well as the basic abiotic factors, so that the relationships that drive the system can develop. With rare exceptions, every natural ecosystem (1) depends on solar energy and (2) recycles matter. In contrast, urban ecosystems depend primarily on fossil fuels while directing matter into landfills.

3.5.1 Matter

Matter is defined as anything that occupies space and has mass — that is, it can be measured where gravity is present. Matter can be changed from one form to another and it can be recycled, but it cannot be created or destroyed. **Energy**, which has no mass and does not occupy space, has the ability to move matter. Like matter, energy can be changed from one form to another, it can be measured, and it cannot be created or destroyed. Unlike matter, energy cannot be recycled.

The basic building blocks of all matter are atoms. Only 93 different kinds of atoms occur in nature, known as the naturally occurring **elements**. Of these, only 17 — called the essential elements — are directly involved in sustaining all plant and animal life on Earth (Table 3.7).

Like blocks, the elements can be combined and disassembled to build different kinds of matter, but the atoms themselves do not change during this process. This constancy of atoms is referred to as the Law of Conservation of Matter, which states, "during an ordinary chemical change, there is no detectable increase or decrease in the quantity of matter." Matter is constantly being recycled, and natural ecosystems do a good job of this through decomposers and detritus feeders, but urban ecosystems do not.

Table 3.7 Essential Elements for Sustaining Life

Carbon (C)	Nitrogen (N)	Manganese (Mn)
Hydrogen (H)	Sulfur (S)	Copper (Cu)
Oxygen (O)	Calcium (Ca)	Chlorine (Cl)
Phosphorus (P)	Iron (Fe)	Molybdenum (Mo)
Potassium (K)	Magnesium (Mg)	Zinc (Zn)
Iodine (I)	Boron (Bo)	

The pneumonic for remembering these elements is: "See (C) Hopkins (HOPKINS) café (Ca Fe) managed (Mg) by (Bo) mine (Mn) cousin (Cu) Clyde Mo Zinc (Cl Mo Zn)."

3.5.1.1 Biogeochemical Cycles (Finite Amount of Substance Made Infinite)

One of the most remarkable functions of ecosystems is to provide a process where a finite amount of matter (e.g., carbon, hydrogen, nitrogen) becomes an infinite resource for all living organisms. This is done by an extremely simple process called recycling. The simplicity of the process becomes evident when one studies how biogeochemical cycles work. First, the actors in the process are always the same, i.e., producers, consumers, decomposing bacteria and fungi, or other microbes that carry out specific functions (Figure 3.6). What is recycled is one of the essential nutrients including nitrogen, carbon, hydrogen, oxygen, phosphorus, and sulfur. Living organisms only borrow nutrients for a period of time, and then release them back into the ecosystem to be picked up by another organism. This is why it is possible that the carbon atoms in your body may have been in the body of a different animal or a plant a few months ago. It is important to understand how matter is processed in an urban when compared to a natural environment.

The human impact on any biogeochemical cycle is to overload it with too much of the raw material, e.g., carbon, nitrogen, or phosphorus. How is this done? In the

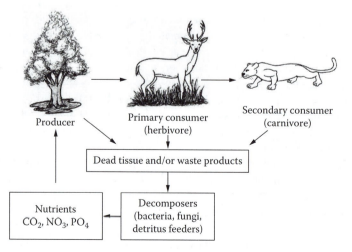

Figure 3.6 Trophic levels and recycling of nutrients.

nitrogen cycle, plants can only accept elemental nitrogen in the form of nitrate (NO_3). This is done by combining three atoms of oxygen with one atom of nitrogen. A particular assemblage of nitrogen-fixing bacteria are responsible for the union of oxygen with nitrogen, or it can be accomplished by lightning or volcanic action. Plants then use this molecular form of nitrogen to make proteins. Human agricultural and lawn care practices require large inputs of nitrate fertilizer — more than required for plant growth. This overindulgence in the use of nitrate fertilizer results in quantities of unused fertilizer being left on the land, which eventually washes into natural waterways. Overfertilizing also puts the nitrogen-fixing bacteria out of work.

The human impact on the carbon cycle is to discharge large quantities of CO_2 into the atmosphere by burning fossil fuels to produce electricity and to run the transportation industry. Plants cannot assimilate the excess CO_2. The excess CO_2 in the atmosphere is suspected of causing climatic change called "global warming," which results in higher than normal ambient temperatures. Recall that temperature is one of the primary abiotic predictors of the type of plant community that will develop in an area (Figure 3.3). Increased atmospheric CO_2 may also be the cause of higher than normal temperatures in urban when compared to rural environments.

Urban wildlife does its part to recycle matter — food wastes are a heavily exploited resource, while birds and rodents often use discarded cloth and paper as nesting materials. Humans, on the other hand, are part of a throwaway culture in which matter is used and then deposited in landfills. The introduction of additional quantities of the essential elements into ecosystems throws the system out of balance and overloads the ability of plants and animals to incorporate these materials. For example, a city with one million human residents consumes an estimated 25,000 tons of water and 2,000 tons of food per day, and produces 50,000 tons of effluent water and 2,000 tons of waste daily (Deelstra 1989). Landfills are the antithesis of functioning ecosystems. They represent a one-way-trip, a use-it-and-lose-it mentality. Given enough time, the matter contained in these mountains of waste will be broken down into other forms, but the process takes much longer than under more natural conditions.

Many city governments have begun to realized the value of landfill garbage and have developed waste-to-energy (WTE) facilities to extract the recyclables and use combustible materials to produce electricity. In fact, several states import large quantities of garbage from elsewhere just to maintain operations at their WTE facilities. Chapter 6 offers a detailed account of how landfill waste or garbage left within urban ecosystems becomes a valuable resource and attractant for wildlife.

Another way in which humans impact natural systems is by introducing elements that are not included in the list of essential elements, such as mercury (Hg), lead (Pb), aluminum (Al), as well as chemical compounds like pesticides (e.g., DDT, PCB) that cannot be assimilated or tolerated by living organisms. Chapter 4 examines the impacts of these selected environmental pollutants on the population dynamics of wildlife populations.

3.5.2 Energy

Of the solar energy that is intercepted by the Earth's atmosphere, only 0.8% is used for photosynthesis (Odum and Barrett 2004). Solar energy absorbed by green plants is transformed into chemical energy by photosynthesis. Photosynthesis allows sunlight energy to become available to almost every aquatic and terrestrial ecosystem.

We owe our lives to the sun...How is it, then, that we feel no gratitude?

Lewis Thomas, Earth Ethics, Summer 1990

While sunlight is the foundation of natural ecosystems, urban environments make little use of this infinite energy supply. Instead, the primary sources of energy for man-made ecosystems are fossil fuels (e.g., coal and gas), finite energy sources that have adverse effects on terrestrial and aquatic ecosystems around the globe.

As was mentioned earlier in this chapter, the fundamental difference between matter and energy is that energy cannot be recycled. No matter what path it takes through the ecosystem, ultimately energy is lost from the system in the form of heat. This fact is in keeping with two natural laws: (1) the First Law of Thermodynamics states that energy can be changed from one form to another, and (2) the Second Law of Thermodynamics states that there will be a loss of energy during each energy conversion. Without a steady supply of energy (kinetic or potential) in the form of sunlight, every biotic component in natural and urban ecosystems would spontaneously move toward increasing **entropy** (disorder, death, and decay).

Imagine you have two wooden desks. One desk is kept inside, protected from factors that cause entropy (and imagine, while you're at it, the amount of energy required to do this!). The other desk is left out in your backyard with no protection from weather, wildlife, children, etc. Can you predict what would happen to the two desks over time? Of course, both desks will eventually succumb to entropy, but the one in the yard will do so much more quickly. In fact, no structural component of natural and urban ecosystems will last forever without some form of energy intervention to sustain its form.

Urban ecosystems are unsustainable in the long term because, in addition to their need for a steady supply of solar energy, they can only function with a continuous new supply of matter. Urban systems quickly decay without heavy subsidies of energy, water, and nutrients in the form of fertilizers provided by urban residents. Urban decay is prevalent in any large metropolitan area where an eroding tax base has left certain areas nearly devoid of the essential goods and services required to sustain the structural and functional integrity of the neighborhood.

3.5.3 Thermodynamics and Conservation of Matter

One way to illustrate how matter and energy are processed in natural ecosystems is to examine what happens with coal when it is ignited. A lump of coal represents a stored form of potential energy resulting from photosynthesis millions of years ago. By applying a catalyst (fire) the potential energy inside the coal is transformed into kinetic energy, as well as the original chemical elements of carbon, hydrogen, oxygen, and sulfur:

$$\text{Coal (CSHO)} + \text{Fire} \rightarrow CO_2\uparrow + SO_x\uparrow + H_2O\uparrow + \text{Ash}\uparrow + \text{Light} + \text{Noise} + \text{Heat}$$

The kinetic energy from burning the lump of coal consists of heat, light, and noise. The thousands of carbon, hydrogen, sulfur, and oxygen atoms contained in this coal sample are transformed by fire into carbon dioxide (CO_2), sulfur oxide (SO_x), water (H_2O), and the incompletely burned residue, referred to as ash. Note that energy has been lost, but matter was conserved, albeit in the form of other chemical compounds. The excess CO_2 is released into the atmosphere and may be picked up by plants in another round of photosynthesis, or the molecule may accumulate and cause the global warming phenomenon mentioned earlier. The SO radical can combine with H_2O in the atmosphere and cause a serious abiotic change called "acid rain." Acid rain is known to cause the destruction of aquatic and terrestrial ecosystems in urban and rural areas. The ash, unless captured, becomes part of the particulate matter in the atmosphere which is believed to cause some of the urban climatic effects illustrated in Table 3.3.

3.6 ECOSYSTEM SERVICES

Ecosystem services are ecological processes that produce, directly or indirectly, goods and services from which humans benefit. Nature provides "life support services" at virtually every scale, and many are free and irreplaceable by technology. The list of ecosystem services is extensive. Norberg (1999) listed 29 specific goods or services provided by nature. These goods and services were categorized according to certain ecological criteria including:

1. Goods or services associated with certain species or a group of similar species and for which the goods or the target of the service is internal to the ecosystem, e.g., foods, timber, pharmaceuticals, pollination, and predators;
2. Services that regulate some exogenous chemical or physical input, i.e., the system itself cannot alter the magnitude of the input, e.g., processes that drive material and energy flows in ecosystems;
3. Organization of biological processes at virtually all scales, i.e., from the way gene sequences are organized to networks of energy and material flows at the level of ecosystems (Norberg 1999).

In addition to housing the genetic library for all life on earth, organisms help to sustain a flow of ecological services that are prerequisites for human economic activities (Folke et al. 1996). There is some survival value for the urban resident in knowing what types of goods and services are provided by nature (Table 3.8), because there are losses or changes in these goods and services during the urbanization process. Losses or changes in ecosystem services (Table 3.9) can have significant negative impacts on the quality of life for humankind on local and global scales. However, awareness of the existence of a particular good or service is not sufficient. "The dynamics of the ecosystem(s) generating and sustaining a good or service, along with links to other systems, to energy, biogeochemical, and hydrological flows, and to human activities have to be addressed" (Limburg and Folke 1999). In addition McDonnell et al. (1997:23) stated that "To understand the ecology of urbanization, the individual components of ecosystem structure, physical

Table 3.8   The Tangible and Intangible Contributions of Nature

GOODS

Breathable air	Building materials (e.g., timber, stone)
Drinkable water	Fuel/energy (e.g., oil, natural gas, wood, nuclear)
Nutrition	Pharmaceuticals
wild foods (terrestrial and aquatic)	Other chemicals and minerals
cultivated foods (terrestrial and aquatic)	Art/ornamental objects
plant and animal species for domestication	Genetic information
Clothing (both plant- and animal-derived)	

SERVICES

Tangible

Intangible

Biotic
  photosynthesis
  pollination
  recycling of nutrients
  decomposition and waste assimilation
  population growth and control
  natural selection
  adaptation
  resistance to invasion of foreign species
  succession
  denitrification
  erosion prevention
  soil generation and preservation
Abiotic
  maintenance of ecosystem structure
  nutrient regeneration/recycling
  water purification
  detoxification/filtering of pollutants
  maintenance of atmospheric composition
  climate moderation
  UV protection by ozone
  mitigation of floods and droughts
  resilience after catastrophic events

Enjoyment/recreational opportunities
Educational opportunities
Aesthetic appreciation
Cultural inspiration
Historic appreciation
Religious inspiration

Source: Folke, C. et al., *Ecological Applications,* 6:1018–1024, 1996; Norberg, J., *Ecological Economics,* 29:183–202, 1999.

and chemical environments, populations, communities, and ecosystems need to be studied in order to appreciate the ecologically important impacts of urban development and change on natural areas."

However, in urban landscapes, humans (*Homo sapiens*) are the dominant species. The human impact on nature's services cannot be fully understood without the integration of ecology, sociology, political agendas and the decision-making process, and natural resource economics. This integration can be explained in terms of how people prioritize their resource and service needs and then act upon them through patterns of consumption, investment, resource use, waste production, and

Table 3.9   Environmentally and Biologically Relevant Effects of Urbanization

Physical and Chemical Environment	Population and Community Characteristics	Ecosystem Structure and Function
Air pollution	Altered assimilation rates	Altered decomposition rates
Hydrological changes	Altered biological cycles	Altered nutrient retention
Local climate change	Altered disturbance regimes	Greater nutrient flux
Soil changes and movement	Altered reproductive rates	Increased habitat patches
Water pollution	Altered succession rates and direction	Habitat fragmentation and simplification
	Altered survival rates	Increased debris dams
	Changed growth rates	Increased sediment loading in streams
	Genetic drift and selection	Loss of biological diversity
	Introduced species	Loss of biological organization
	Landscape fragmentation	Loss of forest understory
	Population size and structure	Loss of plant productivity
	Reduced species diversity	Loss of redundant pathways for nutrient and energy transfer
	Social and behavioral changes	

Source: McDonnell, J. J. and Pickett, S. T. A., *Ecology,* 71:1232–1237, 1990.

disposal decisions — which in turn affect ecosystem energy and mass flow into, through, and from urban ecosystems. In this regard, consideration of the effects of local economics, public policy, law, diversity of stakeholder perspectives, human dimensions, land-use planning decisions, and problem-solving skills become important. This is the conceptual framework we used to develop the remaining chapters in this book.

This book is not just about human and wildlife interactions in the city. The tentacles of urban sprawl extend well beyond the city limits by way of highways and associated structures, power production and transmission, and the mass communication industry. As such, the impacts of urbanization on the exurban environments, habitats, and wildlife need to be part of the story told in this book. Examples of this extended impact are provided in subsequent chapters on green and gray spaces, population dynamics, and urban streams and soils, among others.

PERSPECTIVE ESSAY 3.1: FOR THE LOVE OF LAWNS

Why does our society seem to be in love with the monotonous, sterile lawn? To understand this affinity for turf grass, it can help to examine our culture's history. According to Warren Schultz's book *A Man's Turf: The Perfect Lawn* (1999), small strips of lawn began to show up in formal gardens in Europe during the seventeenth and eighteenth centuries. It was also during this time that vast expanses of short grass began to show up in the European countryside. Originally, these proto-lawns were kept grazed by flocks of sheep. The larger the expanse, the more sheep the landowner had, thereby demonstrating his wealth to the neighbors. Vast expanses of short grass became a symbol of wealth and power — yet another example of the human tendency to one-up the Joneses.

Many of the early U.S. colonists came from this cultural influence; George Washington was one of the first to install a lawn around his home (Schultz 1999). However, the lawn as we know it today didn't become popular until the late nineteenth century. The development of "improved" turf grass varieties later became a goal of the golf industry, and the invention of the lawn mower made an expanse of grass available to the common homeowner.

Today, the lawn is a symbol of territory, a "green moat" around our homes. Schultz (1999) says that a well-kept lawn has come to indicate that the man of the house is powerful, in control of nature, and is taking care of business at home. He also states that a well-kept lawn announces to our neighbors that we are abiding by the rules of society.

If you doubt this is the case, think about the typical urban landscape. It is really quite predictable. Just about every home or business will have a row of evergreen hedges around the perimeter of the building, called "foundation plantings." Unwanted views usually are "screened" with a row of evergreen hedges as well. With the exception of newly constructed developments, there will be large shade trees and some kind of groundcover, turf grass being the most common of these. Often you will see a flower bed of some sort at the base of trees and along walkways, usually containing nonnative annuals and perennials to add color. The landscape is dominated by vast expanses of mowed turf grass. If traditional landscaping were ecologically sensitive, or even neutral, these practices would be a harmless cultural quirk, but the typical suburban lawn and garden creates a myriad of ecological problems.

As mentioned earlier, exotic plant species often are used in these settings. Because they are not native to the area, they tend to require more intensive maintenance to survive (ironically, often these same plants are advertised as "easy care"). Those species that do well can escape the confines of the yard and compete with endemic species. As you'll remember from earlier in this chapter, a region's plants determine the animal species that live there. As native plants are lost, food and shelter resources are lost as well; this can have an adverse effect on resident wildlife.

Maintaining the perfect lawn is ecologically expensive. Exotic plants generally require more moisture than native landscapes, which puts more pressure on community water supplies. Turf grass needs pesticides, herbicides, and fertilizers to achieve the preferred deep green color and lush texture, all of which significantly decrease water quality. According to Schultz (1999), 70 million pounds of chemicals are applied to lawns in the U.S. each year! The majority of these chemicals are washed into local streams and reservoirs by rain and sprinkler systems. The deleterious effects of these chemicals on the population dynamics of aquatic and terrestrial wildlife are discussed in Chapters 4 and 7. A "healthy" lawn endangers more than our water supply. Running a lawn mower for one hour releases as much hydrocarbon into the air as driving a car for 11.5 hours (Schultz 1999).

John M. Davis 2003

PERSPECTIVE ESSAY 3.2: PEOPLE AND WILDLIFE — THE LESSER ANTEATER (*TAMANDUA TETRADACTYLA*)

The animal shown in Figure 3.7 is an anteater and lives in the Amazon basin and the forests of northern South America. It goes by several common names. Its primary food sources are arboreal ants and termites. Lesser anteaters rip open bark with a third toe adapted for this function and use their sticky long tongues to lap up their food.

Given this animal's normal geographic range and food habits, it was quite a surprise to find one hiding under a bush of a residential home in Seward, Nebraska, in September 1968; Seward is several thousand miles north of the anteater's normal habitat. The events that transpired after the discovery of the anteater proved to be interesting, informative, and even a little bizarre.

The story began on a Sunday afternoon when my neighbor, a music teacher, called me, a biology teacher, and told me that there was a female anteater in his backyard. I was skeptical to say the least. I went over to his house and, sure enough, there was an anteater in his backyard!

It was cold outside, so I quickly picked up the anteater to take it to my laboratory at Concordia College. In so doing, I lost the left sleeve of my coat to the animal's strong prehensile third toe. I placed the anteater in a chicken wire cage, only to observe her tear it apart with her front toes. I then placed her in a more secure cage made of steel.

Now I had to figure out what to feed her. I called the Henry Dorley Zoo in Omaha, Nebraska. I told them I had an anteater — I even gave them the scientific name — and explained that I needed to know what to feed it. I also asked if they would be interested in taking the animal off of my hands. They were skeptical of my identification, but they gave the recipe for anteater gruel (which consisted of a mixture of egg yolk, dog food, milk, and a few other ingredients tossed into a blender to form a soup-like concoction — which the anteater subsequently

Figure 3.7   The lesser anteater (*Tamandua tetradactyla*). (Source unknown)

ignored. The Henry Dorley Zoo referred me to the Children's Zoo in Lincoln, Nebraska.

I called the Children's Zoo and told them I had an anteater. They were skeptical of my identification also — even though I was a biology teacher, I did not command a great deal of respect with zoo personnel in the taxonomy department. Nevertheless, they agreed to pick up the animal and I warned them to bring a strong cage. It took about a week for personnel from the Children's Zoo to arrive. When they did, they were quite surprised to find out that I really did have an ANTEATER!

The fact that an anteater was found in Seward, Nebraska, was big news locally and statewide. The local newspaper did a story on her, statewide news services picked up the story, and television stations in Lincoln and Omaha reported on the event. I believe that nearly every child attending elementary school in Seward had a field trip to my lab to see the anteater. I was holding the anteater for the children so they could see how strong she was. I forgot about that third toe and again lost a sleeve on my suit coat.

The nagging question of how the anteater ended up in Seward, Nebraska, all the way from South America remained unanswered. My neighbor's house was just across the street from a new church, and the anteater was found on the same day the building was being dedicated. Many people from all over Nebraska and other states were at the event, so maybe the anteater hitched a ride!

Eventually, the local police department received a call from a lady in Chicago, Illinois, who claimed the anteater was hers. She had left it in her car at the Lincoln airport to fly to Omaha for a one-day meeting. The police asked for her address in Chicago. They then contacted the Chicago police to verify the address — they found a vacant lot! They never heard from the lady again. Anteaters are not on the list of exotic pets desired by humans, so the lady incident just made no sense.

The Children's Zoo adopted the anteater and called her "Connie" after Concordia, which was nice, but Connie died soon after her adoption. An autopsy determined the cause of death to be malnutrition and heavy parasite loads in her intestines.

Readers may assume this story is a bizarre and uncommon tale, but it isn't all that unusual. A student of mine who worked at a wildlife rehabilitation center in Houston, Texas, has many stories to tell of the exotic animals that often ended up at the facility. There seems to be no limit to the appeal that exotic pets have for so-called animal lovers, but the novelty can wear off quickly as unexpected expenses rise and the wild "pets" refuse to settle into domestic bliss. Some of the species that were found by the public and delivered to the wildlife shelter during this student's two-year tenure with the nonprofit organization included: African pygmy hedgehogs (*Atelerix albiventris*), green iguanas (*Iguana iguana*), red-tailed boa constrictors (*Boa constrictor imperator*), Burmese pythons (*Python molurus bivittatus*), a Moluccan cockatoo (*Cacatua moluccensis*), a coatimundi (*Nasua nasua*), and a pet black bear (*Ursus americanus*).

One of three things usually happens when exotic animals are introduced to a new habitat, on purpose or unintentionally. In one case, the animal is completely unsuited for the environmental conditions it finds itself in, and it either dies of starvation, predation, or disease. In the second case, the animal has lost its fear of humans during previous captivity, and may even associate humans with food and shelter, so it may find its way into someone's yard. At this point, the local wildlife rehabilitation center or nature center or zoo is likely to get a call from an astonished citizen who starts the conversation by saying something like, "You're not going to believe this, but there's an anteater in my backyard!"

The last scenario is somewhat less common, but has a greater long-term impact. Some translocated animals find themselves ideally suited to their new habitat, so much so that they may even begin to outcompete some native species. Examples of wild species that have been intentionally or accidentally introduced to North America include everything from the house sparrow (*Passer domesticus*) and the European starling (*Sturnus vulgaris*) to Mediterranean geckos (*Hemidactylus turcicus* Linnaeus) and fire ants (*Solenopsis wagneri*).

Clark E. Adams

REFERENCES

Adams, L. W. (1994). *Urban Wildlife Habitats: a Landscape Perspective*, University of Minnesota Press, Minneapolis.

Begon, M., Harper, J. L., and Townsend, C. R. (1996). *Ecology: Individuals, Populations and Communities*, Blackwell Science, Oxford.

Borman, S. and Korth, R. (1998). *Through the Looking Glass: a Field Guide to Aquatic Plants*, University of Wisconsin-Stevens Point Foundation, Stevens Point.

Brockman, C. F. (1986). *Trees of North America: a Field Guide to the Major Native and Introduced Species North of Mexico*, Golden Press, Racine, WI.

Burger, J. (1999). *Animals in Towns and Cities*, Kendall/Hunt Publishing Company, Dubuque, IA.

Cushman, R. C. and Jones, S. R. (2004). *Peterson Field Guides: the North American Prairie*, Houghton Mifflin, New York.

Davis, J. M. (2003). Urban systems, In *Texas Master Naturalist Statewide Curriculum*, 1st ed., Haggerty, M. M. Ed., Texas Parks & Wildlife Department, College Station.

Deelstra, T. (1989). Can cities survive: solid waste management in urban environments, *AT Source*, 18:21–27.

Folke, C., Holling, C. S., and Perrings, C. (1996). Biological diversity, ecosystems, and the human scale, *Ecological Applications*, 6:1018–1024.

Hope, J. (2000). The geese that came in from the wild, *Audubon*, 102:122–126.

Landry, S. B. (1994). *Peterson First Guides: Urban Wildlife*, Houghton Mifflin Company, New York.

Limburg, K. E. and Folke, C. (1999). The ecology of ecosystem services: introduction to the special issue, *Ecological Economics*, 29:179–182.

McDonnell, M. J. and Pickett, S. T. A. (1990). Ecosystem structure and function along the urban-rural gradients: an unexploited opportunity for ecology, *Ecology*, 71:1232–1237.

McDonnell, M. J., Pickett, S. T. A., Groffman, P., Bohlen, P., Pouyat, R. V., Zipperer, W. C., Parmelee, R. W., Carreiro, M. M., and Medley, K. (1997). Ecosystem processes along an urban-to-rural gradient, *Urban Ecosystems*, 1:21–36.

Norberg, J. (1999). Linking nature's services to ecosystems: an ecological perspective, *Ecological Economics*, 29:183–202.

Odum, E. P. and Barrett, G. W. (2004). *Fundamentals of Ecology*, 5th ed., Thomson Brooks/Cole, Belmont, CA.

Ostfeld, R. S., Keesing, F., Jones, C. G., Canham, C. D., and Lovett, G. M. (1998). Integrative ecology and dynamics of species in oak forests, *Integrative Biology,* 1:178–186.

Ostfeld, R. S., and Keesing, F. (2000). Biodiversity and disease risk: the case of Lyme disease, *Conservation Biology,* 14:722–728.

Pickett, S. T. A., Burch, W. R., Jr., Dalton, S. E., Foresman, T. W., Grove, J. M., and Rowntree, R. A. (1997). A conceptual framework for the study of human ecosystems in urban areas, *Urban Ecosystems,* 1:185–201.

Schultz, W. (1999). *A Man's Turf: The Perfect Lawn,* Clarkson Potter, New York.

Sullivan, R. (2004). *Rats: Observations on the History and Habitat of the City's Most Unwanted Inhabitants,* Bloomsbury, New York.

Wilson, E. O. (1992). *The Diversity of Life.* Belnap Press, Cambridge, MA.

Withers, M. A., Palmer, M. W., Wade, G. L., White, P. S., and Neal, P. R. (1998) Changing patterns in the number of species in North America floras. In Sisk, T. D. (Ed.) *Perspectives on the Land Use History of North America: A Context for Understanding Our Changing Environment.* U.S. Geological Survey, Washington, DC.

Wright, R. T. (2004). *Environmental Science: Toward a Sustainable Future,* 9th ed., Pearson Education, Upper Saddle River, NJ.

Zipperer, W. C., Wu, J., Pouyat, R. V., and Pickett, S. T. A. (2000). The application of ecological principles to urban and urbanizing landscapes, *Ecological Applications,* 10:685–688.

CHAPTER **4**

Principles of Population Dynamics

If any population was miraculously allowed to grow exponentially indefinitely it would eventually expand outward at the speed of light and come to weigh as much as the visible universe.

Author Unknown

CONTENTS

Population dynamics is the study of those concepts that predict or explain an animal's presence, abundance, and concentrations in a particular habitat. This chapter focuses on those principles of population dynamics that moderate species presence, abundance, and concentrations in urban communities. When studying a wildlife population, scientists may ask the following broad questions:

1. Can individuals of the population survive in the particular habitat (i.e., urban ecosystems)?
2. What adaptations does the species have that enhance its survival in the habitat?
3. What is the average density of the population in this type of habitat?

4. What abiotic and/or biotic factors maintain and/or precipitate changes in the average population density over time?

The final section of this chapter examines the impacts of human development on the population dynamics of urban wildlife species. These impacts are discussed in terms of habitat fragmentation (i.e., habitat patches), including the influence of corridors (issues not discussed in Chapter 5), supplemental feeding, animal damage control, and introduction of toxic elements into the urban environment.

4.1 SURVIVAL

When a wildlife population is found in an urban habitat (or any habitat, for that matter), two definitive statements can be made. First, at least some of the individuals that make up the population are able to tolerate the conditions found in the habitat. Second, some individuals within the population are able to acquire enough resources to survive and reproduce. However, reduced predation and supplemental feeding can alter the natural sequence of events listed (section on supplemental feeding below). Before we discuss these statements further, let's review the definitions of *conditions* and *resources* in ecological terms (see Chapter 3). Conditions, as you no doubt remember, are those abiotic factors that vary in space and time; temperature and wind speed are examples of conditions. Resources are those things that are consumed or used by the organisms, such as food and nesting sites (Begon et al. 1996).

Few, if any, vertebrate species are equipped to survive the entire range of conditions found on earth. Consequently, the geographic distributions of species are limited to particular areas. However, many wildlife species that inhabit urban areas usually have large geographic distributions, sometimes close to worldwide distributions (e.g., Norway rat, *Rattus norvegicus*). This pattern indicates that these species have a wide range of tolerance for various abiotic conditions. Ranges of tolerance for different conditions will vary between species. For example, a specific temperature range will predict whether brook trout (*Salvelinus fontinalis)* or the common carp (*Cyprinus carpio*) are likely to be found in an aquatic ecosystem. Additionally, water temperature will predict the amount of dissolved oxygen in the aquatic ecosystem — the lower the temperature, the greater the oxygen concentration.

Trout have a narrow range of tolerance to both water temperature and oxygen concentration; they need cold water and high levels of dissolved oxygen. On the other hand, the common carp has a broad range of tolerance to both temperature and dissolved oxygen. These relations are illustrated in Figure 4.1A and Figure 4.1B.

One can predict the loss of the trout from a stream if the temperature increases or oxygen decreases beyond the trout's narrow range of tolerance for these two variables. A carp, on the other hand, having a broad range of tolerance for extreme changes in temperature and oxygen concentration, is far less affected than the trout and would likely continue to exist in the changed environment. In general, those species that survive well in urban ecosystems are like the common carp in their ability to survive wide ranges and shifts in abiotic conditions.

If an organism can tolerate the abiotic conditions, then it must still be able to obtain enough resources to survive. Further, in order for the population to persist in

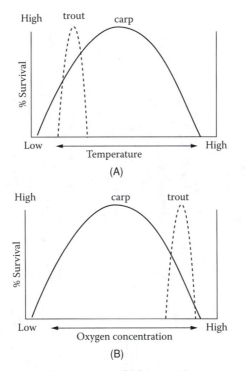

Figure 4.1   (A) Temperature tolerance curve. (B) Oxygen tolerance curve.

the habitat, at least some of its individuals must access enough resources to devote energy to reproduction. Several species of urban wildlife are masters at accessing and using the resources. For example, chimney swifts (*Chaetura pelagica*) are so named because they commonly use chimneys as nesting sites during the summer and as roosting sites during their spring and fall migrations. Populations of these birds likely initially increased as a result of European settlement in North America because they were able to both tolerate the conditions of urban areas and readily use the available resources.

4.2 ADAPTATIONS

Whether an organism lives or dies is based on the types of genetic adaptations it has to enhance its survival and have priority access to resources in order to reproduce. The types of adaptations an organism has are reflected partly in how it looks and how it behaves. For example, the teeth of a coyote and a beaver are adaptations to carnivore and herbivore diets, respectively (Figure 4.2A and Figure 4.2B). The canine and carnassial teeth of the coyote are carnivore adaptations for grabbing and holding prey and then biting off large chunks of meat. The large incisors and molars of the beaver are herbivore adaptations for chiseling off and grinding up woody plant material. The stalking ability of a coyote and the innate nervous twitch of a rabbit are behavioral adaptations for catching prey and avoiding capture, respectively.

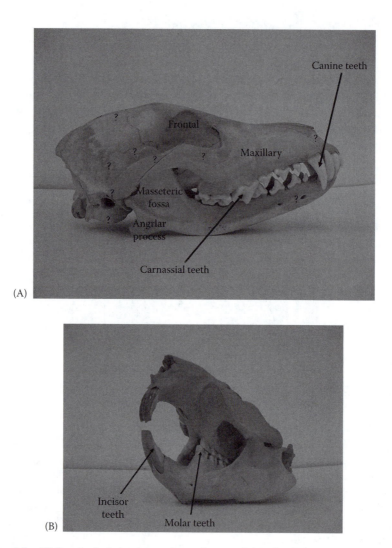

(A)

(B)

Figure 4.2   (A) Coyote skull showing canine and carnassial teeth adaptations for meat-eating. (B) Beaver skull showing large incisors and molars adaptations for eating woody plant material. (C. E. Adams.)

The process of natural selection is illustrated by any animal that survives in its environment and produces viable offspring. In the urban ecosystem, the process of natural selection usually works against species that were present prior to the urbanization process. Recall that the process of urbanization results in a destabilization and simplification of natural ecosystems. Many endemic species are selected against because they are specialists (require special conditions for survival, e.g., food, habitats, plant and animal community interrelationships) that are usually lost during urbanization.

The environment that an organism lives in "tests" its genetic adaptations, determining whether it is fit for that particular environment. Once the minimum criteria for survival have been met, the next challenge is to obtain sufficient resources to reproduce. Reproduction is the process through which genes are introduced into the ecosystem to be tested. Successful reproduction allows the individual the opportunity to pass along its adaptive genetic code to a new generation. In time, the adapted species, by sheer numbers, may force the less adapted species out of the habitat. This scenario is played out constantly and consistently in urban ecosystems.

There are several general adaptations that urban wildlife must have to survive and reproduce in urban environments. These adaptations include the following:

1. *Generalists or Specialists:* Urban wildlife may be generalists able to use a wide assemblage of food sources and shelter, and recognize alternative opportunities for survival in the structure of urban ecosystems. On the other hand, urban wildlife can be fortunate specialists that need and have access to a resource readily available in urban environments; e.g., hummingbirds.

2. *Tolerant of human presence or active when humans are not:* Animals that are successful in the urban ecosystem must become habituated to human activity; in other words, humans are not seen as a threat. In fact, some urban coyotes have grown so familiar with the presence of humans that they will prey on their pets in broad daylight and in the presence of pet owner(s). Urban raccoons are notorious for launching nighttime raids on pet food dishes or garbage cans left unprotected. There are even incidences of cougars stalking and sometimes attempting to prey on urban joggers. White-tailed deer will forage on lawn shrubs and gardens while urban residents are mowing their lawns. It is this tolerance of human presence that allows people to enjoy wildlife viewing in urban/suburban environments.

Urban wildlife have been categorized based on the ability of various species to adapt to urban habitats, or the lack thereof. A taxonomy developed by Blair (1996) was based on the average daily densities of all bird species in a business district, office park, residential area, golf course, open-space recreation area, and natural preserve. Avian species were grouped into "urban exploiters" (three species including pigeons, swifts, and house sparrows) — those species are adept at exploiting the ecosystem changes caused by urban sprawl. This group reaches its highest densities in developed sites and usually represent a very small subset of the world's species.

Another category was the "urban avoiders," which are particularly sensitive to human-induced changes in the landscape. In part, urban avoiders include forest-interior (especially insectivorous) birds that disappear quickly in the initial stages of suburban encroachment. This group reaches their highest densities in natural areas.

A final category was those birds that are "suburban adaptable." This group is able to exploit the additional resources (e.g., ornamental vegetation) that accompany moderate levels of development. This category contained thirty avian species which were found in all or several sites. Suburban adapters — mammals and birds that are mainly adapted to forest edges and open areas — flourish in suburban habitats, especially older subdivisions where ecological succession has produced extensive vegetation. One of the most important traits that separates the three categories is the extent to which species depend on human-subsidized resources to exist in an area (Blair 1996; McKinney 2002).

Finally, there are several characteristics that distinguish urban from rural wildlife communities. A community is defined as all the populations of wildlife living in a particular area. For example, when compared to rural communities, urban wildlife communities tend to:

1. Demonstrate reduced species diversity,
2. Consist of species without a common evolutionary past which have not been found elsewhere in such a combination,
3. Be represented by more exotic or introduced species than native species,
4. Have many generalist species rather than specialists in utilizing the resources in urban habitats,
5. Demonstrate changed behavioral patterns between species and within species groups (Adams 1994; Rebele 1994; Blair 1996; Blair and Launer 1997; Kloor 1999; Alberti 2003).

4.3 DENSITY

In general, species that live in *both* rural and urban habitats (e.g., raccoons) typically reach higher densities in urban areas. However, urban wildlife population densities follow some general patterns seen in other ecosystems. First, populations of species that feed lower on the food chain (e.g., herbivores) are higher in density than populations that feed higher on the food chain (e.g., carnivores). This occurs because energy transfer between the different trophic levels is usually between 2 and 24% efficient (Pauly and Christensen 1995). Consequently, less energy for reproduction is available to the populations that feed at higher trophic levels. However, an urban ecosystem can support higher densities of mice and hawks because of the available levels of energy to each population.

A second pattern of population densities that is similar between urban and natural ecosystems is the dispersion pattern of individuals in the population. Urban populations exhibit what is referred to as clumped dispersion patterns, meaning the individuals making up the population are in closer proximity to each other than would be expected at random. The main factor in determining this dispersion pattern in urban areas is the distribution of resources. As stated before, food sources in urban areas are often clumped, resulting in the attraction of several individuals to a common location. This phenomenon has important consequences for both humans and wildlife. For example, urban residents commonly place multiple hummingbird feeders in close proximity to each other to maximize observations of the delicate hummers. This dispersion of food sources for the hummingbirds is more clumped, richer and consistent than would be found in natural habitats. An unfortunate consequence of this phenomenon is increased fighting between the male hummingbirds for access to these food sources, which can result in serious injuries or even death (Personal observation, Lindsey).

However, it is important to note that urban ecosystems are significantly different from natural ecosystems in at least two respects: the amount of anthropogenic garbage and intentional supplemental feeding that serves as a food source for some wildlife populations. As a result of these artificial food sources, omnivore populations (e.g., opossums, rats, coyotes) and many seed-eating birds probably maintain higher densities in urban areas than in natural ecosystems, although this hypothesis has not been definitively tested.

4.4 FACTORS AFFECTING POPULATION DENSITIES

The actual number of any species of plant or animal within ecosystems is controlled by the opposing factors that either increase or decrease the numbers of individuals in the population. Populations will increase in density by reproduction or via immigration (moving in) of individuals into the population. Populations will decrease in density by mortality or emigration (dispersal or moving out) of individuals from the population. Each species has its own reproductive potential for increase and potential for longevity. As described before, natural selection has shaped different strategies of reproduction and survival for various species. For example, small mammals tend to reproduce at very young ages and produce relatively high numbers of offspring at each reproductive episode. However, this high reproductive potential is offset by relatively short life spans and high mortality rates of the young. In contrast, larger mammals are usually older at sexual maturity and produce fewer offspring at each reproductive episode, but tend to have longer life spans and lower mortality rates of the young. (There are, of course, exceptions to this general rule.)

But what factors control how much reproduction and/or mortality occurs at a given time in each population? The most common examples of factors that limit growth of populations by changing reproductive and/or mortality rates are food supply, nesting/denning sites, predators, diseases, and climate.

In general, urban habitats greatly reduce limiting factors for some species. What draws some wildlife species to urban areas is an abundance of food (a free lunch) and water, alternative shelters, and lack of natural predators. A variety of virtually limitless food sources are presented in the form of garbage, pet foods, bird feeders, pets, fish and amphibians in backyard ponds, gardens, orchards, and landscape plants. In 2001, 52.6 million U.S. residents fed wild birds, and 18.8 million reported feeding other wildlife as well (U.S. Department of the Interior 2001). Urban food resources are consistent, rich, and clumped, which leads to year-round urban wildlife populations. As a result, the numbers of some urban wildlife species have increased at nearly exponential rates (e.g., resident Canada geese, urban white-tailed deer, gulls, pigeons, blackbirds, starlings, rats, and mice). For example, in the early 1960s it was estimated there were only 50,000 giant Canada geese left in all of North America. In 1998, the figure stood at over 2 million in the eastern U.S. alone (see Chapter 12).

Sometimes food is so abundant in urban areas that further supplementing the food sources would not change the numbers of animals in any significant way. For example, Calhoon and Haspel (1989) demonstrated that increasing supplemental food of urban cats in Brooklyn, New York, did not affect numbers of cats. This finding indicated that above a certain level, food was not a limiting factor to an increase in the population. However, other limiting factors might have existed (e.g., climate changes; shortages of water, shelter, mates).

The two population growth patterns (how densities change over time) are illustrated in Figure 4.3. The J-curve demonstrates exponential population growth under optimal conditions. The S-curve shows a population at equilibrium constrained by factors that limit additional growth. Some species of urban wildlife are increasing exponentially with no end in sight (i.e., the J-curve in Figure 4.3). However, limiting

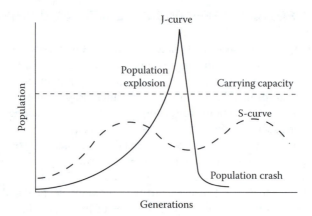

Figure 4.3   Population growth curves.

factors place constraints on population growth, preventing sustained exponential growth.

The J-type growth curve, characteristic of some urban wildlife, represents a population explosion; members of a population produce more offspring than can be adequately supported by the urban ecosystem, also known as the **carrying capacity**. For example, urban white-tailed deer have a propensity to produce more twins than single offspring in urban compared to rural ecosystems (video production titled *Suburban Deer Management*, Media Services, Cornell University). This results in deer populations that cannot find adequate food resources to meet the nutritional needs of all herd members. This results in overgrazing on all plant resources, very small deer from undernourishment or starvation, and high localized deer densities (indicated by high numbers of road kills) and should lead to the massive population crash illustrated in Figure 4.3. However, barring disease, a serious crash is unlikely to happen in the near future, given the steady inputs of nutrients and the protectionist attitudes and behavior by some human residents of urban ecosystems. Disease can have a much more devastating effect on a highly concentrated population within an urban environment.

There are few natural predators for some of the most densely populated species of urban wildlife except the automobile. As pointed out earlier in this book, the few predators that exist may not engage in a normal predator/prey relationship (e.g., urban cats are less likely to prey on urban rats). Coyotes and cougars are the largest predators in urban ecosystems, but they often opt for prey species that are naïve and/or ill-adapted to win the predatory encounter, e.g., pets.

Any population of animals with the above characteristics can potentially become so numerous in urban communities that they can qualify as a "nuisance" species. A nuisance species is defined as "any animal that has a negative impact on human health or economics." For example, the top three bird nuisance species in urban areas were pigeons, blackbirds, and starlings (Fitzwater 1988). Raccoons, coyotes, and skunks were the most common nuisance mammals (Table 2.1). Chapter 12 analyzes the complexity of the pest species management issues associated with resident Canada geese and urban white-tailed deer.

The most significant population controls on urban wildlife are the wildlife damage management programs conducted by state, Federal, and private organizations. The task of confronting wildlife damage problems in urban communities is an overwhelming obligation, resulting in a scenario of winning a few battles, but certainly not the war. As pointed out in Chapter 2, state wildlife agencies have neither the budget, personnel, nor the tradition to engage in the daunting task of urban wildlife management. Most urban wildlife species have been given the opportunity to grow and prosper, and they are!

4.5 EFFECTS OF HABITAT FRAGMENTATION ON POPULATION DYNAMICS

One of the most profound effects of urban development on the population dynamics of urban wildlife is habitat fragmentation that creates islands of habitat or "patches" in urban areas (discussed in detail in Chapter 5). Habitat patches (or islands) vary in size, edge circumference, shape, vegetative makeup, connectivity to other patches (i.e., corridors), and surrounding land use patterns. The selective pressures imposed by habitat patch characteristics results in variations in the presence, abundance, and densities of animals within urbanized areas. In fact, habitat patches affect animal behavior, reproductive patterns, survivability, immigration and emigration or dispersal capabilities, and foraging activity.

From the standpoint of population dynamics, it is reasonable to expect repetitive "boom and bust" (the J-shaped growth pattern, Figure 4.3) cycles of population growth by those species occupying habitat patches if:

1. There are fluctuations in the availability of food, space, and cover, e.g., climatic changes and urban development.
2. There is competition within and between species groups for these resources (increases particularly when population numbers and densities increase beyond the carrying capacity of the patch).
3. The species is small and has a high reproductive rate under any circumstances.
4. The availability of a specific prey species becomes a limiting factor for predators.
5. There are no dispersal corridors at all or safe (e.g., roads and highways) dispersal corridors between patches.

This "boom and bust" growth cycle has been documented by Barko et al. (2003) for white-footed mice (*Peromyscus leucopus*). Ostfeld et al. (1998) used the source-sink model to explain the occurrence of boom and bust growth cycles in habitat patches. They divided patches into those in which birth rates exceed death rates (sources) and those in which death rates exceed birth rates (sinks). In other words, populations that become isolated either decline to extinction or increase rapidly and overshoot the carrying capacity of the habitat patch.

Another aspect of habitat fragmentation in urban areas that influences certain aspects of the population dynamics of resident wildlife was how they used corridors between habitat patches for dispersal and/or foraging activity. Of some note, is the relative frequency of corridor use by large predators including cougars (*Felis concolor*), bobcats (*Lynx rufus*), foxes (*Vulpes vulpes*), and coyotes (*Canis latrans*). Bobcats and coyotes used corridors more as habitat, rather than for travel, and tended to cross over a road rather than use safer underpasses such as culverts (Tigas et al. 2002). Tigas found that both species can persist in urban environments because they can adjust behaviorally to habitat fragmentation and human activities, in part, through temporal and spatial avoidance. Juvenile cougars use urban habitat corridors for dispersal (Beier 1995). Mammalian carnivores generally prosper in the dissected habitats within urban developments. They prosper because they become resource generalists benefiting from the supplemental resources (i.e., garden fruits and vegetables, garbage, direct feeding by humans), and a variety of landscape elements (i.e., forest and grassland patches and corridors), associated with residential development (Crooks 2002; Atwood et al. 2004). One only has to begin paying attention to the newspaper and magazine articles that report human encounters with cougars, coyotes, bears, and other predatory animals to become convinced of how often they do live among us. More often than not, it comes as a total shock to an urban/suburban dweller to learn that they have lost a valued pet to the predatory activities of coyotes. Cougar attacks on hikers, joggers, and cyclists are becoming more frequent and publicized in the print and electronic media. Once predators lose their inherent fear of humans, the rules of the cohabitation game change dramatically.

Tree-lined streets can function as habitat corridors for some species. Birds and tree squirrels are frequent users of tree-lined streets. Blackbirds (*Turdus merula*), woodpigeons (*Columba palumbus*), magpies (*Pica pica*), and starlings (*Sturnus vulgaris*) can feed and reproduce within the tree-lined streets. Fernandez-Juricic (2000) found that out of 24 species that inhabited a local park, 14 were also observed in tree-lined streets. Use of tree-lined streets provides a way for birds to travel from patch to patch while avoiding the inhospitable urban matrix. Squirrels have been observed using power lines and the canopy of tree-lined streets for the same purpose (personal observations, authors). Urban habitats with little ground cover, high dog and cat populations, and auto traffic make movement along the ground especially risky for squirrels (Williamson 1983).

Rights-of-way are managed vegetated areas that line railroad tracks, highways, and power lines. Although usually highly managed by periodic mowing, these greenways consist of herbaceous cover, shrubs, and occasional trees. Small mammals, such as rabbits, shrews, and mice can feed and reproduce in these strips of land. Mammals as large as foxes can survive in rights-of-way, as can many bird species (Transportation Research Board, 2002). In fact, these areas can be managed for the explicit purpose of inviting wildlife.

Constructed corridors between habitat fragments were found to cause increased movement and associated changes in the demography (e.g., age and sex ratios) of meadow voles (*Microtus pennsylvanicaus*). However, there was no clear pattern of detectable differences in vole population size or recruitment in unfragmented, isolated, or nonlinked fragments, and corridor-linked fragments (Coffman et al. 2001).

4.6 EFFECTS OF SUPPLEMENTAL FEEDING
ON POPULATION DYNAMICS

While people enjoy viewing wildlife up close and personal, there are potential deleterious effects of providing supplemental food (i.e., human-derived subsidies, Fedriani et al. 2001) to urban wildlife. These effects on game animals were summarized in 2000 by the Wildlife Management Institute in a booklet titled: "Feeding Wildlife... Just say NO!" by Scot Williamson. In straightforward language and with the use of illustrations, it carefully explains why such feeding programs are invariably costly and rarely beneficial to wildlife in the shorter or longer run. Fedriani et al. (2001) said the impacts of human-derived subsidies on the population dynamics of urban wildlife need to be examined in the contexts of the effect on consumer populations, resource availability, food web, and community dynamics. Any urban animal that is offered an alternative food supply that is free (do not have to hunt for it), abundant, or in some situations "everlasting," and satisfies its daily caloric requirements is going to be affected in several ways. The urban animal may:

1. Switch its foraging activity to concentrate on the subsidized resources (perhaps exclusively),
2. Improve its nutritional condition and gain weight faster because of the lower expenditure in foraging energy (i.e., reduce the size of their home range) required to access the subsidized resource,
3. Begin reproductive activity earlier in life and have larger numbers of offspring more often,
4. Increase its population density in proximity to the location of the subsidized resource,
5. Pick more territorial battles with its own species and other urban animals (including domestic pets) to compete for priority access to the subsidized resource,
6. Survive longer and during climatic conditions that would normally limit survival (e.g., winter),
7. Run a higher risk of predation and disease transmission due to increased species densities at or around the subsidized resource,
8. Become accustomed to the presence of humans — even entering their homes, more often than not, as an invited guest,
9. Rapidly colonize areas were the subsidized resource is available.

Additionally, supplemental feeding affects other ecosystem components. Martinson and Flaspohler (2003) found that aggregations of bark-foraging birds near a feeder resulted in increased predation on nearby bark-dwelling arthropods. The management implication of this study was that birds may increase tree health by consuming leaf-chewing arthropods overwintering in tree bark. So increasing the density of bark-foraging and excavating birds through supplemental feeding may increase predation on certain pest species, thus minimizing the need for chemical alternatives. The use of natural predators for pest control has also been demonstrated by large bat colonies occupying bridges (see Chapter 6, Bridges and Bats).

4.7 EFFECTS OF ANIMAL DAMAGE CONTROL ACTIVITIES
ON POPULATION DYNAMICS

Animal damage control (ADC) activities become necessary as a reaction to urban wildlife management policies and practices that have facilitated rather than prevented pest problems in the first place. ADC efforts target those species that have "crossed the line" in terms of having (1) a negative impact on human health or economics, (2) worn out their welcome, or (3) become an element of fear. Species fitting the first category and their impacts on human health and economics were covered in Chapter 2. Species that have worn out their welcome could be urban deer (i.e., a few hanging around in the neighborhood is tolerable, a few hundred is not). Species that become elements of human fear may be the result of urbanites not knowing what they are dealing with (e.g., an opossum or very big rat), an assumption of danger (e.g., all snakes are poisonous), or knowing that a real potential for danger exists (e.g., predatory animals). Regardless of the reason for the need for ADC activities, there are consequences of these activities that affect the population dynamics of targeted species.

In order to understand the effects of ADC activities on the population dynamics of selected species, it is necessary to reexamine the techniques and strategies used in ADC (Conover 2002). At the risk of becoming too simplistic, ADC activities are basically designed to remove the "offending" animals permanently or to another location (e.g., translocation) or to discourage the depredative effect through aversive conditioning. The removal of any animal from its population can have several effects on the population dynamics of the species. For example, the population density decreases depending on the number of animals removed. Removal of some members of the population releases more resources for the remaining individuals and may provide opportunity for accelerated immigration. The status of the removed animals within their population hierarchy may be an important factor if the removed individuals had priority access to mates and control of the reproductive activities within the population. This could lead to increased intraspecific competition for the abandoned hierarchical position. Furthermore, sex and age ratios within the population might be altered, particularly if these are variables that predict capture and removal. ADC activities may also have an effect on the predator and prey trophic interrelationships within the urban animal community. Figure 4.4 is an idealized representation of the oscillating cycle of the numbers of predators and prey species in a community. The prey species numbers are high because the predator species is low in the community. However, since there is an abundance of prey species for the few predators, predation increases. The predator's numbers increase over time because of an abundant food supply, causing the prey numbers to decrease over time. Now the cycle shows predator numbers high and prey numbers low. The prey species will again increase because the predator is at a point in the cycle where few prey species exist, causing a decrease in predator numbers. The application of the predatory and prey cycle to ADC activities can be explained as follows. If predators are the primary targeted species group, and ADC activities are successful in the near elimination of predators, one can certainly expect the prey species to increase— perhaps dramatically.

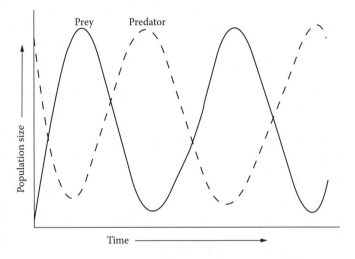

Prey   Predator

Population size

Time

Figure 4.4 Idealized oscillating population growth patterns of predator and prey.

Coyotes rank second in importance in the number of calls received by Wildlife Services for public assistance (see Table 2.1). Coyotes are one of the most "controlled" species across their range, often with high percentages of local populations removed annually. Although coyote population numbers seem to recover rapidly from these control efforts, the genetic consequences have not previously been explored (Williams et al. 2003). Speculation existed concerning the degree to which inbreeding by subdominant males (often offspring of removed dominant males) would affect the gene pool of the population. Williams et al. (2003) found that various aspects of coyote social structure and dispersal patterns adequately maintained genetic variation and promoted genetic homogeneity over relatively small geographic scales during periods of locally aggressive removal. In other words, control efforts tended to stabilize rather than disrupt various aspects of the population dynamics of this species. A more extensive treatment of the coyote as a species of special management concern is given in Chapter 11.

4.8 EFFECTS OF ENVIRONMENTAL POLLUTANTS ON WILDLIFE POPULATION DYNAMICS

Chemicals have replaced bacteria and viruses as the main threat to health. The diseases we're beginning to see as the major causes of death in the latter part of this century and into the 21st century are diseases of chemical origin (for humans and wildlife, author's emphasis).

Dick Irwin, toxicologist at Texas A&M Universities

Some of the most insidious materials to humans and wildlife today are the witches' brew of toxic chemicals that are present in the soil, air, and water. In fact, humans have probably produced and released, directly or indirectly, every conceivable toxicant possible into the natural ecosystem. There is an endless list of toxic chemicals, but they can be categorized into hazardous materials (HAZMATS), pesticides, herbicides, rodenticides, heavy metals, industrial discharges, radioactive wastes, and pharmaceutical products. All of these materials represent pollutants of "type," which means that living organisms do not know what to do with them once they enter the ecosystem. They do not represent the 17 essential nutrients for life (see Chapter 3), so organisms cannot process them into normal bodily functions once they have been assimilated. The inability to process "unnatural" chemicals, more often than not, leads to death (immediately or eventually), cancer, abnormal thyroid function, decreased fertility, decreased hatching success, demasculinization and feminization of males, and alteration of immune function among others. Susceptibility to environmental toxicants is more noticeable in certain organisms (e.g., fish and amphibians), which may be due to the highly specialized genetic programs they need to survive in their natural surroundings. Frogs and other amphibians have been vanishing worldwide over the last few decades (Cone 2005).

The literature on environmental toxicology is extensive and cannot be given a full treatment in this text. Rather, our purpose is to raise the readers' level of awareness concerning the existence of environmental toxicants and alert them to what one might call the "hidden dangers" that now exist in higher concentrations in urban rather than rural environments. Our treatment of the impacts of environmental toxicants on the population dynamics of urban wildlife will focus on selected pesticides and pharmaceuticals.

Urban runoff (point and nonpoint) is a significant pollutant source. Point sources include industrial waste discharges into the air and water or on the soil, and municipal wastewater treatment plant discharges into aquatic habitats. Nonpoint sources consist primarily of stormwater runoff from impervious surface covers (e.g., concrete pavements). Some of the more urban-specific environmental toxicants are pharmaceutical, personal care, and common household products. These toxicants enter the environment from hospitals, pharmaceutical manufacturing plants, and household wastes contained in the wastewater treatment stream. Few, if any, wastewater treatment facilities can extract these toxicants from the water that discharges them into streams and rivers. Manure, used as fertilizer (e.g., applied to urban gardens) may contain veterinary pharmaceuticals, which directly contaminate soils and runoff into streams and rivers after heavy rains (Glaser, 2004 and Ternes et al. 2004).

Without question, environmental toxicants have significant negative impacts on the population dynamics of various species of wildlife. For example, with the advent of modern pesticides (i.e., DDT) in 1943, came a marked decline in the populations of many birds. The decline of the American robin (*Turdus migratorius*) was linked in the early 1950s to DDT spraying for Dutch Elm disease. A recurring spring phenomenon during the 1960s was the delivery of dead or dying robins to my biology classroom by my students (personal observation, C.E. Adams). DDT in the soil was incorporated into and accumulated in the tissue of earthworms (voracious detritus feeders), which are a mainstay in a robin's daily diet. The ingestion of one too many contaminated earthworms constituted a lethal DDT dose for the robin and its chicks. In this case, the impact of DDT on the population

dynamic of urban wildlife was to adversely influence avian survival and reproductive success. The ability for DDT to bioaccumulate in the food chain put those
organisms at the top of the food chain (e.g., fish-eating birds and mammals) at
the highest risk of adverse effects on their reproductive capability (see Perspective
Essay 6.1, The Peregrine Story).

Another impact on the population dynamics of urban and other wildlife caused
by pharmaceutical toxicants is the development of hermaphroditic (intersexes: individuals with gonads containing both female and male reproductive tissue) members
of a population, e.g., fish, frogs, alligators (Cone 2005; Gibson et al 2005). It should
not be a surprise to anyone that the hermaphroditic condition in organisms that were
once discretely male or female will cause problems in reproduction and possibly a
population crash, e.g., cricket frogs (*Acris crepitans*), once abundant, declined dramatically around Chicago (Cone 2005). Unfortunately, there do not seem to be any
enforceable mechanisms in place, at this time, to control the release of pharmaceutical toxicants into natural ecosystems. However, media reports on the deleterious
effects of these toxicants on various wildlife species is increasing, which may
increase public awareness and action to alter existing conditions.

REFERENCES

Adams, L. W. (1994). Urban wildlife habitats: a landscape perspective, University of Minnesota Press, Minneapolis, MN.

Alberti, M., Marzluff, J. M., Shulenberger, E., Bradley, G., Ryan, C., and Zumbrunnen, C.
(2003). Integrating humans into ecology: opportunities and challenges for studying
urban ecosystems, *BioScience,* 53:1169–1179.

Atwood, T. C., Weeks, H. P., and Gehring, T. M. (2004). Spatial ecology of coyotes along a
suburban-to-rural gradient, *Journal of Wildlife Management,* 68:1000–1009.

Barko, V. A., Feldhamer, G. A., Nicholson, M. C., and Davie, D. K. (2003). Urban habitat:
a determinant of white-footed mouse (*Peromyscus leucopus*) abundance in southern
Illinois, *Southeastern Naturalist,* 2(3):369–376.

Beier, P. (1995). Dispersal of juvenile cougars in fragmented habitat, *Journal of Wildlife
Management,* 59: 228–237.

Begon, M., Harper, J. L., and Townsend, C. R. (1996). *Ecology: Individuals, Populations and
Communities,* Blackwell Science, Oxford.

Blair, R. B. (1996). Land use and avian species diversity along an urban gradient, *Ecological
Applications,* 6:506–519.

Blair, R. B. and Launer, A. E. (1997). Butterfly diversity and human land use: species
assemblages along an urban gradient, *Biological Conservation,* 80:113–125.

Calhoon, R. E. and Haspel, C. (1989). Urban cat populations compared by season, subhabitat,
and supplemental feeding, *Journal of Animal Ecology,* 58:321–328.

Coffman, C. J., Nicols, J. D., and Pollock, K. H. (2001). Population dynamics of *Microtus
pennsylvanicus* in corridor-linked patches, *Oikos,* 93:3–21.

Cone, M. (2005). Hermaphrodite frogs linked to pesticide use, *Los Angeles Times,* March 2.

Conover, M. (2002). *Resolving Human-Wildlife Conflicts: the Science of Wildlife Damage
Management,* Lewis Publishers, Boca Raton, FL.

Crooks, K. R. (2002). Relative sensitivities of mammalian carnivores to habitat fragmentation,
Conservation Biology, 16(2):488–502.

Fedriani, J. M., Fuller, T. K., and Sauvajot, R. M. (2001). Does availability of anthropogenic food enhance densities of omnivorous mammals? An example with coyotes in southern California, *Ecography,* 24:325–331.

Fernandez-Juricic, E. (2000). Avifaunal use of wooded streets in an urban landscape, *Conservation Biology,* 14: 513–521.

Fitzwater, W. D. (1988). Solutions to urban bird problems, *Proceedings of the Vertebrate Pest Control Conference,* Crabb, A. C. and Marsh, R. E., Eds., University of California, Davis. 13:254–259.

Gibson, R., Smith, M. D., Spary, C. J., Tyler, C. R., and Hill, E. M. (2005). Mixtures of estrogenic contaminants in bile of fish exposed to wastewater treatment works effluents, *Environmental Science and Technology,* 39:2461–2471.

Glaser, A. (2004). The ubiquitous Triclosan: a common antibacterial agent exposed, *Pesticides and You,* 24:12–17.

Kloor, K. (1999). A surprising tale of life in the city, *Science,* 286:663.

Martinson, T. J. and Flaspohler, D. J. (2003). Winter bird feeding and localized predation on simulated bark-dwelling arthropods, *Wildlife Society Bulletin,* 31:510–516.

McKinney, M. L. (2002). Urbanization, biodiversity, and conservation, *Bioscience,* 52:883–890.

Ostfeld, R. S., Keesing, F., Jones, C. G., Canham, C. D., and Lovett, G. M. (1998). Integrative ecology and the dynamics of species in oak forests, *Integrative Biology,* 1:178–186.

Pauly, D. and Christensen, V. (1995). Primary production required to sustain global fisheries, *Nature,* 374:255–257.

Rebele, F. (1994). Urban ecology and special features of urban ecosystems, *Global Ecology and Biogeography Letters,* 4:173–187.

Ternes, T. H., Joss, A., and Siegrist, H. (2004). Scrutinizing pharmaceuticals and personal care products in wastewater treatment, *Environmental Science and Technology,* October:393–399.

Tigas, L. A., Van Vuren, D. H., and Sauvajot, R. M. (2002). Behavioral responses of bobcats and coyotes to habitat fragmentation and corridors in an urban environment, *Biological Conservation,* 108:299–306.

Transportation Research Board (2002). *Interactions between Roadways and Wildlife Ecology: A Synthesis of Highway Practice,* National Cooperative Highway Research Program, Washington, DC.

U.S. Department of the Interior, Fish and Wildlife Service and U. S. Department of Commerce, Bureau of the Census. 2001. National Survey of Fishing, Hunting, and Wildlife-Associated Recreation. United States Government Printing Office, Washington, DC.

Williams, C. L., Blejwas, K., Johnston, J. J., and Jaeger, M. M. (2003). Temporal genetic variation in a coyote (*Canis latrans*) population experiencing high turnover, *Journal of Mammalogy,* 84: 177–184.

Williamson, R. D. (1983). Identification of urban habitat components which affect eastern gray squirrel abundance, *Urban Ecology,* 7:345–356.

SECTION III

Urban Habitats and Hazards

CHAPTER 5

Special Habitat Considerations: Green Spaces

Environments can, by the judicious use of those tools employed in gardening or landscaping or farming, be built to order with assurance of attracting the desired bird.

Aldo Leopold 1933

CONTENTS

This chapter focuses on the structural features of urbanization that are recognized by selected wildlife species as alternative habitats that provide food, water, shelter, and protection from the elements and predators. The structural features of urbanization (Table 1 in McDonnell and Pickett 1990) can be placed into two categories. One category we have labeled as green spaces (e.g., plants are the dominant cover type) classified as either remnant, successional, or managed habitats. The second category we have labeled as gray spaces (e.g., the dominant cover type consists of concrete, asphalt, or bare ground) which include airports, buildings, highways and roads with associated bridges and underpasses, and landfills. Examples of green spaces are discussed in this chapter in the context of the ecological inter-relationships that resulted in the presence of urban wildlife species. Gray spaces are reviewed in Chapter 6.

5.1 GREEN SPACES

Developing a list of potential green spaces in the urban landscape is a daunting task. Even more challenging is constructing a biological or ecological classification of green spaces. Categorization of urban green spaces by land use (i.e., cemetery, park, golf course) is artificial and not reflective of ecological and biological features of the habitat. For example, housing subdivisions can exhibit different ecological characteristics (e.g., tree canopy height, species diversity, number of exotics, nutrient enrichment) depending upon the socioeconomic status of the homeowners. For example, exotic rodents such as house mice (*Mus musculus*) and Norway rats (*Rattus norvegicus*) occurred more often in habitats close to poor neighborhoods than in wealthier neighborhoods (Huckstep 1996 in Nilon and Paris 1997). Additionally, green spaces will reflect the features of the region in which they are found. For example, a golf course in west Texas will be structurally different from a golf course in western North Carolina.

Further evidence that urban green spaces should not be classified based on land use comes from urban wildlife studies. For example, Hostetler and Knowles-Yanez (2003) found that land use was not an adequate predictor of bird distributions in the Phoenix metropolitan area. Rather, the authors hypothesized that birds in urban areas most likely selected habitat based on biological features such as habitat structure (e.g., what kinds of trees are available). It is therefore warranted to classify urban green spaces based on ecological features rather than a human-derived construct such as land use. We have organized urban green spaces into three broad habitat types: remnant, successional, and managed habitats. This organization follows closely the urban habitat classification scheme described by Nilon and Paris (1997).

5.2 REMNANT HABITAT PATCHES

Remnant habitat patches (RHPs) are sites that have not been cleared or heavily managed by humans and contain species that are typical to the geographic region. RHPs can be found in urban areas because their particular habitat features may impede development. For example, houses may be built outside of a riparian habitat because of frequent flooding. RHPs may also occur simply because development has not progressed enough to remove these habitats. Other times, humans will intentionally leave these habitats "undisturbed." Many residential developers in the eastern U.S. will leave patches of forest interspersed within and between housing units (Figure 5.1).

As urbanization encroaches upon native habitats in different directions, RHPs become like islands embedded in the matrix of the urban/suburban landscape. While there is usually no direct intentional manipulation or management of the RHP, the type of land use surrounding the habitat (i.e., matrix) has real impacts on the numbers and types of wildlife found therein. The wildlife that typically inhabits an RHP is a function of the patch size, proximity to other natural vegetation or habitats, and the land use of the surrounding matrix (Schiller and Horn 1997; Hennings and Edge 2003; Melles et al. 2003).

Figure 5.1   Remnant habitat patches in close proximity to housing development in Houston, Texas. (C. E. Adams)

Scientists have long hypothesized that the species richness (number of species) found in a given area is positively correlated with the area size (Krebs 2001). Researchers have found this generalization to hold true for birds that live in habitat islands within the urban landscape (Fernandez-Juricic and Jokimaki 2001). For example, Soulé et al. (1988) found that size of chaparral habitat patches in San Diego County, California, was positively correlated with the numbers of bird species. Crooks et al. (2001) later studied the same habitat patches and found that 30 populations of birds had gone extinct and only 12 colonizations had occurred since the original study. Extinctions were more likely to occur in small habitat patches. Additionally, colonization of the patches was positively correlated with size.

Theoretically, recolonization of RHPs will be dependent on the proximity of those patches to other natural habitats (MacArthur and Wilson 1967). These larger, natural habitats will serve as sources of wildlife that can potentially colonize the RHPs within the urban/suburban landscape (Marzluff and Ewing 2001). The farther the RHP is from the source habitat, the less likely the species will be able to disperse to the RHP. This is especially true for species that are poor dispersers such as small mammals (Dickman and Doncaster 1989), amphibians, or soil invertebrates. Isolation of a habitat patch can occur in two fundamental ways. First, and most obviously, long distances between patches will produce an isolating effect. Therefore, scientists commonly recommend that corridors or "stepping stones" should be conserved in the urban landscape to facilitate movement between patches (Marzluff and Ewing 2001). Fernandez-Juricic (2000) found wooded streets that connect urban parks in Madrid, Spain served as intermediate habitats for approximately half of the bird species usually found in the parks.

Second, the matrix surrounding the RHP can effectively isolate a patch if it is incompatible with dispersal of organisms (Jules and Shahani 2003). Imagine a gray squirrel dispersing one half mile across a residential development between forest fragments. It likely will confront some obstacles to its dispersal such as dogs and fifth graders with BB guns. However, another squirrel dispersing one half mile between forest fragments across major highways and parking lots will certainly face a greater probability of death. It is important to realize that the matrix does not influence different species similarly. The matrix acts as a filter that allows the passage of some species and prevents the dispersal of others. As a consequence, the composition of species in these isolated RHPs will be different than in patches that are connected to other natural habitats.

Crooks (2002) found that mesopredators (middle-sized predators such as domestic cats and raccoons) were unaffected by habitat fragmentation in an urban area. In contrast, other predators such as bobcats, badgers, long-tailed weasels, coyotes, and mountain lions were less likely to be found in small, isolated RHPs within the urban matrix. Crooks suggested that part of the reason predators such as raccoons and domestic cats are not affected by fragmentation is because they can move through the urban matrix easily. The other predators were sensitive to the urban matrix and were in effect stranded within the RHPs. As these populations went extinct, they could not be recolonized because of the urban matrix. Bolger et al. (1997a) used the same habitat patches to study the effects of urban fragmentation on native rodent species. They demonstrated that small, isolated RHPs could not support viable populations of native rodents. Additionally, there was no evidence of recolonization in these patches after extinction of the local populations, leading the researchers to hypothesize that the urban matrix, even at short distances, was impenetrable by the rodents.

A study by Barko et al. (2003) was designed to assess the habitat features that determine the abundance of white-footed mice in bottomland forest patches. Interestingly, they found the highest abundances in patches that were surrounded by urban land use in comparison to patches surrounded by upland forest habitat. While this finding could be interpreted to mean that these patches were the most suitable to white-footed mice populations, the researchers hypothesized that the urban matrix prevented emigration from the patches thereby leading to an abnormally high population density.

Edge is defined as the area where two different habitat patches meet (Smith and Smith 2001). Volumes have been written on the effects (both negative and positive) of edge habitats on various species. It is not warranted to review here all the literature related to edge effects. However, it is important to note that edge can either reduce or enhance wildlife populations (Yahner 1988) within urban RHPs. For example, Bolger et al. (1997b) categorized breeding birds in coastal Southern California based on the species' sensitivities to the edge between natural shrub habitats and urban development. Edge/fragmentation-enhanced species including the house finch (*Carpodacus mexicana*), northern mockingbird (*Mimus polyglottos*), lesser goldfinch (*Spinus psaltria*), and Anna's hummingbird (*Calypte anna*) readily adapted to the presence of humans and the consequential changes in habitat. The researchers found that populations of these species were actually higher in closer proximity to urban development than to natural habitat patches. Edge/fragmentation-insensitive species such as California quail (*Callipepla californica*), California thrasher (*Toxostoma redivivum*), rufous-sided towhee (*Pipilo erythropthalmus*), wrentit (*Chamaea fasciata*),

Figure 5.2   Edge habitat in suburban area can result from juxtaposition of remnant habitat patches (woods) and managed habitat patches (lawn). (C. E. Adams)

Bewick's wren (*Thyomanes bewickii*), California towhee (*Pipilo crissalis*), California gnatcatcher (*Polioptila californica*), scrub jay (*Aphelocoma coerulescens*), common bushitit (*Psaltriparus minimus*), and mourning dove (*Zenaida macroura*) showed no variation in their high abundance across the fragmented landscape. Edge/fragmentation-reduced species including the black-chinned sparrow (*Spizella atrogularies*), sage sparrow (*Amphispiza belli*), lark sparrow (*Chondestes grammacus*), rufous-crowned sparrow (*Aimophila ruficeps*), Costa's hummingbird (*Calypte costae*), and western meadowlark (*Sturnella neglecta*) showed marked variations in their abundance pattern across the landscape. Higher abundances were found in larger natural patches than in highly fragmented patches and near edges. The edge between a patch and the surrounding matrix is sometimes a combination of features of both habitats, gently changing from one into the other. Other times the edge is abrupt, as is the case in many urban habitats (Figure 5.2). Not only does the urban matrix present a dispersal barrier for many species, but also the edge area between remnant patches and the urban matrix can effectively reduce the size of the RHP for species that are sensitive to urban habitats.

In a study of Neotropical migrant bird species, Friesen et al. (1995) found, as in other studies previously mentioned in this chapter, that size of the habitat fragment correlated with species richness and population abundances. However, they also found that diversity and abundance decreased with increasing development in the matrix regardless of the patch size. In fact, a small (4-ha) woodlot with no surrounding development could support a richer bird community than a larger (25-ha) woodlot surrounded by high housing density. Bock et al. (2002) found that native rodents in Boulder, Colorado avoided the edges between grasslands and suburbs. Mechanisms that result in the sensitivity of some species to edge are poorly understood, although hypotheses suggest that edge habitats attract more avian nest predators such as blue jays (*Cyanocitta cristata*) and crows (*Corvus sp.*), and more mammalian mesopredators like raccoons, domestic cats, and striped skunks (*Mephitis mephitis*), which can reduce prey abundance and diversity. Additionally, nest parasites such as brown-headed cowbirds (*Molothrus ater*) can more easily access nests close to the edge (Paton 1994). Negative human disturbances are also more likely to occur at the edges of RHPs. Zipperer (2002) showed that remnant forest patches in urban areas were subject to several disturbances including dumping of trash and yard wastes, tree removal, and erosion.

5.3 SUCCESSIONAL HABITAT PATCHES

Successional habitat patches (SHPs) are sites that were previously cleared or managed by humans but have subsequently been abandoned. A vacant lot is an example of a successional habitat. We use the term successional because these habitats are no longer under direct management by humans and are subject to the natural process of succession. Succession is defined as the changes in the community structure over time following a disturbance (Ricklefs and Miller 1999). By community structure, we mean both the biological and physical structure. The biological structure of a community is made up of the different types of organisms and the population sizes of those organisms.

An example of physical structure would be the vertical stratification of a forest (Smith and Smith 2001). If you have walked through different types of forests you have probably observed that some forests have several vertical layers such as the herbaceous layer (wildflowers and grasses), the shrub layer (small shrubs and very young trees), the understory tree layer (small trees), and the canopy layer (the largest trees). Other forests such as an urban wooded park that is maintained by mowing or a southern pine forest that burns periodically however will have only a canopy layer and the herbaceous layer.

The disturbance in an urban or suburban habitat that initiates succession is typically removal or clearing of all or part of the biological community. We represent a disturbance every week when we mow our lawn. Because of this weekly disturbance to our lawns by mowing, succession cannot proceed. However, if we stopped mowing our lawns, succession would bring about significant changes in the species of plants that live on our property. For example, the lawns of residents living in the eastern U.S. would eventually revert back to a forest habitat. A previously disturbed site near Washington D.C. developed into an oak-beech forest in approximately 70 years (Ricklefs and Miller 1999).

How does succession proceed? A successional habitat patch can be thought of as a prime real estate spot. Different species are vying for access to the patch and competing with other species once they arrive. The first species to arrive to an SHP are those that have excellent dispersal abilities. For example, many grasses and other small herbaceous plants have light seeds that are carried great distances on wind currents. Have you ever blown the seeds of a dandelion on a warm summer day? Dandelions are examples of fantastic dispersers. These same species are usually adapted to disturbed conditions and initially flourish following a disturbance event. Because these species have invested their limited energy into dispersal and rapid reproduction, they are usually poor competitors with other species. Consequently, over time, other species will replace them within the habitat patch. Additionally, some early colonizers can significantly change the abiotic or nonliving conditions of the habitat. This change can either make it easier or more difficult for other species to invade. For example, black locust trees in the eastern U.S. enrich the soil with nitrogen. As a result of their presence, other tree species that otherwise would not have been able to cope with the nitrogen-poor soil are able to colonize and survive. Other species however, have the opposite effect. Many exotic plants such as kudzu (*Pueraria montana*) or Japanese honeysuckle (*Lonicera japonica*) will change the physical structure (e.g., reduction of light) of a community in such a way that other species cannot invade or survive.

The rate at which succession occurs and the composition of the SHP will be a function of number and proximity of source species available for colonization (Dzwonko and Loster 1997). SHPs in urban areas in the eastern U.S. sometimes undergo succession at a relatively slow pace. One hypothesis suggests seed dispersal of woody species in these patches is limited because of low avian activity. Robinson and Handel (1993) tested this hypothesis by planting woody species to attract avian seed dispersers in an abandoned urban landfill. They predicted that by providing perching habitat for the birds, a positive feedback would result that would lead to more woody species recruitment. The plantings did attract avian seed dispersers which introduced 20 new plant species to the area. Another factor that has been found to influence forest habitat succession is the quantity of leaf litter on the forest floor. Kostel-Hughes et al. (1998) demonstrated that forest floor leaf litter was higher in density, depth, and mass with increased distance from New York City. Small-seeded species were more likely than larger-seeded species to germinate in shallow leaf litter leading the authors to conclude that physical changes to the forest floor from urbanization could change the trajectory of succession in urban forests. Additionally, small-scale disturbances by humans were identified as important factors in successional forest habitats in urban areas in Syracuse, New York (Zipperer 2002).

It should be obvious that, if the plant community of a given area changes over time, the animal community will change as well. Unfortunately little is known about how animal communities change over time in urban SHPs. From rural studies, we know that different wildlife species are adapted to the different stages of succession. For example, typical bird species of early successional habitats in the eastern deciduous forests include the grasshopper sparrow (*Ammodramus savannarum*). As habitats become shrubbier, species such as yellow-breasted chats (*Icteria virens*) (Thompson and Nolan 1973), cardinals (*Cardinalis cardinalis*), and indigo buntings (*Passerina cyanea*) become more abundant. Typical birds of mid-successional forests include summer tanagers (*Piranga rubra*) and wood thrushes (*Hylocichla mustelina*). Black-throated green warblers (*Dendroica virens*) and ovenbirds (*Seiurus aurocapillus*) prefer more mature forests (Johnston and Odum 1956; Kricher 1973; Holmes and Sherry 2001).

5.4 MANAGED HABITAT PATCHES

Managed habitat patches (MHPs) are sites under direct and intense management by humans. Examples of urban/suburban MHPs are cemeteries (see Perspective Essay 5.1), parks, golf courses (Figure 5.3), and residential yards. The physical and biological structure of MHPs is a result of several synergistic factors including culture, history, economics, ecology, and regional flora and fauna. Additionally, as with the remnant and successional habitat patches, MHPs are also affected by their sizes, proximity to source habitats, and surrounding matrix. MHPs can mimic a broad range of "natural" habitat types such as deciduous forests, desert scrub, or grasslands. Additionally, many incorporate a diversity of habitats that may include both terrestrial and aquatic communities. For example, In the Midwest U.S., cemeteries sometimes contain fragments of remnant grassland communities (Stowe et al. 2001). This results in part from the fact that many urban cemeteries were originally established on the outskirts of towns and were subsequently incorporated into the cities as they grew (Gilbert 1989). A growing trend among homeowners is the construction of ponds or water gardens in their yards, and every golf course will contain one or several small ponds or lakes.

Probably the most common habitat feature of many MHPs is trees, both deciduous and coniferous. Homeowners often plant trees for aesthetic or functional purposes. Most urban cemeteries possess "a structural framework of trees planted when they were first laid out" (Gilbert 1989), but the types of trees planted varies widely according to geographic region and the time period in which the cemetery was founded. Those established around the turn of the nineteenth century or a few decades earlier were often meant to conform to Romantic or Victorian aesthetic standards

Figure 5.3   **(See color insert following page 276)** Golf courses have the potential to provide habitat needs for several species of wildlife including these deer. (Ken Hammond/ USDA)

and often functioned secondarily as arboretums (Gilbert 1989). Consequently, many of these cemeteries will contain a mixture of exotic, ornamental, and popular native tree species. For example, Woodlawn Cemetery in Toledo, Ohio, founded in 1876, contains the exotics European larch (*Larix decidua*), European beech (*Fagus sylvatica*), and star magnolia (*Magnolia stellata*), as well as the natives red maple (*Acer rubrum*), tulip poplar (*Liriodendron tulipifera*), Ohio buckeye (*Aesculus glabra*), and the now-rare American elm (*Ulmus americana*). Various showy trees and shrubs are also very common, including magnolias (*Magnolia sp.*), Japanese maples (*Acer palmatum*), dogwoods (*Cornus sp.*), redbuds (*Cercis sp.*), crepe myrtles (*Lagerstroemia spp.*), hydrangeas (*Hydrangea spp.*), fruit trees, and willows (*Salix sp.*) (Woodlawn Cemetery 2001–2003). Coniferous trees are also common features of MHPs. For homeowners, coniferous tree species provide shelter and aesthetics year round. Common coniferous species planted in urban MHPs include but are not limited to Douglas-fir (*Pseudotsuga menziesii*), Norway spruce (*Picea abies*), blue spruce (*Picea pungens*), white pine (*Pinus strobus*), Scotch pine (*Pinus sylvestris*), and various cedars and junipers.

Trees planted in urban MHPs often reach very large sizes because they do not have to compete as much for resources as in a natural forest habitat. American Forests, a nonprofit citizen's conservation organization, maintains a registry of the largest trees in each state. Several state tree records occur in urban MHPs such as cemeteries (Figure 5.4) and residential yards (American Forests 2004). Large trees provide important habitat for several species of wildlife. For example, Sharp-shinned hawks (*Accipiter striatus*) in and around Montreal used mature coniferous trees for nesting (Coleman et al. 2002). Another study (Mager and Nelson 2001) found red bats (*Lasiurus borealis*) prefer roosting in the foliage of trees with broad crowns and

Figure 5.4 Many urban cemeteries are home to large, old trees which can serve as important nesting and denning sites for wildlife. (John M. Davis/Texas Parks and Wildlife Department)

large leaves in an urban park in Charleston, Illinois. These roosting sites are similar to sites selected by bats in natural habitats. The researchers noted that the urban habitats such as parks were important for bats because they provided large trees interspersed with open habitats for foraging. In fact, the researchers had the most success in trapping bats close to edges of forests/fields and to streetlights. They hypothesized that the bats could find a high concentrated diversity of insect prey in these areas. Tree-trimming practices and natural senescence often result in the formation of cavities, which many types of birds and mammals will use for nesting, roosting, and/or shelter. Several species of birds use tree cavities for nesting including bluebirds, woodpeckers, owls, ducks, flycatchers, and chickadees. Cemeteries in west Texas and the panhandle are good roosting and nesting places for owls due to the high density of large conifers in otherwise relatively treeless regions (Ardath Lawson, personal observation).

Other studies show that the MHPs provide relatively easy access to resources needed for some species of wildlife. For example, studies (Boal and Mannan 1998; Mannan and Boal 2000) of Cooper's hawks (*Accipiter cooperii*) in Tucson, Arizona, showed that the hawks nested in groves of large exotic trees within residential or recreational areas. The authors hypothesized that the hawks selected urban areas for nesting because of the availability of water and prey (doves). As a result of the high density of doves in these areas, the hawks did not need to forage far from their nests, as they would normally do in rural habitats.

Studies have shown that MHPs such as urban cemeteries and golf courses hold rich avian diversity. Thomas and Dixon (1973) discovered 95 different species of birds frequenting the Boston cemeteries. Thirty-four of the species used cemetery habitats for breeding and nesting. In addition to the blue jays, robins (*Turdus migratorius*) and starlings (*Sturnus vulgaris*) commonly found in urban habitats, they also discovered a number of birds not typically found within densely populated cities, including sharp-shinned hawks, yellow-shafted flickers (*Colaptes auratus*), belted kingfishers (*Ceryle alcyon*), bobwhites (*Colinus virginianus*), ring-necked pheasants (*Phasianus colchicus*), and black-billed cuckoos (*Coccyzus erythropthalmus*). This last species seems especially surprising, given that it is assumed to be an inhabitant of dense, undisturbed forests (Alsop 2001). Lussenhop (1977) conducted a study in central Chicago and found that, instead of urban birds spilling over into the more diverse cemetery habitat, rural birds from the surrounding nonurbanized areas seemed to be making the most use of the cemetery. Species such as rock doves (*Columba livia*), sparrows (*Passer domesticus*), and starlings were not found in higher densities than surrounding urban habitats. However, higher numbers of native, typically rural, birds such as red-headed woodpeckers (*Melanerpes erythrocephalus*), eastern wood-pewees (*Contopus virens*), red-winged blackbirds (*Agelaius phoeniceus*), and indigo buntings were found living and nesting within the cemetery than would have been expected in an urban area generally.

As stated previously, some MHPs will contain remnant habitat patches and, consequently, have been identified as biological repositories within the urban landscape (Barrett and Barrett 2001). Golf courses in Kent, England, contain uncommon habitat types such as heathland and dune and slack communities and rare populations of orchids, broomrape, and mosses (Green and Marshall 1987). The threatened Big Cypress fox squirrel (*Sciurus niger avicennia*) can be found on golf courses in Florida and have provided researchers with valuable information regarding this normally secretive species (Jodice and Humphrey 1992). Harrison (1981) found that cemeteries in Macon, Georgia, were biologically diverse. Abundance of leaf litter invertebrates from the cemeteries was comparable to the samples taken from undisturbed forests in the region. Additionally the cemeteries were as diverse in avian species as a comparison study site, the Ocmulgee National Monument. However, it should be noted that the composition of the bird communities between the cemeteries and the Monument were different. Twenty species were unique to the National Monument and 14 species were unique to the cemeteries.

Despite growing interest in urban MHPs as habitats for wildlife, little is known about the differences in life history characteristics between rural and urban populations. Thompson (1977, 1978) made several observations about a gray squirrel (*Sciurus carolinensis*) population living in an urban cemetery. First, urban squirrels reached sexual maturity at a significantly older age than did rural squirrels (Longley 1963). The mechanism resulting in this disparity is unclear but may be a result of sampling differences. Breeding season and number of young born per litter were equal between urban and rural squirrels; however, survival of the young was higher among urban squirrels. Dykstra et al. (2000) found equal reproductive rates between suburban and rural red-shouldered hawks. However, Boal and Mannan (1999) found that, although urban Cooper's hawks exhibited larger clutch sizes, the nestling mortality rates were higher (50%) compared to their rural counterparts (5%). The primary cause of nestling death in urban areas was the parasitic disease trichomoniasis.

Other studies have demonstrated that behavioral differences exist between rural and urban populations of species. For example, great tits *(Parus major)* sing at higher frequencies in noisy urban areas than do individuals in quieter areas (Slabbekoorn and Peet 2003). Estes and Mannan (2003) found that male Cooper's hawks in urban areas delivered significantly more prey biomass to the nest than did rural males. Additionally, the urban males were more likely to deliver prey directly to the nest, and urban females rejected deliveries more often than did rural females. Finally, both rural males and females vocalized more often than did urban individuals. The authors attributed these contrasting nesting behaviors to the differences in prey abundance between rural and urban locations. Moreover, tolerance levels of some birds, as indicated by their flushing distances, increases in areas with high human visitation (Fernandez-Juricic and Jokimaki 2001).

While it appears that urban MHPs offer a wealth of resources for several species of wildlife, we believe that a word of caution is warranted. Urban habitats, whether remnant fragments of natural communities or highly managed patches, cannot support the same level of biodiversity found in native biological communities. Additionally, they will support a vastly different composition of species, with rarer, more specialized species being absent. Formation of urban habitats selects for those species that are adapted to the specific biological and physical structure of urban environments and to a constant stream of anthropogenic disturbances. The high rate of visitors to urban parks in Madrid, Spain, reduced the avian richness of those sites (Fernandez-Juricic and Jokimaki 2001).

For example, the physical and biological structure of MHPs that have trees as the dominant cover are radically different from natural forests. First, vertical layers are removed, which results in the absence of escape cover for small vertebrates or nesting habitats for shrub and ground-nesting birds (Zalweski 1994; Blair 1996; Marzluff and Ewing 2001; Livingston et al. 2003). In fact, the existence of open lawns in MHPs dramatically reduces the species diversity of those habitats. Additionally, maintenance of lawns is ecological homicide because it requires the use of pesticides, fertilizers, and the burning of fossil fuels, all contributing factors to the loss of diversity and function of ecosystems.

Exotic species (including domestic pets) can have dramatic effects on the species composition of MHPs. Many golf courses and residential yards are dominated by exotic plantings. While these exotic plants provide additional physical structure, they are usually not preferred by native wildlife species. Many researchers (Germaine et al. 1998; Green and Baker 2003) have emphasized the use of native plants in restoring avian communities within urban areas. Domestic pets, primarily domestic cats, are significant factors determining the distribution and abundance of some wildlife species within the urban landscape (Soule et al. 1988; Lepczyk et al. 2003). Baker et al. (2003) found that wood mice (*Apodemus sylvaticus*) abundance in residential gardens in Bristol, England, was negatively correlated with the presence of cats, and Churcher and Lawton (1987) found that cats played a major role in the population dynamics of urban sparrows.

PERSPECTIVE ESSAY 5.1: BIRDS IN TEXAS CEMETERIES

I was interested in learning more about the types of birds that frequented cemeteries in this state, so I solicited the TexBirds mailing list for information on bird sightings in Texas cemeteries, as well as for information on the cemeteries themselves. From the birders who responded to this request for information, I received a large amount of information concerning the bird species within Texas that frequently utilize urban cemeteries either year round or as a stop on their migration route during the spring and fall.

Much of this data was collected in two cemeteries in Dallas and Houston, Texas. The 33-acre Greenwood Cemetery in Dallas (Figure 5.4) was founded in 1850 and contains a large number of mature trees, mostly white pine, with a scattering of juniper and small amount of crepe myrtle serving as the understory. This cemetery was birded frequently from 1977 to 1979 by Homer Klonis of Dallas, primarily during the spring and fall migratory periods. Klonis observed 67 different bird species (Table 5.1). Most sightings were common urban birds (e.g., species 1 and 2, Table 5.1). A number of typically rural birds were also observed, including little blue herons (*Egretta caerulea*), a great egret (*Ardea alba*), a sharp-shinned hawk, common nighthawks (*Chordeiles minor*), a summer tanager (*Piranga rubra*), a yellow-billed cuckoo (*Coccyzus americanus*), and a loggerhead shrike (*Lanius ludovicianus*). Many Neotropical migrants were also spotted, most notably a wide variety of different warblers and sparrows that were passing through.

Table 5.1   Total Bird Observations by Sight in Greenwood Cemetery, Dallas, TX, from April 1977 through September 1979

Species (Common Names)	Total Number Observed
Starling	>706
House Sparrow	142
Chipping Sparrow	131
Blue Jay	>94
Northern Mockingbird	>92
Northern Cardinal	>65
Mourning Dove	64
American Robin	64
Nashville Warbler	62
Great-tailed Grackle	>50
Common Grackle	>45
Chimney Swift	>36
American Crow	>34
Ruby-crowned Kinglet	>28
Brown-headed Cowbird	21
Brown Thrasher	>19
Cedar Waxwing	17
Lincoln's Sparrow	16
Empidonax spp.	14
Black-and-white Warbler	12
Field Sparrow	>11
Great Crested Flycatcher	10
Northern Flicker	9
Brown Creeper	9
Little Blue Heron	8
Black-throated Green Warbler	8
American Redstart	7
Junco	7
Eastern Phoebe	6
Red-breasted Nuthatch	6
Solitary Vireo	6
Bay-breasted Warbler	6
Swainson's Thrush	5
Yellow-rumped Warbler	5
Baltimore Oriole	5
Red-winged Blackbird	5
Red-headed Woodpecker	4
Carolina Chickadee	>4
Tennessee Warbler	4
Common Nighthawk	3
Red-bellied Woodpecker	3
Hairy Woodpecker	3
Great Horned Owl	2
Downy Woodpecker	2
Yellow Warbler	2
Blackburnian Warbler	2
Wilson's Warbler	2
Spotted Towhee	2
Song Sparrow	2

(Continued)

Table 5.1   (*Continued*)

Species (Common Names)	Total Number Observed
Sharp-shinned Hawk	1
Yellow-billed Cuckoo	1
Yellow-bellied Sapsucker	1
Eastern Kingbird	1
Scissor-tailed flycatcher	1
Tufted Titmouse	1
Loggerhead Shrike	1
Carolina Wren	1
Bewick's Wren	1
Golden-crowned Kinglet	1
Blue-gray Gnatcatcher	1
Orange-crowned Warbler	1
Magnolia Warbler	1
Canada Warbler	1
White-crowned Sparrow	1
White-throated Sparrow	1
Summer Tanager	1

Note: Most observations occurred during the spring and fall migratory periods.
Data courtesy of Homer Klonis of Dallas, TX.

The second cemetery for which a significant amount of data was collected was Forest Park Cemetery in west Houston, a 115-acre area surrounded by a mixture of apartments and homes on the west side and office buildings on the north. The vegetation within this cemetery consists of sparsely-planted trees with a few flowering shrubs; although live oak (*Quercus virginiana*) is the dominant tree species, water oak (*Quercus nigra*), hackberry (*Celtis occidentalis*), redbud (*Cercis canadensis*), eastern red cedar (*Juniperus virginiana*), chestnut oak (*Quercus prinus*), Chinese tallow (*Sapium sebiferum*), willows, palm trees, and various conifers are also represented. There were also a few open areas consisting entirely of unmown grass; these sections are most likely slated for future development. Data were collected by Harry Elliott of Houston who observed 29 different bird species on two days in March (Table 5.2). Again, urbanized species (e.g., 1 to 9, Table 2) were present in sizeable numbers, but migrants and species less commonly found in typical urban habitats were also observed, including loggerhead shrikes, purple martins, various warblers, and a large hawk, possibly a Cooper's hawk.

Several birders contributed information on other cemeteries throughout the state and some of the more interesting avian species that could be found in them (Table 5.3). The city cemetery in Weslaco, Texas, was especially well-known for its sightings of rare or unusual subtropical birds (including the infrequently observed blue mockingbird). In comparison, one could find large numbers of shorebirds such as curlews and godwits in cemeteries in Galveston, Texas. Cemeteries in West Texas and the panhandle were cited as being good roosting and nesting places for owls, due to the high density of large conifers in otherwise relatively treeless regions.

Ardath Lawson

Table 5.2 Total Bird Observations by Sight and Song in Forest Park Cemetery, Houston, TX, in March 2003

Species (Common Names)	Total Number Observed
Yellow-rumped Warbler	175
American Robin	150
Red-winged Blackbird	150
Mourning Dove	45
Northern Mockingbird	27
European Starling	23
Rock Dove	20
Great-tailed Grackle	>15
House Sparrow	15
Blue-gray Gnatcatcher	13
White-winged Dove	12
Common Grackle	>10
Northern Cardinal	>9
Killdeer	>8
Blue Jay	8
Ruby-crowned Kinglet	8
American Crow	6
Purple Martin	>5
Loggerhead Shrike	5
Red-bellied Woodpecker	4
Accipiters (probably Cooper's Hawk)	2
White-eyed Vireo	>2
Carolina Wren	>2
Eurasian Collared Dove	>1
Inca Dove	>1
Yellow-bellied Sapsucker	>1
Canada Warbler	>1
Orange-crowned Warbler	1
Eastern Meadowlark	1

Note: Numbers of some birds are estimates.

Data courtesy of Harry Elliott, Houston, TX.

Table 5.3 Other Bird Species Observed in Texas Cemeteries

Cemetery	Species (Common Names)
Weslaco City Cemetery, Welsaco, Hidalgo Co.	Green Parakeet
	Red-crowned Parrot
	Black-bellied Whistling Duck
	White-tailed Kite
	Great Kiskadee
	Green Jay
	Blue Mockingbird
Oakwood Cemetery, Comanche, Comanche Co.	Mississippi Kite
Unknown, Galveston Co.	Plain Chachalaca
	Long-billed Curlew
	Veery
	American Goldfinch

(Continued)

Table 5.3 (*Continued*)

Cemetery	Species (Common Names)
Various West Texas Cemeteries	Marbled Godwit
	Barn Owl
	Eastern Screech Owl
	Great Horned Owl
	Long-eared Owl
State Cemetery, Austin, Travis Co.	Golden-fronted Woodpecker
Old Fairview Cemetery, Bastrop, Bastrop Co.	Black-throated Gray Warbler

REFERENCES

Alsop, F. J. (2001). *Birds of North America: Eastern Region,* DK Publishing, New York.

American Forests (2004). National Register of Big Trees, Accessed at http://www.american-forests.org/resources/bigtrees/.

Baker, P. J., Ansell, R. J., Dodds, P. A. A., Webber, C. E., and Harris, S. (2003). Factors affecting the distribution of small mammals in an urban area, *Mammal Review,* 33(1):95–100.

Barko, V. A., Feldhamer, G. A., Nicholson, M. C., and Davie, D. K. (2003). Urban habitat: a determinant of white-footed mouse (*Peromyscus leucopus*) abundance in southern Illinois, *Southeastern Naturalist,* 2(3):369–376.

Barrett, G. W. and Barrett, T. L. (2001). Cemeteries as repositories of natural and cultural diversity, *Conservation Biology,* 15(6):1820–1824.

Blair, R. B. (1996). Land use and avian species diversity along an urban gradient, *Ecological Applications,* 6(2):506–519.

Boal, C. W. and Mannan, R. W. (1998). Nest-site selection by Cooper's hawks in an urban environment, *Journal of Wildlife Management,* 62(3):864–871.

Boal, C. W. and Mannan, R. W. (1999). Comparative breeding ecology of Cooper's hawks in urban and exurban areas of southern Arizona, *Journal of Wildlife Management,* 63(1):77–84.

Bock, C. E., Vierling, K. T., Haire, S. L., Boone, J. D., and Merkle, W. W. (2002). Patterns of rodent abundance on open-space grasslands in relation to suburban edges, *Conservation Biology,* 16(6):1653–1658.

Bolger, D. T., Alberts, A. C., Sauvajot, R. M., Potenza, P., McCalvin, C., Tran, D., Mazzoni, S., and Soulé, M. E. (1997a). Response of rodents to habitat fragmentation in coastal southern California, *Ecological Applications,* 7(2):552–563.

Bolger, D. T., Scott, T. A., and Rotenberry, J. T. (1997b). Breeding bird abundance in an urbanizing landscape in coastal southern California, *Conservation Biology,* 11(2):406–421.

Churcher, P. B. and Lawton, J. H. (1987). Predation by domestic cats in an English village, *Journal of Zoology,* 212:439–455.

Coleman, J. L., Bird, D. M., and Jacobs, E. A. (2002). Habitat use and productivity of sharp-shinned hawks nesting in an urban area, *Wilson Bulletin,* 114(4):467–473.

Crooks, K. R. (2002). Relative sensitivities of mammalian carnivores to habitat fragmentation, *Conservation Biology,* 16(2):488–502.

Crooks, K. R., Suarez, A. V., Bolger, D. T., and Soule, M. E. (2001). Extinction and colonization of birds on habitat islands, *Conservation Biology,* 15(1):159–172.

Dickman, C. R. and Doncaster, C. P. (1989). The ecology of small mammals in urban habitats, II. Demography and dispersal, *Journal of Animal Ecology,* 58:119–129.

Dykstra, C. R., Hays, J. L., Daniel, F. B., and Simon, M. M. (2000). Nest site selection and productivity of suburban red-shouldered hawks in southern Ohio, *The Condor,* 102:401–408.

Dzwonko, Z. and Loster, S. (1997). Effects of dominant trees and anthropogenic disturbances on species richness and floristic composition of secondary communities in southern Poland, *Journal of Applied Ecology,* 34:861–870.

Estes, W. A. and Mannan, R. W. (2003). Feeding behavior of Cooper's hawks at urban and rural nests in southeastern Arizona, *The Condor,* 105:07–116.

Fernandez-Juricic, E. (2000). Avifaunal use of wooded streets in an urban landscape, *Conservation Biology,* 14(2):513–521.

Fernandez-Juricic, E. and Jokimaki, J. (2001). A habitat island approach to conserving birds in urban landscapes: case studies from southern and northern Europe, *Biodiversity and Conservation,* 10:2023–2043.

Friesen, L. E., Eagles, P. F. J., and Mackey, R. J. (1995). Effects of residential development on forest-dwelling neotropical migrant songbirds, *Conservation Biology,* 9(6):1408–1414.

Germaine, S. S., Rosenstock, S. S., Schweinsburg, R. E., and Richardson, W. S. (1998). Relationships among breeding birds, habitat, and residential development in greater Tucson, Arizona, *Ecological Applications,* 8(3):680–691.

Gilbert, O. L. (1989). *The Ecology of Urban Habitats,* Chapman and Hall, New York.

Green, B. H. and Marshall, I. C. (1987). An assessment of the role of golf courses in Kent, England, in protecting wildlife and landscapes, *Landscape and Urban Planning,* 14:143–154.

Green, D. M. and Baker, M. G. (2003). Urbanization impacts on habitat and bird communities in a Sonoran desert ecosystem, *Landscape and Urban Planning,* 63:225–239.

Harrison, J. O. (1981). Older urban cemeteries as potential wildlife sanctuaries, *Georgia Journal of Science,* 39:117–126.

Hennings, L. A. and Edge, W. D. (2003). Riparian bird community structure in Portland, Oregon: habitat, urbanization, and spatial scale patterns, *The Condor,* 105:288–302.

Holmes, R. T. and Sherry, T. W. (2001). Thirty-year bird population trends in an unfragmented temperate deciduous forest: importance of habitat change, *The Auk,* 118(3):589–609.

Hostetler, M. and Knowles-Yanez, K. (2003). Land use, scale, and bird distributions in the Phoenix metropolitan area, *Landscape and Urban Planning,* 62:55–68.

Jodice, P. G.R. and Humphrey, S. R. (1992). Activity and diet of an urban population of Big Cypress fox squirrels, *Journal of Wildlife Management,* 56(4):685–692.

Johnston, D. W. and Odum, E. P. (1956). Breeding bird populations in relation to plant succession on the Piedmont of Georgia, *Ecology,* 37:50–62.

Jules, E. S. and Shahani, P. (2003). A broader ecological context to habitat fragmentation: Why matrix habitat is more important than we thought, *Journal of Vegetation Science,* 14:459–464.

Kostel-Hughes, F., Young, T. P., and Carreiro, M. M. (1998). Forest leaf litter quantity and seedling occurrence along an urban-rural gradient, *Urban Ecosystems,* 2:263–278.

Krebs, C. J. (2001). *Ecology: the Experimental Analysis of Distribution and Abundance,* Benjamin Cummings, Menlo Park, CA.

Kricher, J. C. (1973). Summer bird species diversity in relation to secondary succession on the New Jersey Piedmont, *American Midland Naturalist,* 89(1):121–137.

Lepczyk, C. A., Mertig, A. G., and Liu, J., (2003). Landowners and cat predation across rural-to-urban landscapes, Biological Conservation, 115:191–201.

Livingston, M., Shaw, W. W., and Harris, L. K. (2003). A model for assessing wildlife habitats in urban landscapes of eastern Pima County, Arizona (USA). *Landscape and Urban Planning,* 64:131–144.

Longley, W. H. (1963). Minnesota gray and fox squirrels, *American Midland Naturalist,* 69(1):82–98.

Lussenhop, J. (1977). Urban cemeteries as bird refuges, *Condor,* 709:456–461.

MacArthur, R. and Wilson, E. O. (1967). *The Theory of Island Biogeography,* Princeton University Press, Princeton, NJ.

Mager, K. J. and Nelson, T. A. (2001). Roost-site selection by eastern red bats (*Lasiurus borealis*). *American Midland Naturalist,* 145:120–126.

Mannan, R. W. and Boal, C. W. (2000). Home range characteristics of male Cooper's Hawks in an urban environment, *Wilson Bulletin,* 112(1):21–27.

Marzluff, J. M. and Ewing, K. (2001). Restoration of fragmented landscapes for the conservation of birds: a general framework and specific recommendations for urbanizing landscapes, *Restoration Ecology,* 9(3):280–292.

McDonnell, M. J. and Pickett, S. T. A. (1990). Ecosystem structure and function along urban-rural gradients: an unexploited opportunity for ecology, *Ecology,* 71(4):1232–1237.

Melles, S., Glenn, S., and Martin, K. (2003). Urban bird diversity and landscape complexity: Species-environment associations along a multiscale habitat gradient, *Conservation Ecology,* 7(1):5, online at http://www.consecol.org./vol7/iss1/art5.

Nilon, C. H. and Paris, R. C. (1997). Terrestrial vertebrates in urban ecosystems: Developing hypotheses for the Gwynns Falls watershed in Baltimore, Maryland, *Urban Ecosystems,* 1:247–257.

Paton, P. W. C. (1994). The effect of edge on avian nest success: how strong is the evidence? *Conservation Biology,* 8:17–26.

Ricklefs, R. E. and Miller, G. L. (1999). *Ecology,* W. H. Freeman, New York.

Robinson, G. R. and Handel, S. N. (1993). Forest restoration on a closed landfill: rapid addition of new species by bird dispersal, *Conservation Biology,* 7(2):271–278.

Schiller, A. and Horn, S. P. (1997). Wildlife conservation in urban greenways of the mid-southeastern United States, *Urban Ecosystems,* 1:103–116.

Slabbekoorn, H. and Peet, M. (2003). Birds sing at a higher pitch in urban noise, *Nature,* 424:267.

Smith, R. L. and Smith, T. M. (2001). *Ecology and Field Biology,* Benjamin Cummings, Menlo Park, CA.

Soulé, M. E., Bolger, D. T., Alberts, A. C., Wright, J., Sorice, M., and Hill, S. (1988). Reconstructed dynamics of rapid extinctions of chaparral-requiring birds in urban habitat islands, *Conservation Biology,* 2(1):75–92.

Stowe, J. P., Jr., Schmidt, E. V., and Green, D. (2001). Toxic burials: the final insult, *Conservation Biology,* 15(6):1817–1819.

Thomas, J. W. and Dixon, R. A. (1973). Cemetery ecology, *Natural History,* 82(3):61–67.

Thompson, C. F. and Nolan, V., Jr. (1973). Population biology of the yellow-breasted chat (*Icteria virens*) in southern Indiana, *Ecological Monographs,* 43(2):145–171.

Thompson, D. C. (1977). The social system of the grey squirrel, *Behaviour,* 64:305–328.

Thompson, D. C. (1978). Regulation of a northern grey squirrel (*Sciurus carolinensis*) population, *Ecology,* 59(4):708–715.

Woodlawn Cemetery, (2001–2003). Historic Woodlawn Cemetery, http://www.historic-woodlawn.com/index.html.

Yahner, R. H. (1988). Changes in wildlife communities near edges, *Conservation Biology,* 2(4):333–339.

Zalewski, A. (1994). A comparative study of breeding bird populations and associated landscape character, Torun, Poland, *Landscape and Urban Planning,* 29:31–41.

Zipperer, W. C. (2002). Species composition and structure of regenerated and remnant forest patches within an urban landscape, *Urban Ecosystems,* 6:271–290.

CHAPTER 6

Special Habitat Considerations: Gray Spaces

What sort of coyote has Los Angeles created? It's a creature that will jump over chainlink for a bowl of Alpo. It's an animal that can learn and remember which storm-sewer channels lead to which golf courses...which dumpsters behind which supermarkets are likely to be overflowing with old vegetables and delightfully rancid fish... It has eaten from the Tree of Forbidden Knowledge, and it recalls fondly the taste of Fifi and Mr. Boots.

David Quammen

The thing scared me to death,

**New Yorker who encountered a wild turkey on the balcony of
his 28th-floor apartment on West 70th Street**

CONTENTS

6.1 INTRODUCTION

Over the past decade, several ecologists have recognized the need to apply the science of ecology to urban environments. "Urbanization can be characterized as an increase in human habitation, coupled with increased per capita energy and resource consumption and extensive modification of the landscape, creating a system that does not depend principally on local natural resources to persist" (McDonnell and Pickett 1990:1231). Their list of structural features unique to urbanization included dwellings, factories, office buildings, warehouses, roads, pipelines, power lines, railroads, channelized stream beds, reservoirs, sewage disposal facilities, landfills, and airports.

Lost natural habitats are replaced by four types of altered habitat that become progressively more common toward the urban core. The four types of habitat are presented below in terms of increasing habitability to most native species and decreasing proportion of coverage toward the urban core.

1. Built habitat: buildings and sealed surfaces, such as roads and parking lots cover over 80% of central urban area
2. Managed vegetation: residential, commercial, and other regularly maintained green spaces
3. Ruderal vegetation: empty lots, abandoned farmland, and other green space that is cleared but not managed
4. Natural remnant vegetation: remaining islands of original vegetation (usually subject to substantial nonnative plant invasion (McKinney 2002)

Habitat types 2, 3, and 4 were discussed extensively in Chapter 5. This chapter examines how various wildlife species utilize the "built" habitats within urban ecosystems. Our synthesis of the literature demonstrated that wildlife attempts to occupy nearly every "nook and cranny" in urban structures. This should not be surprising given the remarkable flexibility of urban adapters (Chapter 4) in taking advantage of the new structures provided in urban settings.

6.2 BUILDINGS, WINDOWS, AND TOWERS

Wildlife is most noticed by urban residents when the animal appears in or in close proximity to the areas where we live, work, or recreate. The famous cliché "build it and they will come" is correctly applied to the response of wild animals to the urban structural features of homes, high-rise office buildings, warehouses, and athletic stadiums, among others. Some wild animals are also selective in terms of what part of the building they prefer (e.g., roof, attic, walls, basement, inside, or outside); others are generalists and will occupy any available spaces to which they have access.

6.2.1 Buildings

Wild animals occupy urban buildings because they have been invited by the structural design, are opportunistic in seeking shelter, or need an area to rest (e.g., hibernate) or raise young. More often than not, humans who occupy the buildings adopted by wild animals do not even know they are present until the telltale signs of fecal droppings, structural damage from gnawing, smells, or sounds tip them off concerning the animal's presence. The assemblage of vertebrate animals that have accepted urban structures and features associated with theses structures (e.g., lawns and gardens) as alternative substrates to conduct their life cycle activities include 204 bird, 50 mammal, and 41 amphibian or reptile species (http://www.enature.com/).

On the other hand, birds and buildings are a lethal combination. In Chicago, experts estimate that any single, tall building could be killing 2,000 birds a year during peak migration (Figure 6.1). From 1968 to 1998, more than 26,000 migrating

Figure 6.1 Some experts estimate up to 100 million birds die each year in collisions caused by the artificial lights of tall buildings, communication towers, and airports. This photo shows a sample of birds collected beneath high-rise buildings in Toronto during one migration season. (Mark Jackson and Fatal Light Awareness Program)

birds died crashing into a single building along the Chicago lakefront. Birds migrate at night using the stars as a navigational tool and often following a corridor along a body of water such as Lake Michigan. Lighted buildings can disorient them, attracting them to their deaths. Bird biologists speculate that, from a vantage point over the lake, the low, dark mass of McCormick Place may appear to be a cluster of trees. It might look, to birds, like a haven of food and shelter (De Vore 1998).

6.2.2 Windows

On the other hand, human-built structures often pose serious hazards to birds (Figure 6.2). Windows on residential homes account for at least one hundred million bird fatalities (migrants and resident birds) each year in the U.S. Plate glass collisions account for 34% of all avian mortality, second only to hunting (42%). Collisions occur during all seasons, all times of day, and with windows facing any direction (Klem 1991). Approximately 25% (225/917) of the avian species in the U.S. and Canada have been documented striking windows. The 20 most reported species are listed in Table 6.1.

There seems to be a greater vulnerability for those species whose activities occur on or near the ground, such as several species of thrushes, wood warblers, and finches. The ovenbird was the Neotropical migrant reported most often as a window kill. Birds fail to see windows as barriers in their flight patterns and are vulnerable to window strikes wherever they coexist. Any factor (e.g., bird feeders) that increases the density

Figure 6.2   This immature Northern goshawk (*Accipiter gentillis*) is just one of millions of avian fatalities caused each year when birds mistake reflective windows for open sky. (Fatal Light Awareness Program, 2002)

Table 6.1 Species Most Frequently Reported Striking Windows in the U.S. and Canada

American robin (*Turdus migratorius*)	White-throated sparrow (*Zonotrichia albicollis*)
Dark-eyed junco (*Junco hyemalis*)	Ruby-throated hummingbird (*Archilochus colubris*)
Cedar waxwing (*Bombycilla cedrorum*)	Tennessee warbler (*Vermivora peregrine*)
Ovenbird (*Seiurus aurocapillus*)	Yellow-bellied sapsucker (*Sphvrapicus varius*)
Swainson's thrush (*Catharus ustulalus*)	Purple finch (*Carpodacus purpureus*)
Northern flicker (*Colaptes auratus*)	Common yellowthroat (*Geothlvpis trichas*)
Hermit thrush (*Catharus guttatus*)	Rose-breasted grosbeak (*Pheucticus ludovicianus*)
Yellow-rumped Warbler (*Dendroica coronata*)	Gray catbird (*Dumatella carolinensis*)
Northern cardinal (*Cardinalis cardinalis*)	Wood thrush (*Hvlocichla mustalina*)
Evening grosbeak (*Coccothraustes vespertinus*)	Indigo bunting (*Passerina cyanea*)

Source: Klem, D., Jr., *Wilson Bulletin,* 101:606–620, 1989.

of birds near windows will account for strike frequency (Klem 1989). The effects of window strikes for birds ranged from no visible damage, to being knocked unconscious, to fractured bones and superficial and internal bleeding (Klem 1990).

Suggested methods to prevent bird window strikes involve decreasing bird density near the window and/or increasing window visibility for birds. For example, decreasing bird density can be accomplished by moving feeders and bird baths 20 to 30 feet from windows. Window visibility can be enhanced using window screens, using interior vertical blinds or leaded glass decorations, placing cutout decals or vertical strips on the windows, and covering the window with soap or planting shade trees outside windows to reduce reflection (Klem 1991).

6.2.3 Towers

Communication towers, wind turbines, smoke stacks, and high-rise buildings cause bird and bat mortalities. Banks (1979) estimated that 1.2 million birds per year were killed by communications towers across the U.S. (Figure 6.3, top). The Federal Aviation Administration (FAA) tracks the number of towers across the continent to monitor aviation hazards. Generally, once a tower reaches 200 feet or higher, the FAA considers it a potential aviation hazard. As of November 2, 1998, the FAA's Digital Obstacle File listed 39,530 towers, in height classes ranging from 200 to over 1,000 feet high, distributed across the lower 48 states (for benefits to birds, see Perspective Essay 6.1). The real total is actually higher, because towers that are close together often get lumped as one aviation obstruction. Since the 1990s, the birth of the cell phone and Personal Communication Service industry has accelerated tower construction to over 5000 new towers per year. Considering the greater proliferation of new towers, the annual estimated bird mortality at communication towers could be over 5 million birds per year (http://www.towerkill.com/issues).

It is difficult to obtain an accurate count of bird mortality at towers because of scavengers that consume the dead birds before they can be identified and the lack

Figure 6.3   There are many lethal bird and bat encounters with communication towers and wind
turbines that extend anywhere from 200 to over 1,000 feet high. (United States
Department of Energy and John M. Davis/Texas Parks and Wildlife Department)

of long-term longitudinal studies at specific sites. One study summarized the data
collected over a 37-year period at a television tower in Nashville, TN. The study
identified the overall numbers and species of birds that were killed (Table 6.2).

Table 6.2 Bird Fatalities Collected at the WMSV Television Tower: 1960–1997

Rank	Species	Number	Rank	Species	Number
1	Ovenbird	4362	11	Blackburnian Warbler	337
2	Tennessee Warbler	3579	12	Gray Catbird	328
3	Magnolia Warbler	1992	13	Yellow-breasted Chat	227
4	Red-eyed Vireo	1618	14	Philadelphia Vireo	205
5	Black-and-white Warbler	1177	15	Northern Waterthrush	203
6	Chestnut-sided Warbler	953	16	Palm Warbler	192
7	Bay-breasted Warbler	855	17	Indigo Bunting	164
8	American Redstart	555	18	Kentucky Warbler	160
9	Black-throated Green Warbler	367	19	Rose-breasted Grosbeak	146
10	Common Yellowthroat	357	20	Yellow-rumped Warbler	131

Source: Nehring, J. D., Middle Tennessee State University, Murfreesboro, TN, 1998.

Shire et al. (2000) provided a summary of 47 studies of bird kills at communication towers which included 184,797 birds of 230 different species (approximately one quarter of the number of species in the U.S.). There was an 80% correlation between the top 20 species listed by Nehring (1998) and Shire et al. (2000). The majority (92%) of bird kills at towers are migratory, predominantly or frequently at night. Of note was the prevalence of ovenbirds as tower victims as well as window victims.

Birds die by direct impact with the tower or its guy wires or with other birds flying in circles around a lighted tower. Most fatalities occur when a cloud cover prevents moonlight, and birds' ability to use their nighttime navigation systems. Birds attempt to use the tower lights as a reference point during flight and fly in circles around the tower until they run into something or fall to the ground in exhaustion. At this time, there are no management strategies available to mitigate bird mortality at communication towers. Perhaps the day will come when the towers are no longer needed to facilitate the mass communication needs of the public. Even then, it is doubtful that the towers would be dismantled very quickly.

Wind power development is a relatively new technology to produce electrical energy, which could replace less environmentally friendly technologies such as coal-fired generating plants. Now there is a growing body of evidence associating the deaths of birds and bats with collisions into 340-foot-high wind turbines (Figure 6.3, bottom) Of the 33,000 bird fatalities reported in the references used by Erickson et al. (2001), 34% were diurnal raptors, 32% protected passerines, 14% nonprotected birds, 9% owls, and 4% waterbirds and/or waterfowl. Rugge (2001) recorded 316 bird fatalities, of which 55% (174) were raptors (more than half were red-tailed hawks, *Buteo jamaicensis*). Factors that influenced raptor fatalities at wind turbines included:

1. Geographic and habitat locations (i.e., topographic relief),
2. Season and types of available prey species,
3. Wind conditions that promoted flight characteristics (e.g., soaring or kiting),
4. Flight behavior while foraging.

Hoover and Morrison (2005) provided some proactive management strategies to reduce raptor fatalities at wind turbines. They suggested that, prior to installation of new wind turbines, site assessments should determine whether topographical and/or weather conditions may produce dangerous conditions for foraging raptors.

Bat mortality at wind turbines has received considerable recent attention due to a large bat kill at a West Virginia wind farm in 2003 (Williams 2003 as cited by Johnson et al. 2004). Although speculative, bat fatalities at wind turbines may be associated with migration. It is largely unknown why bats collide with turbines, but there is some speculation concerning whether the bats turn off their echolocation abilities during migration. No research has been done on the sound wave effects of wind turbines traveling at 100 miles per hour, and the possible effect on the echolocation capabilities of migratory bats.

An ongoing study by Johnson et al. (2003 and 2004) and Johnson and Strickland (2003) identified six different species of bats that collided with turbines (Table 6.3). Most of the fatalities were comprised of three species of tree bats that migrate long distances and do not hibernate, including the hoary bat, eastern red bat, and silver-haired bat. As with bird fatalities at communication towers, one would expect that geographic and habitat locations and season may be factors that influenced species compositions of bat fatalities at wind turbines. However, the species composition was consistent in Tennessee, Wisconsin, Washington, Oregon, Colorado, Wyoming, Pennsylvania, California, Minnesota, and West Virginia (Table 6.3). Additionally, the highest numbers of bat fatalities occurred at three wind plants including Buffalo Ridge, MN ($N = 420$), Backbone Mountain, WV ($N = 242$), and Foote Creek Rim, WY ($N = 135$). There may be some relationship between number of bats found and the level of investigation by researchers at the various wind plants. Some bats could not be identified because of heavy insect scavenging. There is some irony in the phenomenon of insect-eating bats being eaten by insects. No bat fatalities classified as threatened or endangered species have been documented at wind plants (Johnson and Strickland 2003).

Table 6.3   Number and Total Proportion of Bats Collected at the Wind Towers in 10 Different States from 1998 to 2002

Species	Number of carcasses	Percent of identified fatalities
Hoary bat (*Lasiurus cinereus*)	601	41
Red bat (*Lasiurus borealis*)	441	30
Eastern pipistrelle (*Pipistrellus subflavus*)	142	10
Silver-haired bat (*Lasionycteris noctivagans*)	113	8
Little brown bat (*Myotis lucifugus*)	109	7
Unidentified	25	2
Big brown bat (*Eptesicus fuscus*)	22	1
Northern long-eared bat (*Myotis septentrionalis*)	7	<0.5
Mexican free-tailed bat (*Tadarida brasiliensis*)	1	—
Long-eared myotis bat (*Myotis evotis*)	1	—
Total	1,462	100

Source: Johnson, G. D. et al., *American Midland Naturalist,* 150:332–342, 2003; Johnson, G. D. et al., *Wildlife Society Bulletin,* 32:1278–1288, 2004; Johnson, G. D. and Strickland, M. D., Western Ecosystems Technology, Inc. Cheyenne, WY, 2003.

6.3 ROADS AND HIGHWAYS

Urban sprawl could not have occurred without the Highway Trust Fund (HTF), created by the Highway Revenue Act of 1956 (Pub. L. 84–627) primarily to ensure a dependable source of financing for the National System of Interstate and Defense Highways and also as the source of funding for the remainder of the Federal-aid Highway Program. The HTF provided each state with the resources to build a massive interstate highway system throughout the U.S.

One of the most widespread forms of modifications of the natural landscape during the past century has been 3.9 million miles of road construction. Trombulak and Frissell (1999) estimated an average loss of 11.8 million acres of land and water bodies that formerly supported plants, animals, and other organisms. With the continuing growth in the size of highways (e.g., number of lanes) and higher traffic volumes, there is a growing threat to wildlife that affects a wider range of wildlife species and presents an almost impassable barrier for many species of reptiles, amphibians, and small mammals (Jackson 2000). At first, Forman (2000) estimated a "road-effect zone" (area affected ecologically by roads and associated vehicular traffic) that encompassed 19% of the total area of the continental U.S. In a more recent publication, Forman et al. (2003) estimated that the area covered by roads, roadsides, and medians was equivalent to 1% of the land base of the U.S., a surface area equivalent to the state of South Carolina. The long-term consequences of highway construction include:

1. Animal mortality (i.e., roadkills) and secondary effect on carrion feeders,
2. Loss and change of habitat and biological communities,
3. Habitat fragmentation and secondary effect on dispersal and vagility and/or isolation of selected species,
4. Degradation of habitat quality,
5. Increase in human exploitation through poaching and hunting,
6. Disruption of social structure,
7. Reduced access to vital habitats,
8. Population fragmentation and isolation,
9. Disruption of dispersal processes that maintain gene flow within species populations,
10. The need to develop structures that mitigate the impacts of highways on wildlife populations (Spellerberg 1998; Jackson 2000).

Some of these consequences are discussed in more detail below.

6.3.1 Animal Mortality

All wildlife species have the basic need to find adequate food and water, shelter, and mates. The movement necessary to fulfill these basic needs is usually what compels an animal to cross the extremely inhospitable landscape of a highway (Jacobson 2002). The highways of America are littered with road-killed animals (Figure 6.4). The victims include a variety of different species and numbers of each species (Table 6.4).

Figure 6.4   Road-killed wildlife. (John M. Davis/Texas Parks and Wildlife Department)

Clearly, the census of road-killed animals shows a preponderance of large ungulates (mostly white-tailed deer, *Odocoileus virginianus*). Deer roadkills increased in 26 states from 1982 through 1991. The national deer roadkill data for 1991 conservatively totaled at least 726,000 deer (92% fatal), causing $1.1 billion in property

Table 6.4   Vertebrate Animals and Numbers Listed as Traffic Victims in 19 Journal Articles

Vertebrate Animals	Numbers	Vertebrate Animals	Numbers	Vertebrate Animals	Numbers
Frogs/ Toads	1,129	Cottontail	14,401	Porcupine	1
Salamanders	100	Coyote	933	B.T. prairie dog	1
Lizards	95	Deer	90,150	Raccoon	4,037
Snakes	957	Domestic cat	429	Rat—cotton	56
Turtles	223	cow	2	norway	17
Doves/ Pigeons	97	dog	98	kangaroo	316
Domestic Fowl	266	pig	10	Skunk	3,353
Grouse/ Quail	94	Franklins ground squirrel	28	Tree Squirrel	1,683
Pheasant	11,351	13-lined ground squirrel	320	Weasel	2
Raptors	214	Jackrabbit	1,782	Woodchuck	8
Songbirds	2,403	Mice	11		
Waterfowl	15	Mink	1		
Armadillo	50	Mole	5		
Badger	617	Muskrat	1,256		
Chipmunk	8	Opossum	2,517		

Source: Adams, C. E., *American Biology Teacher,* 45:256–261, 1983.

damage (avg. $1,577/accident), 29,000 human injuries, and 211 human fatalities (Conover et al. 1995; Romin and Bissonette 1996).

However, deer are not the only victims of highway traffic. A road survey of small vertebrate road-killed animals in Canada by Clevenger et al. (2003) identified 677 animals (56 different species) including 313 mammals (18 species), 316 birds (36 species) and 48 amphibians (two species). Vehicular collisions were the cause of at least 50% of bobcat and coyote mortalities (Tigas et al. 2002). A survey of road-killed snakes in Arizona by Rosen and Lowe (1994) included 368 snakes, which included two species of conservation interest. They estimated that from tens to hundreds of millions of snakes have been killed by automobiles in the U.S.

In eastern Texas road mortality may have caused the loss of the timber rattlesnake (*Crotalus horridus*) populations and large snake populations (Jackson 2000). Mumme et al. (2000) examined the impact of road mortality on the demographic makeup of the Florida scrub jay (*Aphelocoma coerulescens*), a threatened species. Nearly half of the reported deaths of the endangered Florida panther (*Puma concolor coryi*) are from collisions with vehicles (Foster and Humphrey 1995). Traffic mortality had a significant negative effect on the local densities of frogs and toads (Fahrig et al. 1995).

Roads and automobiles are relatively recent environmental variables in the evolutionary history of all wildlife animals. As such, there is no genetic program that enables wildlife to deal with a potentially lethal encounter with automobile traffic. For example, some raptors (e.g., owls) swoop low to the ground to capture prey. Swooping is not an adaptive trait for survival when done over a busy interstate highway. Most animals are not accustomed to the light intensity of automobile headlights, which causes them to freeze in place rather than move out of the path of an approaching vehicle at night. Some animals cannot get across the road fast enough (e.g., turtles, amphibians, and small mammals). Armadillos (*Dasypus novemcinctus*) have the strange habit of jumping straight up when a car passes over them which is why road-killed armadillos usually display massive trauma to the back. The classic "playing possum" response to danger is one reason opossums are well-represented among roadkill victims (Jacobson 2002).

The presence of a road may modify an animal's behavior through home range shifts and altered movement patterns, reproductive success, escape responses, and physiological states (Trombulak and Frissell 1999). On the other hand, both turkey vultures (*Cathartes aura*) and black vultures (*Coragyps atratus*) preferentially establish home ranges in areas with greater road densities (Coleman and Fraser 1989), probably because of the increase in carrion from roadkills.

6.3.2 Overpasses, Underpasses, and Escape Routes

Highway construction affects wildlife through the direct loss and fragmentation of habitat and by disrupting animal movement and dispersal (e.g., migratory routes and home range activities). From a landscape ecology perspective, highways have the potential to undermine ecological processes through the fragmentation of wildlife populations, restriction of wildlife movements, and the disruption of gene flow and metapopulation dynamics (Jackson and Griffin 2000). In order to mitigate highway impacts on wildlife, alternative structural designs have been used to provide safe passage for large and small animals over or under the highway or to provide escape routes (e.g., fencing) that direct wildlife away from the highway or toward under- or overpasses (Transportation Research Board 2002; Figure 6.5).

The Wildlife Crossings Toolkit (www.wildlifecrossings.info) an online information source developed by the USDA Forest Service, was designed for professional wildlife biologists and engineers faced with integrating the highway infrastructure and wildlife resources. The Toolkit is a searchable database of case histories of

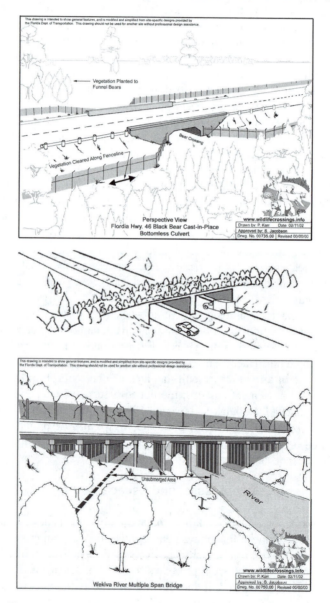

Figure 6.5   Three examples of wildlife crossing designs. (Provided by USDA Forest Service in their Wildlife Crossings Toolkit.)

mitigation measures, and articles on decreasing wildlife mortality and increasing animals' ability to cross highways. Professional wildlife biologists and engineers can use the Toolkit to creatively solve challenges associated with highways.

There is an International Conference on Ecology and Transportation (ICOET). Conducted every two years, ICOET is designed to address the broad range of ecological issues related to surface transportation development, providing the most current research information and best practices in the areas of wildlife, fisheries, wetlands, water quality, overall ecosystems management, and related policy issues. ICOET is a multidisciplinary, interagency supported event, administered by the Center for Transportation and the Environment (http://www.icoet.net). The first and second conferences were held in 2001 and 2003, respectively. The third conference is scheduled to be held in 2005. The website contains additional details concerning these conferences. Proceedings are produced after each conference that contain reports on research conducted to determine the impacts of the highway infrastructure on wildlife and their habitats.

6.3.3 Structural Design Considerations

Bridge overpasses, underpasses, and fencing are designed to provide safe passage for a large assemblage of vertebrates, including amphibians, reptiles, and mammals or whole faunal communities within a region. In addition, many of the bridge designs provide other benefits to some animals, including nesting, resting, brood-rearing, and hibernation sites. The bridge structures also provide alternative dispersal sites for large and small animals, particularly when young mammals are leaving parental home ranges and seeking to establish their own territories.

It is important to maintain connectivity between the island populations of some animals created by the habitat fragmentation caused by highway construction. Connectivity is important in terms of reversing (1) loss through dispersal, (2) chance extinctions, and (3) the effects of inbreeding such as genetic drift and the loss of genetic variability. However, the potential negative consequences of increased connectivity include: (1) disease or parasite transmission into a population that may have been disease free (e.g., chronic wasting disease in deer), (2) allowing exotic or other competitors to enter a habitat, and (3) outbreeding depression in some species that may have adapted to isolation and could suffer the negative effects from the introduction of distantly related genetic material (McKelvey et al. 2002).

In some cases, bridge design changes were serendipitous for some animals (see Chapter 8). For example, a design change in the reconstruction of the multiple-span bridge over the Colorado River in Austin, Texas, created a wider crevice size between the bridge's expansion joints. New spaces between the expansion joints were three-quarters to four inches wide and about 16 inches deep. Mexican free-tailed bats found the new spaces irresistible and moved in soon after bridge construction was finished (Murphy 1990). Crevice width and depth are important bridge design considerations for bats. Textured surfaces on concrete and wood bridges are important features for day and night roosting by bats and nest building by swallows. The rough textured surfaces and temperature range inside box culvert bridges make ideal hibernacula for some bat species (Walker et al. 1996). Jackson and Griffin (2000) suggested the following general considerations for wildlife crossing structures designed to mitigate highway impacts on wildlife:

1. Place structures in areas of known migratory and dispersal routes of selected species and/or areas of high highway mortality.
2. Size of the underpass to be relative to the width of the highway — bigger is better.
3. Consider lighting — some species are hesitant to enter underpasses that lack sufficient ambient light — but avoid artificially lit areas.
4. Consider moisture conditions; wet substrates are important for some amphibian species.
5. Consider sustained temperature regimes and air flow.
6. Consider traffic noise, as it is a problem for some mammals sensitive to human disturbance.
7. Consider a substrate that provides tactile security.
8. Design approaches (covered or clear) that provide visual security for the design species.
9. Include fencing to help guide animals to the passage system and prevent them from circumventing the system.
10. Consider placement away from a high level of human disturbance.
11. Consider potential interactions among species (e.g., predator and prey) at crossing sites (Jackson and Griffin 2000; Transportation Research Board 2002).

The Highway Bridge Replacement and Rehabilitation Program (HBRRP) is authorized by the federal Transportation Equity Act for the 21st Century (TEA21). The purpose of the program is to replace or rehabilitate public highway bridges over waterways, other topographical barriers, other highways, or railroads when the state and the Federal Highway Administration determine that a bridge is significantly important and is unsafe because of structural deficiencies, physical deterioration, or functional obsolescence (http://www.fhwa.dot.gov/bridge/hbrrp.htm). This program offers a unique and timely opportunity to rebuild, reconstruct, or refurbish bridges to accommodate wildlife using designs such as those illustrated in the Wildlife Crossings Toolkit (Figure 6.5). These accommodations could be the types of structures that now allow safe passages for wildlife. If these accommodations are not part of the HBRRP, the opportunity to provide them will be missed until the next bridge replacement cycle 50 to 70 years in the future (personal communication, S.L. Jacobson, Wildlife Biologist, USDA Forest Service).

6.3.4 Bridges, Birds, and Bats

Some animals have taken advantage of the alternative nesting, roosting, and dispersal opportunities provided by various highway structures (e.g., bridges, underpasses, overpasses, and culverts). For example, in the past 100 to 150 years, cliff swallows (*Hirundo pyrrhonota*), have expanded their range across the Great Plains and into eastern North America, a range expansion coincident with the widespread construction of highway culverts, bridges, and buildings that provide abundant alternative nesting sites (Brown and Brown 1995). Originally nesting primarily in caves, the

Figure 6.6   Swallows nesting under a highway bridge. (Jeffrey S. Pippen)

barn swallow (*Hirundo rustica*) has almost completely converted to breeding under the eaves of or inside artificial structures such as buildings and bridges (Brown and Brown 1999). Since the mid 1980s, the cave swallow (*Hirundo fulva*) has undergone a dramatic range expansion in Texas and also colonized south Florida. In each of these cases, invasion of new territory has been facilitated by the adoption of bridges and culverts for nesting, with new colonies often springing up along highways (West 1995; Figure 6.6).

Human structures may often be superior to natural nest sites in important ways: predators may be less likely to gain access to nests, nest substrates may be superior for the long-term attachment of nests, and the thermal environments may be more favorable. By adopting human structures for nest sites, barn swallows are able to take advantage of localized food sources not otherwise profitably exploited (Speich et al. 1986). Other birds that have occupied bridges (personal communication R.J. Reynolds, Virginia Dept. of Game and Inland Fisheries) include the eastern phoebe (*Sayornis phoebe*), northern rough-winged swallow (*Stelgidopteryx serripennis*), pigeon (*Columba livia*), and osprey (*Pandion haliaetus*). Some birds (e.g., phoebes) will occupy the nests built by other birds under bridges rather than build their own nest.

Many different species of bats have adapted to the various structural designs of highway bridges for day and night roosting, migratory rest stops, brood rearing, and as hibernacula during the winter (Lewis 1994; Keeley 1995; Walker et al. 1996; Adam and Hays 2000; Kiser et al. 2001). Bridges used as night roosts may serve several functions, including conservation of energy (thermoregulation), protection from predators, and locations for information transfer, social interaction, and consumption and digestion of prey (Adam and Hayes 2000:402). The assemblage of bat species (Table 6.5) that use bridges includes 24 of the 45 U.S. species of bats, and another 13 species are likely to do so.

The diversity of bat species that occupy bridges is impressive as well as the numbers of some species. For example, the Congress Avenue bridge in Austin, TX, is the summer home for 1.5 million Mexican free-tailed bats. The spectacular

Table 6.5   Bat Species Known to Occupy Highway Bridges in the U.S.

Common Names	Scientific Names
Big brown bat	*Eptesicus fuscus*
Big–free-tailed bat	*Tadarida molossa*
California leaf-nosed bat	*Macrotus californicus*
California myotis	*Myotis californicus*
Cave myotis	*Myotis velifer*
Eastern long-eared myotis	*Myotis evotis*
Eastern pipistrelle	*Pipistrellus subflavus*
Evening bat	*Nycticeius humeralis*
Fringed myotis	*Myotis thysanodes*
Gray myotis	*Myotis grisescens*
Indiana bats	*Myotis sodalis*
Little brown myotis	*Myotis lucifigus*
Long-legged myotis	*Myotis volans*
Mexican free-tailed bats	*Tadarida brasiliensis*
Mexican long-tongued bat	*Leptonycteris nivalis*
Northern long-eared Myotis	*Myotis septentrionalis*
Pallid bats	*Antrozous pallidus*
Rafinesque's big-eared bat	*Plecotus rafinesquii*
Silver-haired bat	*Lasionycteris noctivagans*
Southeastern myotis	*Myotis austroriparius*
Western or Townsend's big-eared bat	*Plecotus townsendi*
Western pipistrelle	*Pipistrellus hesperus*
Western small-footed myotis	*Myotis subulatus*
Yuma myotis	*Myotis yumanensis*

Source: Keeley, B. W. and Tuttle, M. D., Bat Conservation International, Austin, TX, 1999 (Copyright Bat Conservation International, www.batcon.org. Used with permission.)

evening emergence of the bats has become a wildlife-watching opportunity for Austin residents and tourists alike. Crowds of several hundred are usual. A public education campaign by Bat Conservation International (BCI) changed public opinion regarding the ecological importance (consume 10,000 to 30,000 pounds of insects nightly) of the Congress Avenue bridge bat colony, and corrected misinformation concerning bat natural history (Murphy 1990). The economic impact to the community is discussed in Chapter 8. There is now a collaborative effort between BCI and the Texas Department of Transportation to develop bat-friendly bridge designs.

6.4 LANDFILLS, DUMPSTERS, AND GARBAGE CANS

6.4.1 Organic Waste Accumulations: A Concept Unique to Urban Ecosystems

Examples of urban organic wastes would include garbage or municipal solid waste (MSW), yard clippings and leaves, and human, pet, and urban wildlife excrements. Estimates of the total volume of various organic wastes within urban communities include the following.

As mentioned in Chapter 3, a city with one million inhabitants is estimated to consume 25,000 tons (t) of water and 2,000 t of food per day and produces 50,000 t of effluent water and 2,000 t of waste material daily (Deelstra 1989).

Approximately 231.9 million tons of municipal solid waste (MSW) was generated in the U.S. in 2000. This volume of MSW translates into about 4.51 pounds per person per day (Environmental Protection Agency, U.S. EPA 2002). Piles of MSW on city streets are invitations for a free meal for rats, cats, raccoons, coyotes, and cockroaches.

The volume of pet wastes can only be estimated, given the large varieties of pet types and sizes in urban communities. Nearly 1.4 billion tons of animal manure are produced annually in the U.S. The New Jersey Department of Health estimated that there are over 500,000 dogs in the state. Adding in cats and other smaller pets with dogs results in a significant volume of waste daily. The EPA estimated that for watersheds of up to 20 square miles draining to small coastal bays, 2 or 3 days of droppings from a population of about 100 dogs would contribute enough bacteria and nutrients to temporarily close a bay to swimming and shell fishing. In the Four Mile Run watershed in Northern Virginia, a dog population of 11,400 is estimated to contribute about 5,000 pounds of solid waste every day and has been identified as a major contributor of bacteria to the stream. A single gram of dog feces can contain 23 million fecal coliform bacteria. A general lack of public recognition about the water quality and health consequences of dog waste calls for improved watershed education efforts (USEPA 1993).

There are thousands of Canada geese in some communities (e.g., Minneapolis and St. Paul, MN). Canada geese eat and excrete at a high frequency daily. For example, the number of droppings per goose may range from 28 to 92 per day each weighing from 1.17 to 1.9 grams. Geese are not very concerned where they defecate, whether on lawns, golf courses, athletic fields, park lawns, swimming pools, or city lakes. Goose droppings are organic and require bacterial decomposition, which in lakes increases biological oxygen demand, reducing oxygen levels. The nitrogen and phosphorus in goose manure represents another form of fertilizer. The amount of nitrogen/goose/year ranges from 1.15 to 3.11 pounds. The amount of phosphorus/goose/year ranges from 0.36 to 1.41 pounds. These fertilizer loads will cause eutrophication in city water ways (The Water Line 1997). The resident Canada goose problem is covered in more detail in Chapter 12.

Overall, the average family throws away 1.28 pounds of food per day, for an annual total of 470 pounds per household, or 14% of all food brought into the house (unpublished data, Tim Jones, Bureau of Applied Research in Anthropology, University of Arizona).There is no estimate of the food discarded by restaurants and grocery stores.

The process of removing organic wastes from urban communities is a primary consideration in city planning, because there is so much of it. There is so much waste because urban residents, their pets, and urban wildlife are concentrated in a relatively small area (e.g., 5000 humans/sq. mile) and do not know how to or do not want to recycle the various forms of matter they encounter on a day-to-day basis. Most of it can be recycled.

There are many advanced forms of technologies to recycle urban organic wastes, but cost and convention prevent these technologies from being implemented in most urban communities. Conventional technologies generally involve a process of "filing by piling" (i.e., nothing is actually done with the wastes; they are just allowed to accumulate in different piles of matter straining the ability of local ecosystems to assimilate them).

Garbage accumulation in urban environments has had a profound effect on the legendary predator-prey relationships between cats and rats. Childs (1991) found that inner city rats grow faster, reproduce earlier, and have many more offspring than parkland rats. He observed cat predation only on rats 7 ounces or less, too young to contribute to the rat recruitment rate in the inner city. By weight, at least 30% of inner city garbage contains edible material for rats and cats. Some inner city rats can grow as large as a pound and one half. Inner city cats were more likely to be observed feeding side-by-side with the inner city rat on the same nutrient-rich garbage resource rather than preying on the rat as a food source. In even more bizarre change in cat vs. rat behavior was the observation of rats feeding on young cats (Sullivan 2004).

The habitats adjacent to landfills, dumpsters, and garbage cans will predict, in part, the type of wildlife that will exploit the organic resources (e.g., food). The list of wildlife attracted to residential and/or landfill garbage is impressive. The cast of garbage consumers includes gulls (*Laurus* spp.), vultures (*Cathartes aura* and *Coragyps atratus*), bears (*Ursus* spp.), rats (*Rattus norvegicus*), free-ranging cats (*Felis catus*) and dogs (*Canis familiaris*), coyotes (*Canis latrans*), deer (*Odocoileus* spp.), raccoons (*Procyon lotor*), opossums (*Didelphis marsupialis*), bobcats (*Felis rufus*), and ravens (*Corvus corax*). Predatory hawks and owls appear at landfills to hunt rodents, birds, and young cats (Eberhard 1954; Horton et al. 1983; Childs 1991; Belant et al. 1995, 1998; Manley and Williams 1998; Slate et al. 2000; Restani et al. 2001; Tigas et al. 2002; Williams 2002; Hutchings 2003).

An understanding of the full assemblage of wildlife species attracted to concentrations of human food waste is limited by several factors. First, research on wildlife attracted to garbage sources does not have a lot of prestige in the scientific community, so it is difficult to find investigators and the financial resources to conduct the research. Second, merely identifying what species utilize garbage is not as important a research objective as determining the impact of this type of feeding behavior on other aspects of the animals "normal" existence. For example, Childs (1991) observed three large rats and four cats feeding side-by-side from an overturned trash can. Feral cats can convert to scavenging rather than predation to obtain food. However, a well-fed cat will still kill other wildlife species if the opportunity presents itself (Hutchings 2003). Williams (2003) reported on animal addiction to human food leading to loss of interest in natural foods and digestive disorders. Bears can become aggressive panhandlers for human food (Manley and Williams 1998). At landfill sites, gull and raven numbers increased beyond the carrying capacity of their natural habitats (Belant et al. 1995; Restani et al. 2001). Estimates of gulls using landfills ranged from several hundred to 50,000, but the actual number of gulls may be under-estimated (Slate et al. 2000; Belant 1997). Finally, research on animal utilization of garbage as a food source would fall under an urban wildlife management paradigm, which introduces a complex set of problems resulting from frequent contacts with people during the research process (Van Druff et al. 1996).

6.4.2 Factors That Promote the Presence of Wildlife at Landfills

The classic image of urban wildlife is raccoons emptying a garbage can in an urban area (Figure 6.7). One of the most critical problems to solve in urban areas is household garbage disposal. There are three basic steps in the garbage disposal process, beginning with the residential collection in garbage cans, followed by municipal collection, and finally deposition at a landfill site (i.e., city dump). The first and last steps provide food accumulations for wildlife. As mentioned earlier, the average family throws away an enormous amount of food daily and annually. Excluding the categories of liquids, slop, and other, the food group frequencies, by weight, were grain (20%), meats (16%), fruits (30%), vegetables (32%), and fats (2%) (unpublished data, Tim Jones, Bureau of Applied Research in Anthropology, University of Arizona). This is just the amount and types of food that end up in the garbage can. It does not include food that is fed to pets, goes down garbage disposals, is composted, or is fed directly to wildlife in the neighborhood. Nor does it include food discarded by restaurants or grocery stores. Nevertheless, urban residents provide a buffet of food resources that attract many species of wildlife to landfills.

6.4.3 Standards Used in Landfill Siting

Few city and county politicians want to be in office when the job of siting a new landfill becomes necessary. The task can become embroiled in conflicts related to land acquisition negotiations, consideration of environmental impacts, public dissent if the new landfill is too close to their property and will adversely affect property values or quality of life, and changes in land-use patterns. The landfill siting process gives priority consideration to location concerning proximity to floodplains or areas that have critical habitats, historical/archeological features, and wetlands. The siting is also evaluated in terms of the required setback distances to navigable waters, state and federal highways, public parks, airports, and water supply wells. Other important considerations in the siting process are proposed landfill life and disposal capacity,

Figure 6.7   Raccoon emptying a garbage can. (USDA photo)

municipalities and industries to be served, anticipated waste types, characteristics and amount of waste to be handled, and regional geotechnical characteristics of proposed location (http://www.dnr.state.wi.us/org/aw/wm/solid/landfill/siting.htm).

The landfill siting process contains many environmental concerns not found in the airport siting process discussed later in this chapter. As pointed out above, several environmental impacts need to be investigated before landfill construction can take place. Of special note is the impact of the landfill on the hydrology of the area, because of the high potential of leachate generation and groundwater contamination. However, both neglect to include impacts on surrounding habitats and endemic plants and animals. Like airports, landfills are normally built on the outer fringe of urban or suburban communities, which results in the most direct and influential impact on the natural ecosystem and wildlife of the area.

6.4.4 Types of Habitats Found in and around Landfills

As with airports, the types of ecosystems in and around landfills are a function of the dominant ecosystem in the region of construction. Landfills (or open dumps) can be found wherever there is a concentration of humans, and this includes nearly all terrestrial ecosystems. Depending on the ecosystem where the landfill is located, different wildlife species will be attracted to the food sources that are available at landfills. For example, polar bears (*Ursus maritimus*) in Alaska, black bears (*Ursus americanus*) in the northeastern U.S., gulls (*Laurus* sp.) along coastal states, and rat (*Rattus rattus*) throughout the U.S. represent ecological equivalents for landfill feeding in their geographical regions.

6.4.5 Human–Wildlife Conflicts at Landfills

Anytime wild animals associate human presence with a food supply and become food-conditioned to the resource, there will be human–wildlife conflicts. These conflicts can range from the aggravation of having to pick up loose garbage spread across your lawn by a raccoon to being seriously mauled by a bear. The most publicized human–wildlife conflicts that occur because of the association of humans, wildlife, and unnatural food sources are airplane strikes (covered later in this chapter) and bear attacks, particularly if the bear is a female with cubs. Slate et al. (2000) listed several direct human–wildlife conflicts with gulls that included safety concerns for equipment operators because of reduced visibility as gulls move in large flocks at the working face, inability to control gulls as potential disease vectors, corrosion of equipment, and increased bacterial counts from gull fecal material (Figure 6.8).

Belant (1997) listed the types of human–wildlife conflicts that occurred when gulls leave the landfill and move into the urban community. For example, roof-nesting gulls harassed maintenance personnel, deposited fecal wastes on roofs and nearby vehicles, plugged up roof drainage systems with debris, were noisy, and caused structural damage to buildings. Gull transmission of bacteria that cause enteric disease in humans is a growing concern with large colonies of roof nesters. In a survey of municipalities in the U.S. regarding vertebrate pests, gulls were ranked

Figure 6.8   Gulls at a landfill. (Jeffrey S. Pippen)

as the ninth by Fitzwater (1988) and the fourth most frequently occurring nuisance species in the Middle Atlantic States (see Table 2.1).

Three events have led to an ever increasing number of human–bear conflicts. The first event was the urban/suburban sprawl into traditional bear habitat, forcing a daily coexistence. The second event was the bears' attraction to unnatural foods (e.g., residential or landfill garbage) to which bears became food-conditioned. Bears obtain food from humans by scavenging garbage at dumps, raiding garbage containers, camps, or buildings, stealing food directly from people, or being given handouts by people. Bears that feed on garbage are larger and tend to have higher reproductive rates (Stringham 1989). The third event, in concert with the second, was bear habituation to the presence of humans. Human and marauding bear conflicts are of particular importance because of the potential of physical or mortal danger to both (Rogers 1989; Peine 2001).

6.4.6 Wildlife Management at Landfills

Wildlife management at landfills can begin by addressing the root cause of the problem rather than the symptoms. Addressing the root cause requires creative siting away from transportation centers (e.g., airports), wildlife refuges, wetlands, major flyways of migratory birds, and sensitive habitats. The next requirement is the utilization of modern landfill technologies that reduce the exposure of municipal solid waste (MSW) to wildlife in the surrounding habitats. For example, most modern landfills compact the MSW and deposit it in refuse cells that are covered with 6 to 8 inches of fill dirt.

Other technologies that reduce or eliminate altogether the exposure of MSW to wildlife are waste-to-energy (WTE) facilities that separate the MSW into recoverable and combustible categories. Overall, about 80% of the MSW is burned for energy, 12% is recovered, and 8% is put in landfills (Wright 2004:498).

Many communities are turning to nontraditional waste management facilities such as yard-waste compost facilities, construction and demolition landfills, and trash transfer stations. Gabrey (1997) found that nontraditional waste-management facilities do not appear to attract birds or small mammals at higher than background levels and would not pose a significant nuisance problem to the community or be a hazard to aircraft if located near airports. A key to wildlife management at landfills is waste reduction, particularly food wastes. Food waste is primarily (70%) liquid and can be converted into a dry pulpy material using a dehydrating machine (http://www.wastereductionsystems.net/Food_Waste_Reduction.html). Composting has been an extremely successful food waste reduction option in many communities (http://www.ilsr.org/recycling/wrrs/food/food.html). Composting of food wastes can be done at home using a garbage can as the composting apparatus (see Perspective Essay 6.2). Finally, and probably most importantly, wildlife management at landfills begins at home with the individual in terms of how much he or she purchases, prepares, and consumes food. In general, the majority of affluent urbanites could purchase, prepare, and consume LESS food.

The management alternatives to control gull populations at landfills included nonlethal pyrotechnics, high grass management on gull loafing areas, overhead wires on active and capped portion of landfills, gull distress calls, bird of prey kites and predator-eye or other balloons, repellants, and shooting gulls (Slate 2000). None of these alternatives can be expected to have any long-term effects, given the lack of control over gull reproductive success, limitations imposed by the Migratory Bird Treaty Act, the continued presence of food waste at landfills, and the ineffectiveness of natural controls (e.g., using birds of prey in large gull populations.) In fact, unless the primary factor that attracts any animal to garbage dumps or landfills is removed, control strategies will only provide short-term reductions in animal numbers.

The management of bears at landfills presents a unique set of problems, given its human value as a charismatic species (see Chapter 11). The valuation forces influencing management policies for nuisance bears are extremely varied and usually invoke strong emotions (Peine 2001). Consider how the bear has captured the human emotion for the species in terms of children's books (e.g., Winnie the Pooh), Smokey the Bear, and teddy bears. Some people value the bear for its parts, a picture, a glimpse, as spiritual deities, to attract customers, or as a symbol of identity of place at communities or schools. Nevertheless, nuisance bears have to be dealt with when they have crossed the line and have a negative impact on human health or economics (Schmidt 1997). Bear management strategies included garbage control, public education on the link between human waste-handling and bear problems, physical and chemical aversive conditioning, trap and translocate, and killing the bear (Ternent and Garshelis 1999; Peine 2001; Clark et al. 2002). Unlike gulls, bears are less likely to rebound from or even survive management programs that do not provide alternative habitats where humans do not exist.

6.5 AIRPORTS

Recall the discussion of food chains in Chapter 3 and how the abiotic conditions of an area predict the dominant plant communities, which, in turn, predict the dominant animal communities (Figure 3.3). Different wildlife species would be present in and around an airport, given its proximity to wetland, grassland, tropical forests, or desert ecosystems. On a smaller scale, grasslands attract rodents that, in turn, attract birds of prey. Other factors that promote the presence of wildlife in and around airports are the presence of wildlife attractants, including wetlands, agriculture, wastewater treatment facilities, wildlife refuges, landfills, and aquaculture activities.

For example, agriculture and landfills are two attractants that provide abundant alternative food sources for selected species that will probably multiply beyond the carrying capacity of their natural environment. Gallaher (2003) reported a correlation between wildlife attractants and aircraft strikes with wildlife. Airports often are situated in outlying areas surrounded by woods, agricultural fields, and early successional habitats. In addition, landing fields are planted with grasses and forbs. These habitat conditions provide prime grazing locations for deer (Wright et al. 1998).

6.5.1 Standards for Airport Siting and Zoning

The development, movement, or expansion of a public or private airport is not a trivial matter. Decisions have to be made regarding location (siting), zoning rules have to be reviewed or written, and forums for public input need to be established. An examination of several siting and zoning ordinances across the country revealed certain standards that were common to all. The common siting standards included airport construction considerations, proximity to and required clearance heights over other transportation networks and waterways, and scheduled public hearings. Common zoning considerations dealt with identifying the existence of natural and man-made obstructions to air navigation.

The specific hazards to landing and taking-off, including trees, towers, and particulate (e.g., smoke) air pollutants, have to be identified. Height of the landing field above sea level was another common consideration. Compatibility with existing land uses and future county and/or regional planning commission projects was a common ingredient in zoning ordinances. In general, airport zoning ordinances focused on anything that reduced the size of the area for landing and taking-off without mention of how various wildlife species might have an impact on this consideration (e.g., deer on the runway).

Public hearings on airport siting or zoning usually focused on impacts on property values, quality of life changes, and the "Not In My Backyard" (NIMBY) syndrome (i.e., public distrust of the political agenda). There were no ordinances that addressed some important ecological considerations including endangered or threatened species, flyways of migratory waterfowl, and potential impacts on critical habitats. Environmental impact statements (EIS) submitted to the U.S. Fish and Wildlife Service become a critical part of the airport siting and zoning planning process when ecological issues need to be addressed.

Due to the real and increasing danger of animal/airplane strikes at airports and liability concerns, the Federal Aviation Administration (1997) published an advisory circular (AC No: 150/5200–33) titled, "Hazardous Wildlife Attractants on or Near Airports." The circular specifies the types of land uses, adjacent to or on airport property, that attract birds and other types of wildlife that are potentially hazardous to aircraft. Types of land uses specifically mentioned were landfills, wastewater treatment facilities, water retention ponds, wetlands, agricultural practices, aqua-culture activities, and surface mining operations, among others. The siting criteria of the circular specify the minimum distance of the airport operations from the above types of land uses.

6.5.2 Types of Habitats Found in and around Airports

The types of ecosystems in and around airports are a function of the dominant ecosystem in the region of construction. Airports have been built in every ecosystem type, including hot and cold deserts, grasslands, temperate deciduous forests, sub-tropical and tropical forests, and coastal wetlands. Airports are normally built on the outer fringe of urban or suburban communities, which results in the most direct and influential impact on the natural ecosystem and wildlife of the area. For example, in some ecosystem types, compliance with the standard zoning and siting ordinances requires dramatic habitat alterations including clear-cutting, draining wetlands, landscape alterations, removal of endemic plants and animals, introduc-tions of exotic plants and animals, and the structural changes associated with building the airport.

In order for the human activities associated with air travel to take place, natural habitats need to be simplified to accommodate their conversion into urban airports. For this to take place, the natural ecosystem succession processes are altered or delayed on all or parts of the airport property. There are alternative airport devel-opment designs that focus on sustainable resource use in and around airports (Dalzell 2004).

6.5.3 Wildlife Species Attracted to Airport Habitats

Wetlands can be found in the vicinity of numerous airports, because wetlands are left undeveloped and may provide for airport approaches involving less risk to the public than approaches over developed areas. The dominant avifauna in wetland habitats include migratory waterfowl, wading birds, and shorebirds (Kennamer 1999). Birds can also be drawn to airports by open short grass and warm pavement where they find security from predators and humans and a place to rest, loaf, and feed (MacKinnon 1997).

Baker and Brooks (1981) identified three dominant habitat types at an airport including short grass, agricultural, and old field succession (e.g., abandoned farm land). The three habitat types harbored an assemblage of rodents, rabbits, and birds, which were the prey species for resident raptor (e.g., hawks and owls) populations. A high ratio of edge-to-airport area including a scrub community, sugarcane field, and golf course and resort supported over 16 different avian species (Linnell et al. 1996).

Mowed and unmowed airport grasslands provided habitat for 22 and 29 differ-ent vertebrate species, respectively (Barras et al. 2000). Small mammal abundance was greater in unmowed than mowed areas. Twice as many raptors were observed in unmowed areas than mowed areas. The authors concluded that maintaining mowed vegetation on the entire airport will decrease the rodent population and, in turn, the raptor population. However, some grassland habitats in or around airports represent the only remaining habitat for some threatened or endangered bird species (Rossi-Linderme and Hoppy 2000). Airport grasslands are recognized as substitute habitats, lost to urban development and agriculture, for six species of grassland birds (Kershner and Bollinger 1996).

A 2.5-pound herring gull, struck by an aircraft traveling 290 miles per hour, has an impact of 34 tons (Rossi-Linderme and Hoppy 2000)

6.5.4 Human–Wildlife Interactions at Airports

From 1990 to 2000, 33,000 bird and about 500 deer strikes to civil aircraft were reported in the U.S. (Figure 6.9). These estimates are based on 20% air strike reporting rate. Dolbeer et al. (2000:39) stated that "aggressive programs by natural resource and environmental agencies during the past 50 years (e.g., expansion of wildlife refuge system), coupled with land-use changes, have resulted in dramatic increases in populations of many wildlife species that are a threat to aviation." He demonstrated these population increases in terms of Canada geese (*Branta canadensis*), ring-billed

Figure 6.9   Birds surround aircraft as they take off and land. (USDA APHIS Wildlife Services)

gulls (*Larus delawarensis*), red-tailed hawks (*Buteo jamaicensis*), turkey vultures (*Cathartes aura*), and white-tailed deer (*Odocoileus virginianus*).

Cleary et al. (2003) provided a detailed summary of the degree to which these same species are involved in strikes with aircraft. For example, waterfowl, raptors, gulls, doves, and deer caused the highest number of strikes resulting in substantial aircraft damage. Migratory waterfowl and deer caused the highest number of strikes resulting in human injury or death. Over a 13-year period, reported losses from bird and mammal strikes totaled 364,626 hours of aircraft downtime, and $170.9 million in monetary losses. White-tailed deer, vultures, and geese were ranked 1, 2, and 3, respectively, as the most hazardous species groups to aviation (Dolbeer et al. 2000).

6.5.5 Wildlife Management Priorities at Airports

Airport wildlife management plans center on minimizing aircraft strikes and, in some cases, preserving habitat for threatened or endangered species. The body of literature on animal control strategies has been amply reported by Conover (2002). In general, the strategies for controlling the numbers of each species hazardous to aircraft flight depend on the (1) offending species, (2) legal ramifications concerning the species in question and some control options (see Chapter 10), (3) the desired level of population control, (4) impact on other species, (5) cost effectiveness, and (6) public reactions.

Management strategies can be site-specific or can be conducted at the landscape-level. The standard selection of management techniques are nonlethal, which includes habitat management, removal of food sources, frightening devices, exclusion, repellents, and capture and translocation. Lethal control techniques include nest/egg disturbance, toxicants, shooting, and predators (Belant 1997).

Sometimes wildlife management at airports may present a paradox of priorities. There are cases when the wildlife management plan to mitigate air strikes are in direct conflict with the management plan to sustain populations of threatened and endangered species (Rossi-Linderme and Hoppy 2000). The Bird Aircraft Strike Hazard (BASH) Team concluded that the best way to reduce the number of birds roosting in and around the base was to reduce the attractiveness of the airfield, which would include mowing the surrounding grasslands. However, these grasslands were considered the premier breeding habitat for the upland sandpiper (*Bartramina longicauda*) and grasshopper sparrow (*Ammodramus savannarum*). Partnering with other Federal agencies and working with local and regional groups resulted in a tentative airfield management compromise.

PERSPECTIVE ESSAY 6.1: THE PEREGRINE STORY

On a positive note, high-rise buildings and power plants, bridges and overpasses, power plant smoke stacks, and other structures have provided alternative nesting and brood rearing sites for the peregrine falcon (*Falco peregrine*) (Figure 6.10). The primary reason for species extinction is habitat loss or change (pollution). The falcon was one of many predatory birds that were adversely affected by the indiscriminant use of DDT and PCBs from 1940 to 1970. Predatory birds are at the top of the food

Figure 6.10   Peregrine falcon with chicks. (USFWS photo)

chain and, therefore, will receive the highest concentrations of each chemical due to ever-increasing concentrations as the chemical moves up the food chain; a process called biomagnification. The chemicals accumulated in the body fat of female falcons, which resulted in a decreased ability to metabolize calcium carbonate. The shell gland, which is part of the reproductive tract of the female, could not deposit enough calcium in the egg shell. This resulted in cracked eggs when the female attempted to incubate them. The loss of falcon chicks on a regular basis soon caused a precipitous decline in falcon numbers throughout the U.S. By the 1970s, none remained in the eastern U.S. In 1970, the U.S. Fish and Wildlife Service listed the peregrine falcon as endangered under the Endangered Species Conservation Act of 1969.

Once the use of DDT and PCBs were banned in the 1972, a program to reintroduce and propagate the peregrine falcon began in New York City. The structural and biotic features of large cities are ideal for falcon reintroduction. For example, high-rise buildings contain nesting ledges similar to those found on rocky cliffs in the falcon's natural environment. Furthermore, there is an abundant prey base for falcons in cities. Cade et al. (1996) reported 104 different bird species as falcon food, and Bell et al. (1996) reported 34 species. Common peregrine falcon food species were pigeons (rock doves), northern flicker, blue jay, mourning dove, European starlings, and American robins.

The technique scientists used to help bring back the peregrines is called "hacking." Hacking is a delicate method of releasing birds back into the wild. Young falcons, still unable to fly, are placed in a specially designed box at the site from which they are to be released. This location is known as the hack site.

Site attendants care for the birds 24 hours a day. Peep holes are built into the box, so the attendants can watch the peregrines without disturbing them. Food is provided through a chute, again eliminating any view of humans. By keeping actual interactions with the falcons to a minimum, it insures the birds are kept wild and not imprinted on people. The front of the box, which is covered with bars, allows the young birds to become accustomed to their surroundings, while at the same time, providing them with protection.

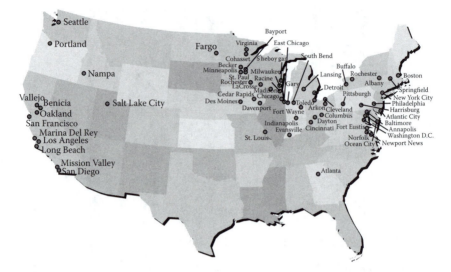

Figure 6.11 Cities where peregrine falcons live. (L. Causey and C. E. Adams)

The box is opened after a week. The young falcons are then 6 weeks of age and ready to take their first flight. Peregrines can instinctively learn how to fly and hunt without the aid of an adult! They will spend the next 3 weeks learning how to fly and the following 3 weeks learning how to hunt. However, during this time they continue to return to the hack site for food. As the immature peregrines become better at hunting, they will wander further from the site (http://www.gyrfalcons.co.uk/hacking.htm).

The reintroduction of peregrine falcons has occurred in many cities across North America (Cade 1996; Figure 6.11). The reintroductions have been so successful in increasing the number of breeding pairs that the bird was removed from the Federal list of threatened and endangered species in 1999. The USFWS has now initiated a monitoring plan for peregrine falcons in 40 states beginning in 2003 and ending in 2015.

PERSPECTIVE ESSAY 6.2: HOME COMPOSTING ON A SMALL SCALE

If a residential area contains some yard space, a simple composting apparatus for food scraps left over from meal preparation and nonconsumption can be utilized for home composting. The system diagrammed in Figure 6.12 only requires the purchase of a 30-gallon plastic garbage can. Holes are punched in the bottom of the can to allow liquid to escape. The can is buried in the ground allowing enough room to replace the lid (Figure 6.12a). Nearly every kind of kitchen scrap can be thrown into the garbage can composters, including plant and animal material and egg shells, but it is not designed for grass clippings and leaves (Figure 6.12b). One author used such a device and added anything organic that would have gone down the garbage disposal to the composter. Some may not want to add animal material because of

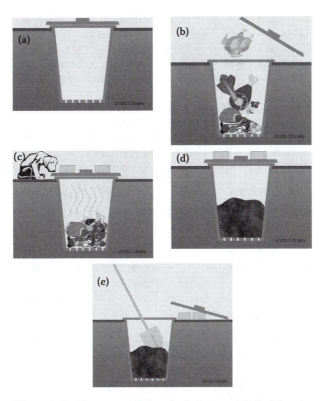

Figure 6.12 Home composting. (L. Causey and C. E. Adams)

fear of disease and/or local ordinance restrictions. The organic material will begin to decompose on its own (Figure 6.12c). In addition, flies will lay their eggs on the organic material, facilitating a more rapid decomposition process as the larvae convert the detritus into humus. Odors are produced during the process of decomposition that will attract pets and other animals. Weights on the top of the lid will prevent their access. Over time, depending on climate, the organic material in the composter will turn into a black humus (Figure 6.12d). This humus can retrieved from the composter and can be used as a mulching material on gardens, flower beds, and other areas that require a rich organic base (Figure 6.12e).

REFERENCES

Adam, M. D. and Hayes, J. P. (2000). Use of bridges as night roost by bats in the Oregon coast range, *Journal of Mammalogy,* 81:402–407.

Adams, C. E. (1983). Road-killed animals as resources for ecological studies, *American Biology Teacher,* 45:256–261.

Baker, J. A. and Brooks, R. J. (1981). Raptor and vole populations at an airport, *Journal of Wildlife Management,* 45:390–396.

Banks, R. C. (1979). Human related mortality of birds in the United States, Special Scientific Report 215, U.S. Fish and Wildlife Service, Washington, DC.

Barras, S. C., Dolbeer, R., Chipman, R. B., Bernhardt, G. E., and Carrara, M. S. (2000). Bird and small mammal use of mowed and unmowed vegetation at John F. Kennedy International airport, 1998 to 1999, in *Proceedings Vertebrate Pest Conference,* Salmon, T. P. and Crabb, A. C., Eds., University of California, Davis, 19:31–36.

Belant, J. L. (1997). Gulls in urban environments: landscape-level management to reduce conflict, *Landscape and Urban Planning,* 38:245–258.

Belant, J. L., Ickes, S. K., and Seamans, T. W. (1998). Importance of landfills to urban-nesting herring and ring-billed gulls, *Landscape and Urban Planning,* 43:11–19.

Belant, J. L., Seamans, T. W., Gabrey, S. W., and Dolbeer, R. A. (1995). Abundance of gulls and other birds at landfills in northern Ohio, *American Midland Naturalist,* 134:30–40.

Bell, D. A., Gregoire, D. P., and Walton, B. J. (1996). Bridge use by peregrine falcons in the San Francisco Bay area, in *Raptors in Human Landscapes,* Bird, D. M., Varland, D. E., and Negro, J. J., Eds., Academic Press, London, pp. 14–23.

Brown, C. R. and Brown, M. B. (1995). Cliff swallows (*Hirundo pyrrhonota*), in *The Birds of North America,* No. 149, Poole, A. and Gill, F., Eds., The Birds of North America Inc., Philadelphia, PA.

Brown, C. R. and Brown, M. B. (1999). Barn swallows (*Hirundo rustica*), in *The Birds of North America,* No. 452, Poole, A. and Gill, F., Eds., The Birds of North America Inc., Philadelphia, PA.

Cade, T. J., Martell, M., Redig, P., Septon, G., and Tordoff, H. (1996). Pergrine falcons in urban North America, in *Raptors in Human Landscapes,* Bird, D. M., Varland, D. E., and Negro, J. J., Eds., Academic Press, London, pp. 3–13.

Childs, J. E. (1991). And the cat shall lie down with the rat, *Natural History,* June:16–19.

Clark, J. E., van Manen, F. T., and Pelton, M. R. (2002). Correlates of success for on-site releases of nuisance black bears in Great Smoky Mountains National Park, *Wildlife Society Bulletin,* 30:104–111.

Cleary, E. C., Dolbeer, R. A., and Wright, S. E. (2003). Wildlife strikes to civil aircraft in the United States 1990–2002. U.S. Department of Transportation, Federal Aviation Administration, Serial Report Number 9, Washington, DC, http://wildlife.pr.erau.edu/Bash90-02b.pdf.

Clevenger, A. P., Chruszcz, B., and Gunson K. E. (2003). Spatial patterns and factors influencing small vertebrate fauna road-kill aggregations. *Biological Conservation,* 109:15–26.

Coleman, J. S. and Fraser, J. D. (1989). Habitat use and home ranges of Black and Turkey vultures, *Journal of Wildlife Management,* 53:782–792.

Conover, M. (2002). *Resolving Human-Wildlife Conflicts: The Science of Wildlife Damage Management,* Lewis Publishers, Boca Raton, FL.

Conover, M. R., Pitt, W. C., Kessler, K. K., DuBow, T. J., and Sanborn, W. A. (1995). Review of human injuries, illnesses, and economic losses caused by wildlife in the United States, *Wildlife Society Bulletin,* 23:407–414.

Dalzell, S. (2004). Airport sustainable design. Massport (http://www.aci-na.org/docs/43-Stewart%20Dalzell%20Airport%20Sustainable%20Design.ppt#256,1,Airport Sustainable Design).

De Vore, S. (1998). Birds and buildings: a lethal combo, *Chicago Wilderness Magazine,* Spring Issue, p. 2.

Deelstra, T. (1989). Can cities survive: solid waste management in urban environments, *AT Source,* 18(2): 21–27.

Dolbeer, R. A., Wright, S. E., and Cleary, E. C. (2000). Ranking the hazard level of wildlife species to aviation, *Wildlife Society Bulletin,* 28:372–378.

Eberhard, T. (1954). Food habits of Pennsylvania house cats, *Journal of Wildlife Management,* 18: 284–286.

Environmental Protection Agency. (2002). *Municipal Solid Waste in the United States: 2000 Facts and Figures.* Office of Solid Waste and Emergency Response. Washington, DC.

Erickson, W. P., Johnson, G. D., Strickland, M. D., Young, D. P., Jr., Sernka, K. J., and Good, R. E. (2001). *Avian Collisions with Wind Turbines: a Summary of Existing Studies and Comparisons to Other Sources of Avian Collision Mortality in the United States,* National Wind Coordinating Committee, Washington, DC.

Fahrig, L., Pedlar, J. H., Pope, S. E., Taylor, P. D., and Wegner, J. F. (1995). Effect of road traffic on amphibian density, *Biological Conservation,* 73:177–182.

Federal Aviation Administration. (1997). Hazardous wildlife attractants on or near airports. Advisory Circular No. 150/5200-33. U.S. Department of Transportation. Washington, DC.

Fitzwater, W. D. (1988). Solutions to urban bird problems, in *Proceedings Vertebrate Pest Conference,* Crabb, A. C. and Marsh, R. E., Eds., University of California, Davis, 13:254–259.

Forman, R. T. T. (2000). Estimate of the area affected ecologically by the road system in the United States, *Conservation Biology,* 14: 31–35.

Forman, R. T. T., Sperling, D., Bissonette, J., Clevenger, A. P., Cutshall, C. D., Dale, V. H., Fahrig, L., France, R., Goldman, C. R., Heanue, K., Jones, J. A., Swanson, F. J., Turrentine, T., and Winter T. C. (2002). Road ecology: science and solutions, Island Press, Washington, DC.

Foster, M. L. and Humphrey, S. R. (1995). Use of highway under passes by Florida panthers and other wildlife, *Wildlife Society Bulletin,* 23:95–100.

Gabrey, S.W. (1997). Bird and small mammal abundance at four types of waste-management facilities in northeast Ohio. *Landscape and Urban Planning,* 37:223–233.

Gallaher, P. A. (2003). The airport environment and its effect on wildlife damage to aircraft, M.S. Thesis, Embry-Riddle Aeronautical University, Daytona Beach, FL.

Hoover, S. L. and Morrison, M. L. (2005). Behavior of red-tailed hawks in wind turbine development, *Journal of Wildlife Management,* 69:150–159.

Horton, N., Brough, T., and Rochard, J. B. A. (1983). The importance of refuse tips to gulls wintering in an inland area of south-east England, *Journal of Applied Ecology,* 20:751–765.

Hutchings, S. (2003). The diet of feral house cats (*Felis catus*) at a regional rubbish tip, Victoria, *Wildlife Research,* 30:103–110.

Jackson, S. D. (2000). Overview of transportation impacts on wildlife movements and populations, In *Wildlife and Highways: Seeking Solutions to an Ecological and Socio-Economic Dilemma,* Messmer, T. A. and West, B., Eds., The Wildlife Society, Bethesda, MD, pp. 7–20.

Jackson, S. D. and Griffin, C. R. (2000). A strategy for mitigating highway impacts on wildlife, in *Wildlife and Highways: Seeking Solutions to an Ecological and Socio-Economic Dilemma,* Messmer, T. A. and West, B., Eds., The Wildlife Society, Bethesda, MD, pp. 143–159.

Jacobson, S. (2002). Using wildlife behavioral traits to design effective crossing structures, Wildlife Crossings Toolkit, http://www.wildlifecrossings.info/.

Johnson, G. D., Erickson, W. P., Strickland, M. D., Shepherd, M. F., and Shepherd, D. A. (2003). Mortality of bats at a large-scale wind power development at Buffalo Ridge, Minnesota, *American Midland Naturalist,* 150:332–342.

Johnson, G. D., Perlik, M. K., Erickson, W. P., and Strickland, M. D. (2004). Bat activity, composition, and collision mortality at a large wind plant in Minnesota, *Wildlife Society Bulletin,* 32:1278–1288.

Johnson, G. D. and Strickland, M. D. (2003). Biological assessment for the federally endangered Indiana bat (*Myotis sodalist*) and Virginia big-eared bat (*Corynorhinus townsendii virginianus*), Western Ecosystems Technology, Inc. Cheyenne, WY.

Keeley, B. W. (1995). Why do bats use this bridge but not that one? *Bats,* 13:18.

Keeley, B. W. and Tuttle, M. D. (1999). *Bats in American Bridges,* Bat Conservation International, Austin, TX.

Kennamer, R. (1999). Wetlands, birds, and airports, http://www.uga.edu/srel/Graphics/Airports_birds.pdf.

Kershner, E. L. and Bollinger, E. K. (1996). Reproductive success of grassland birds at east-central Illinois airports, *American Midland Naturalist,* 136:358–366.

Kiser, J. D., MacGregor, J. R., Bryan, H. D., and Howard, A. (2001). The use of concrete bridges as night roosts by Indiana bats in south central Indiana, *Bat Research News,* 42:33.

Klem, D., Jr. (1989). Bird-window collisions, *Wilson Bulletin,* 101:606–620.

Klem, D., Jr. (1990). Bird injuries, cause of death, and recuperation from collisions with windows, *Journal of Field Ornithology,* 6:115–119.

Klem, D., Jr. (1991). Glass and bird kills: an overview and suggested planning and design methods of preventing a fatal hazard, in *Wildlife Conservation in Metropolitan Environments,* Adams, L. W. and Leedy, D. L., Eds., National Institute for Urban Wildlife, Symp. Ser. 2, Columbia, MD, pp. 99–104.

Lewis, S. E. (1994). Night roosting ecology of pallid bats (*Antrozous pallidus*) in Oregon, *American Midland Naturalist,* 132:219–226.

Linnell, M. A., Conover, M. R., and Ohashi, T. J. (1996). Analysis of bird strikes at a tropical airport, *Journal of Wildlife Management,* 60:935–945.

MacKinnon, B. (1997). Airport wildlife management: beyond airport boundaries, Airport Wildlife Management Bulletin No. 20, Transport Canada, Ottawa, ON.

Manley, T. and Williams, J. (1998). Bearproofing sold waste containers for grizzly and black bears in Lake and City counties, *Intermountain Journal of Science,* 4:101.

McDonnell, M. J. and Pickett, S. T. A. (1990). Ecosystem structure and function along urban-rural gradients: an unexploited opportunity for ecology, *Ecology,* 71:1232–1237.

McKelvey, K. S., Schwartz, M. K., and Ruggiero, L. F. (2002). Why is connectivity important for wildlife conservation, http://www.wildlifecrossings.info/.

McKinney, M. L. (2002). Urbanization, biodiversity, and conservation, *Bioscience,* 52:883–890.

Mumme, R. L., Shoech, S. J., Woolfenden, G. E., and Fitzpatrick, J. W. (2000). Life and death in the fast lane: demographic consequences of road mortality in the Florida scrub-jay. *Conservation Biology,* 14:501–512.

Murphy, M. (1990). The bats at the bridge, *Bats,* 8:7.

Nehring, J. D. (1998). Assessment of avian population change using migration casualty data from a television tower in Nashville, TN. M.Sc. Thesis, Middle Tennessee State University, Murfreesboro, TN.

Piene, J. D. (2001). Nuisance bears in communities, *Human Dimensions,* 6:223–237.

Restani, M., Marzluff, J. M., and Yates, R. E. (2001). Effects of anthropogenic food sources on movement, survivorship, and sociality of common ravens in the Arctic, *Condor,* 103:399–404.

Rogers, L. L. (1989). Black bears, people, and garbage dumps in Minnesota, in *Bear-People Conflicts: Proceedings of a Symposium on Management Strategies,* Bromley, M., Ed., Northwest Territories Department of Renewable Resources, 6–10 April 1987, Yellowknife, Canada, pp. 43–46.

Romin, L. A. and Bissonette, J. A. (1996). Deer-vehicle collisions: status of state monitoring activities and mitigation efforts, *Journal of Wildlife Management,* 24:276–283.

Rosen, P. C., and Lowe, C. H. (1994). Highway mortality of snakes in the sonoran desert of southern Arizona. *Biological Conservation,* 68:143–148.

Rossi-Linerme, G. and Hoppy, B. K. (2000). Natural resource management and the bird aircraft strike hazard at Westover air reserve base, Massachusetts, in *Proceedings Vertebrate Pest Conference,* Salmon, T. P. and Crabb, A. C., Eds., University of California, Davis, 19:63–67.

Rugge, L. (2001). An avian risk behavior and mortality assessment at the Altamont Wind Resource Area in Livermore, California, Thesis, California State University, Sacramento.

Schmidt, R. (1997). Drawing the line for wildlife, *Wildlife Control Technology,* December: 6–7.

Shire, G. G., Brown, K., and Winegrad, G. (2000). Communication towers: a deadly hazard to birds, American Bird Conservancy, Washington, DC.

Slate, D., McConnel, J., Barden, M., Chipman, R., Janicke, J., and Bently, C. (2000). Controlling gulls at landfills, in *Proceedings Vertebrate Pest Conference,* Salmon, T. P. and Crabb, A. C., Eds., University of California Davis, Vol. 19, pp. 6–76.

Speich, S. M., Jones, H. L., and Benedict, E. M. (1986). Review of the natural nesting of the Barn Swallow in North America, *American Midland Naturalist,* 115:248–254.

Spellerberg, I. F. (1998). Ecological effects of roads and traffic: a literature review, *Global Ecology and Biogeography Letters,* 7:317–333.

Stringham, S. F. (1989). Demographic consequences of bears eating garbage at dumps: an overview, in *Bear-People Conflicts: Proceedings of a Symposium on Management Strategies,* Bromley, M., Ed., Northwest Territories Department of Renewable Resources, 6–10 April 1987, Yellowknife, Canada, pp. 35–42.

Sullivan, R. (2004). *Rats: Observations on the History and Habitat of the City's Most Unwanted Inhabitants,* Bloomsbury, New York.

Ternent, M. A. and Garshelis, D. L. (1999). Taste-aversion conditioning to reduce nuisance activity by black bears in a Minnesota military reservation, *Wildlife Society Bulletin,* 27:720–728.

Tigas, L. A., Van Vuren, D. H., and Sauvajot, R. M. (2002). Behavioral responses of bobcats and coyotes to habitat fragmentation and corridors in an urban environment, *Biological Conservation,* 108:299–306.

Transportation Research Board (2002). *Interactions between Roadways and Wildlife Ecology: A Synthesis of Highway Practice,* National Cooperative Highway Research Program, Washington, DC.

Trombulak, S. C. and Frissell, C. A. (1999). Review of ecological effects of roads on terrestrial and aquatic communities, *Conservation Biology,* 14:18–30.

United States Environmental Protection Agency (USEPA). (1993). *Guidance Specifying Management Measures for Sources of Nonpoint Pollution in Coastal Waters.* US EPA, Office of Water, Washington, DC.

VanDruff, L. W., Bolen, E. G., and San Julian, G. J. (1996). Management of urban wildlife, in *Research and Management Techniques for Wildlife and Habitats,* fifth ed., rev., Bookhout, T. A., Ed., The Wildlife Society, Bethesda, MD, pp. 507–530.

Walker, C. W., Sandel, J. K., Honeycutt, R. L., and Adams, C. E. (1996). Winter utilization of box culverts by vespertilionid bats in Southeast Texas, *Texas Journal of Science,* 48:166–168.

West, S. (1995). Cave swallows (*Hirundo fulva*), in *The Birds of North America,* No. 141, Poole, A. and Gill, F., Eds., The Birds of North America Inc., Philadelphia, PA.

Williams, D. (2002). Conspicuous consumption, *National Parks,* 76:40–45.

Wright, R. T. (2004). *Environmental Science: Toward a Sustainable Future.* 9th ed.. Pearson Education, Upper Saddle River, NJ.

Wright, S. E., Dolbeer, R. A., and Montoney, A. J. (1998). Deer on airports: an accident waiting to happen, in *Proceedings Vertebrate Pest Conference,* Baker, R. O. and Crabb, A. C., Eds., University of California, Davis, 18:90–95.

CHAPTER 7

Special Habitat Considerations: Urban Streams and Soils

When the well is dry, we know the worth of water.

Benjamin Franklin, *Poor Richard's Almanac*

CONTENTS

The two resources that are most critical in supporting all life on Earth are fertile soil and adequate water. Both fertile soil and filtered water are renewable resources that, with proper management, can be maintained, used, and reused indefinitely. But, without proper management, soil and water can be depleted or rendered useless. Indeed, there is much evidence to suggest the many ancient extinct civilizations came to their downfall as a result of depleting their soil and/or water resources. In this regard is important to note that from 1990 to 2000 over 1.4 million acres/year of farmland was converted to nonfarm uses in the U.S. alone. Urban sprawl accounted for 1 million acres/year of the total loss of farmland. This chapter provides an overview of the essential information one needs to know in order to understand the critical ecological interrelationships between soil and water that must be sustained.

7.1 WATER

If there is one resource that urban residents take for granted, it is water. Freshwater habitats are among the most altered ecosystems on Earth, because they are the ultimate sumps (low areas that receive drainage) for watershed pollutants. Their water is increasingly diverted for human use through dams and reservoirs that alter flow regimes and fragment drainages. They are the focus of most human activity, both in large cities and agricultural regions, that degrade water quality, and their biota is subject to intense exploitation (Rahel 2002).

Over 130,000 km (80,730 miles) of streams and rivers in the U.S. are impaired by urbanization (Paul and Meyer 2001). The quality of any given stream is negatively correlated with the amount of urbanization in its surrounding watershed (Miltner et al. 2004). A dominant feature of urbanization is a high level of impervious surface cover (ISC) — parking lots, paved roads, and buildings — that prevents water from infiltrating the soil. The results include an increase in localized flooding, surface runoff, erosion, and channelization. ISC has become an accurate predictor of the varying impacts on urban streams.

7.2 THE WATER CYCLE — NATURE'S FILTER

Not many people know the source of their water supplies and even fewer know how their water supply is tied to the water cycle. The cycle does not produce new water; it simply filters out impurities and other pollutants in existing water. It is important to examine the water cycle if one wants to begin to comprehend the problems associated with urban streams and the drawbacks of poor water management practices.

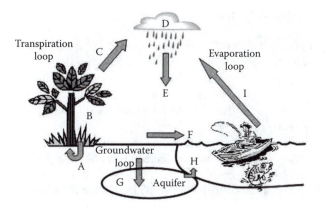

Figure 7.1   Diagram of the water cycle and filters. (C. E. Adams)

The water cycle can be explained in terms of three major filters: (1) the transpiration loop, (2) the groundwater loop, and (3) the evaporation loop. Given the opportunity, the three loops illustrated in Figure 7.1 provide a continuous supply of fresh water for all living things on Earth.

7.2.1 The Transpiration Loop

In the transpiration loop, represented by letters D-E-A-B-C, water infiltrates the soil, but only to a level that allows the plant roots to capture it. Water is filtered as it passes through the plant tissues and out through the leaves, returning to the atmosphere in a process called *evapotranspiration.* The evaporated water condenses into clouds, which eventually produce some form of filtered precipitation, such as rain, sleet, hail or snow. The human impact on this filter, particularly in urban and suburban areas, is the replacement of vegetation with ISC, thus significantly reducing the effect of this filter.

7.2.2 The Groundwater Loop

In the groundwater loop (D-E-G-H-I) water falls to the ground as precipitation, but instead of being completely captured by plant roots, some of the water passes through the soil into an underground water supply called an *aquifer.* In this loop soil acts as the filter. The aquifer water can come back to the surface in the form of seep or artesian flow, or it can be pumped out of the ground by wells. If the water ends up in a surface water impoundment, such as a lake, the water evaporates and reenters the water cycle.

The process of urbanization covers or compacts much of the soil into which precipitation can infiltrate. The presence of pollutants in urban soil may contaminate the water further, rather than purify it, as it recharges into the aquifer. An additional human impact on the groundwater loop is using more aquifer water than can be replaced through the natural recharge rate.

7.2.3 The Evaporation Loop

In the evaporation loop (D-E-F-I) the atmosphere acts as the filter, removing the impurities as water evaporates from streams, rivers, ponds, lakes, and the oceans. A simple demonstration of how this filter works would be to allow a spoonful of saltwater to evaporate. Once all of the water is gone, all that is left behind is salt. This process is similar to that seen in the transpiration filter, but without plants playing an active filtration role.

Human activities, such as manufacturing and burning fossil fuels, release pollutants into the atmosphere that are unaffected by the filtering effect of evaporation. These pollutants include carbon dioxide (CO_2), as well as sulfur oxide (SO) and nitrous oxide (NO) radicals. The latter two pollutants are responsible for the formation of acid rains (see Chapter 3).

7.2.4 Effects of Urbanization on the Water Cycle

The general effects of urbanization on the water cycle are: (1) changing much of the Earth's surface from permeable to impermeable, (2) pollution, and (3) excessive groundwater withdrawal rates (Paul and Meyer 2001). We will examine each of these, as well as a few other factors, in terms of the abiotic and biotic effects of urbanization on stream ecology.

7.3 ABIOTIC EFFECTS OF URBANIZATION ON STREAM ECOLOGY

There is very little published information on the abiotic effects of urbanization on stream ecology (Table 7.1; Paul and Meyer 2001), although there are more data on water chemistry in urban streams than any other aspect of their ecology. *Channelization* (the straightening and/or deepening of a watercourse for purposes of storm-run-off control or ease of navigation) of urban waterways, however, has become an area of growing research interest.

Table 7.1   Deleterious Effects of Varying Levels of ISC on Urban Streams

Effect	Cause
Stream hydrology	Increased surface runoff
	Sedimentation
	Floods and discharge rates
	Decreased lag times
Stream geomorphology	Changing channel width
	Changing channel depth
Stream temperature	Runoff over heated surfaces
	Changes in fish and invertebrate diversity

Source: Paul, M. J. and Meyer, J. L., *Annual Review of Ecology and Systematics,* 32:333–365, 2001.

Table 7.2   Types of Chemicals Found in Elevated Concentrations in Urban Streams and Their Sources

Chemical	Source
Phosphorus	Wastewater and fertilizer runoff from lawns
Nitrate and ammonium	Wastewater and fertilizer runoff from lawns
Metals; Pb, Zn, Cr, Cu, Mn, Ni, Cd, Hg, As, Fe, B, Co, Ag, Sr, Sb, Sc, Mo, Li, Sn	Industrial discharge, brake linings, metal alloys, and accumulations on roads and parking lots.
Pesticides and herbicides	Urban use around homes and work places and lawn and golf course management
Polychlorinated biphenyls (PCBs), polycyclic aromatic hydrocarbons (PAHs), petroleum-based aliphatic hydrocarbons	Storm water runoff from industrial point sources, episodic spills, oil and gasoline spills
Pharmaceuticals: antibiotics, chemotherapeutic drugs, analgesics, narcotics, psychotherapeutic	Hospital effluent and drug flushing down the toilet

Source: Paul, M. J. and Meyer, J. L., *Annual Review of Ecology and Systematics*, 32:333–365, 2001.

7.3.1 Pollutants

Water chemistry is affected by the extent and type of urbanization, amount of waste water treatment (WWT), effluent and/or combine sewer outflows (CSOs), and the extent of storm water drainage area facilitating nonpoint surface (NPS) runoff. In addition, illicit discharges, leaking sewer systems, and faulty septic systems are large and persistent contributors of pollutants to urban streams.

Nearly every chemical used directly or indirectly by urban residents winds up in urban waterways and lakes. Paul and Meyer (2001) provide a detailed analysis of chemical type and source, summarized in Table 7.2. NPS of chemical pollutants are more extensive than point sources, due to the increase in runoff from impervious surface covers (ISC) in urban areas.

Other types of pollutants that enter streams and rivers flowing through urban communities include: fecal wastes from companion animals, fossil fuels such as oil and gasoline, road salts, sediments (i.e., sand, gravel, clay), paper products, and pharmaceuticals. The majority of these pollutants enter the aquatic communities through storm water runoff. This water contains whatever potentially polluting material was deposited on hard surfaces prior to the rain storm.

7.3.2 Stream Channelization

ISC has become an accurate predictor of urban impacts on streams because thresholds of degradation in streams are associated with ISC. For example, as ISC increases, the surface runoff increase proportionately. The rate and flow volume of storm water runoff leads to decreased ground water recharge and alteration of natural drainage patterns through changes in stream bed and bank geomorphology.

Urban storm water management systems are designed to collect and discharge large volumes of water from the community as quickly as possible through the channelization (Figure 7.2). Channelization is necessary because urban communities

Figure 7.2   Stream channelization. (John M. Davis/Texas Parks and Wildlife Department)

consist of large expanses of hard surfaces (concrete or asphalt) that prevent infiltration, increasing surface runoff rates and the potential for flooding. During heavy
rainstorms, huge amounts of water run off from residential lawns, streets and roads,
parking lots, rooftops, and other hard surfaces (nearly 80% of the surface cover at
the urban core). In order to deal with increased surface runoff, many cities have
channelized their streams. Channelization usually involves removing trees and other
vegetation from the streambank, then sculpting the bank into a smooth, straight
channel. This channel may then be seeded with grass or, worse yet, lined with
concrete (Davis 2003).

Converting natural streams into concrete channels accomplishes two things. First,
a concrete channel conveys more water more quickly than does a natural stream.
Second, lining the channel with concrete separates the water from the soil, thus
reducing erosion at a particular site. This may sound like a good thing, but it isn't.
Channelizing a stream fixes one problem, but creates a suite of other problems.

Stream channelization removes the native streamside (riparian) vegetation. This
has several negative effects. First, removal of riparian vegetation eliminates habitat
for a vast assemblage of amphibians, reptiles, birds, and mammals that depend on
that vegetation. Second, replacing the vegetation along urban streams with concrete
eliminates the natural beauty of the stream. Since people prefer to live adjacent to
natural areas (Adams et al. 1984; Lane 1991) property values tend to be higher
where vegetation has not been removed. In Wichita, Kansas, for example, lots
bordering a wetland sold for 50% more than comparable lots away from the natural
area (Ferguson 1998). Natural areas provide natural beauty. Once that natural beauty
is removed, property values are adversely affected.

A third effect of removing the native vegetation along urban streams eliminates
the water cleansing ability of the vegetation. It has been widely documented that

aquatic vegetation locks up excess nutrients and contaminants found in the water. In fact, wetlands planted with aquatic vegetation are currently being used to clean up or polish waste water (APWA 1981; Hammer 1997).

A final effect of removing the vegetation along urban streams is the loss of the cooling effect that shade trees have on the water. Parking lots, rooftops, and streets all become very hot during the summer. As runoff water travels across these surfaces it heats up. This heated water enters urban streams, raising the average temperature of the stream. Water that flows down a shadeless concrete channel does not have a chance to cool down; the result is low dissolved oxygen levels and a reduced capacity for supporting life. The entire biotic community in a stream can be affected by a change of only a few degrees in the water temperature. There are many other specific effects caused by stream channelization, including:

1. *Elimination of pools and riffles.* Natural streams have a normal pattern of deep, slow-moving pools of water and shallow, faster-moving riffles. These habitats support different plant and animal species. When a stream is channelized, these habitats are destroyed along with the plant and animal species that depend upon them.
2. *Reduced organic material.* Most stream systems rely on organic material that is carried in from outside the system to form the basis of the aquatic food web.
3. *Increased water velocity.* The smooth concrete sides to the stream channel increase the velocity of the water in the channel (Riley, 1998).
4. *Reduced or eliminated infiltration.* The concrete lining of the channel does not allow water to infiltrate the soil. Therefore, the amount of water moving through the channel is actually higher than would be found in a natural stream; none is allowed to seep into the ground (Riley, 1998).
5. *Increased downstream flooding and erosion.* Channelized water has to be discharged at some point. When this happens, the area downstream from the channel gets flooded and scoured.
6. *Increased danger for human residents.* Swiftly moving water rushes down these smooth-walled channels. It is not uncommon to hear of people being swept away in these raging torrents (Davis 2003).

7.4 BIOTIC EFFECTS OF URBANIZATION ON STREAM ECOLOGY

The biological and ecological effects of urbanization on urban streams have received even less research attention than the abiotic effects. However, Paul and Meyer (2001) provided an overview of studies conducted on microbes, algae, macrophytes (e.g., submerged aquatic vegetation large enough to be seen easily with the unaided eye), invertebrates, and fish.

There are several deleterious biological and ecological effects of chemical and organic pollutants reaching natural waterways. Stream biota can be lost to pesticide or heavy metal poisoning, osmotic imbalances from excess road salt, suffocation due to sediment clogging their respiratory mechanisms, or loss of oxygen from high biological oxygen demand (BOD) by decomposers. High water temperatures and low saturation of dissolved oxygen are correlated with high relative abundance of introduced species and fish with parasites and diseases.

Fecal coliform and other bacterial pathogens appear in urban waterways from a variety of sources, including human waste treatment plants, companion animals, and high concentrations of urban wildlife (e.g., resident Canada geese discussed in Chapter 12). Bacterial pathogen levels rise in urban streams during and after storms. Some species of these bacteria demonstrated increased antibiotic resistance attributed to genetic transfer and level of metal toxicity in the stream. Nitrifying bacteria densities increase in urban streams due to wastewater treatment plant (WTTP) outflows. Algae species diversity decreases in urban streams, largely due to changes in water chemistry. Bed sediment changes, nutrient enrichment, turbidity, and exotic species introductions result in reduced native macrophyte diversity in urban streams.

Eutrophication: A noticeable ecological effect of organic pollutants on natural waterways is eutrophication. Eutrophic water ways are defined as "nutrient rich" (high concentrations of phosphorus and nitrogen) because eutrophication is caused by large inputs of fertilizer from stormwater drainage, wastewater treatment plants, and combined sewer outflows that cause rapid and extensive algal growth called "blooms." Eutrophic water looks like green pea soup and is incapable of supporting the diversity of aquatic lifeforms that were present before the fertilizers were introduced.

Eutrophic water ways are characterized by low oxygen concentrations, high biochemical oxygen demand (BOD), low light penetration, low biotic diversity, high phosphate and/or nitrate content, elevated water temperatures, high sediment content, and high algae (e.g., single-celled green or blue-green) densities. As more plants and animals die in eutrophic water, more decomposition is required to remove them from the system. Decomposers need oxygen to accomplish the recycling process, which results in less oxygen for other living organisms, which in turn die and require decomposition, which further reduces oxygen levels. Only a few plant and animal species that have a wide oxygen tolerance range can survive in eutrophic waterways.

The aquatic food chain (diagrammed below) begins with phytoplankton, algae attached to hard surfaces and covering the stream bed, and aquatic *macrophytes* (large plants that can be seen without the aid of optical magnification) as the primary producers. The primary consumers are zooplankton and benthic invertebrates that feed on the phytoplankton, attached algae, and macrophytes, as well as each other. Fish larvae feed on the zooplankton, and small fish also eat small benthic invertebrates. The smaller fish are fed on by large invertebrates, and larger fish that are fed on in turn by even larger fish.

Aquatic food chain = phytoplankton → zooplankton → fish larvae
→ small fish → large fish → fish-eating birds

Additionally, fish and invertebrates are fed on by reptiles, birds, and mammals, forming a link in energy and nutrient transfer between aquatic and terrestrial ecosystems. One can sometimes predict the ecological consequences if any one of the trophic levels is stressed. However, this depends on the strength of both direct and indirect interactions among organisms across trophic levels. For example, if some environmental pollutant prevented production of fish larvae, then zooplankton would increase and phytoplankton would decrease, thus removing a primary producer from the system.

Environmental pollutants are accumulated at each trophic level, resulting in very high levels of toxic materials accumulated at the top of the food chain. The consequences of this accumulation were discussed in Chapter 6 as related to causes of the near extinction of the peregrine falcon, a fish-eating bird.

All aspects of aquatic invertebrate habitat are altered by urbanization and are directly correlated with the extent of urban land use. Invertebrate presence declined significantly with increasing ISC. General invertebrate responses were decreased diversity due to toxins, temperature changes, siltation and organic nutrients; decreased abundance due to toxins and siltation; and increased abundance due to inorganic and organic nutrients. Decreases were especially evident for sensitive orders of mayflies (*Ephemeroptera*), stoneflies (*Plecoptera*), and caddisflies (*Tricoptera*). In contrast, the abundance of bloodworms (*Chironomids*), segmented worms (e.g., sludge worms — known as *Oligochaetes*), fly larvae (*Diptera*), and tolerant gastropods (shelled invertebrates) increased. Dominant indicator invertebrates of polluted water are Chironomidae (Diptera) larvae and oligochaete annelids.

Fish communities are indicators of the environmental health of natural waterways. For example, polluted urban streams may contain fish communities that lack intolerant species, include significant numbers of nonnative or introduced fishes, and show low abundance and diversity of the native species. These conditions indicate poor fish habitat and significant ecological disturbances (Paul and Meyer 2001).

The biological and ecological effects of urbanization on fish assemblages (species present and their relative abundance) are given special attention in the chapter because:

- Fish are good indicators of long-term (several years) effects and broad habitat conditions because they are relatively long-lived and mobile.
- Fish communities generally include a range of species that represent a variety of trophic levels (omnivores, herbivores, insectivores, planktivores, and piscivores). They tend to integrate effects of lower trophic levels; thus, fish assemblage structure is reflective of integrated environmental health.
- Fish are at the top of the aquatic food web and are consumed by humans, making them important for assessing contamination.
- Fish are relatively easy to collect and identify to the species level. More common species can be sorted and identified in the field by experienced fisheries professionals and subsequently released unharmed.
- Environmental requirements of more common species are comparatively well known. Life history information is extensive for many game and forage species, and information on fish distributions is readily available.
- Aquatic life uses (e.g., water quality standards) are typically characterized in terms of fisheries (coldwater, cool water, warm water, sport, forage). Monitoring fish provides direct evaluation of "fishability" and "fish propagation," which emphasizes the importance of fish to anglers and commercial fishermen.
- Fish account for nearly half of the endangered vertebrate species and subspecies in the U.S. (Barbour et al. 1999).
- Fish provide many ecosystem services (Table 7.3; Holmlund and Hammer 1999).

Increasing urbanization is related to declines in fish diversity and abundance, and a rise in the relative abundance of tolerant taxa correlates with increasing

Table 7.3   Major Fundamental and Demand-Derived Ecosystem Services Generated by Fish Populations

Fundamental Ecosystem Services

Regulating Services	Linking Services
Regulation of food web dynamics	Linkage within aquatic ecosystems
Recycling of nutrients	Linkage between aquatic and terrestrial ecosystems
Regulation of ecosystem resilience	Transport of nutrients, carbon and minerals
Redistribution of bottom substrates	Transport of energy
Regulation of carbon fluxes from water to atmosphere	Acting as ecological memory: migratory fish
Maintenance of sediment processes: beneficial stream bed alterations	
Maintenance of genetic, species, ecosystem diversity	

Demand-Derived Ecosystem Services

Cultural services	Information services
Production of food	Assessment of ecosystem stress: fish assemblage changes
Aquaculture production	Assessment of ecosystem resilience: fish assemblage changes
Production of medicine: neurotoxin	Revealing evolutionary tracks: co-evolution of fish and other species
Control of hazardous diseases: malaria	Provision of historical information: climatic changes
Control of algae and macrophytes	Provision of scientific and educational information
Reduction of waste: scavenger fish	
Supply of aesthetic values: aquarium industry	
Supply of recreational activities: fishing	

Source: Holmlund, C. M. and Hammer, M., *Ecological Economics,* 29:253–268, 1999.

urbanization. Invasive species increase in the more urbanized reaches of streams. Extensive fish kills are more common in urban streams after storms, given the extensive deposition of pollutants from impervious surface cover runoff. The extirpation of fish species is not uncommon in urban river systems. WTTP effluent selects for fish species that have wide tolerances for changes in dissolved oxygen concentration, temperature, and siltation — recall the discussion of tolerance curves in Chapter 4.

Barbour et al. (1999) summarized a study where several physical and chemical measures of degradation were used to test changes in fish assemblages for Maryland streams. The study found that, out of 270 fish species found in Maryland streams and tested, 8% were tolerant, 62% moderately tolerant, and 30% intolerant of a wide range of chemical and physical changes.

It is helpful to know which species were tolerant to the chemical and physical change, because they can act as predictors of deteriorated stream quality in urban

Table 7.4   Fish Species in Urban Stream Communities Tolerant of Wide Ranges in Chemical and Physical Changes

Common Name	Scientific Name	Trophic Designation[a]
1. Banded killifish	*Fundulus diaphanous*	I
2. Black bullhead	*Amerus melas*	O
3. Blacknose dace	*Rhinichthys atratulus*	G
4. Bluegill	*Lepomis macrochirus*	I
5. Bluntnose minnow	*Pimephales notatus*	O
6. Brown bullhead	*Ameiurus nebulosus*	I
7. Catfish	*Ictalurus spp.*	G
8. Central mudminnow	*Umbra limi*	I
9. Comely shiner	*Notropis amoenus*	I
10. Common carp	*Cyprinus carpio*	O
11. Creek chub	*Semotilus atromaculatus*	G
12. Eastern mudminnow	*Umbra pygmaea*	G
13. Fathead minnow	*Pimephales promelas*	O
14. Golden shiner	*Notemigonus crysoleucas*	O
15. Goldfish	*Cariassius auratus*	O
16. Green sunfish	*Lepomis cyanellus*	I
17. Largescale sucker	*Catostomus macrocheilus*	O
18. Northern squawfish	*Ptychocheilus oregonensis*	P
19. Red shiner	*Cyrpinella lutrensis*	O
20. Redear sunfish	*Lepomis microlophus*	O
21. Reticulate sculpin	*Cottus perplexus*	I
22. Rudd	*Scardinius erythrophthalmus*	O
23. Silver carp	*Hypophthalmichthys molitrix*	O
24. Spotfin chub	*Cyprinella monacha*	I
25. Western mosquito fish	*Gamusia affinis*	O
26. White sucker	*Catostomus commersoni*	O
27. Yellow bullhead	*Ameiurus natalis*	I

[a] Trophic Designations: G = Generalist, plant and animal material; I = Insectivore, opportunistic predator on aquatic insects; O = Omnivore, bottom feeder; P = Piscivore, opportunistic predator on other fish.

Source: Barbour, M. T. et al., U.S. Environmental Protection Agency, Office of Water, Washington, DC, 1999.

areas (Table 7.4; Barbour et al. 1999). For example, the red shiner was found in 50 of 65 sample sites by Marsh-Matthews and Matthews (2000) and accounted for 90% of the fish community in the Chattahoochee River in Atlanta, Georgia (DeVivo 1995 in Paul and Meyer 2001). The red shiner is a popular baitfish and has been introduced into streams and rivers through accidental or intentional releases. The common carp, goldfish, and bullhead catfishes are listed as common, pollution-tolerant species by Rahel (2002). The majority of species listed in Table 7.4 have a trophic classification denoting a broad range of feeding habits characteristic of tolerant and potentially invasive species.

The primary source of invasive exotic fish is the global aquarium industry, which generates $7 billion annually (Holmlund and Hammer 1999). Each year, more than 2000 nonnative fish species, representing nearly 150 million exotic freshwater and marine fishes, are imported into the U.S. for use in the aquarium trade. Dumping aquaria fish into the nearest body of water when they are no longer wanted creates a problem for the native fish species and for ecosystems in general (Phipps 2001).

Each of the 27 species listed in Table 7.4 are also on the freshwater invasive species list provided by the U.S. Geological Survey (http://nas.er.usgs.gov/). The report, titled "Nonindigenous Aquatic Species," includes a list of 672 species, of which 35% are exotics. The remaining 65% on the USGS list are native species to the U.S. that were transplanted outside their native range. *The Field Guide to Freshwater Fishes* by Page and Brooks (1991) lists 790 freshwater species native to the U.S. The USGS list suggests that 435 of the 790 native species (55%) have become invasive species! How did this happen, and what are some of the ecological impacts?

Rahel (2002) discussed the occurrence and ecological impacts of invasive fish species in terms of biotic homogenization. Biotic homogenization is the increased similarity of biotas over time caused by the replacement of native species with nonindigenous species, usually as a result of introductions by humans. In human-created habitats, endemic species typically are replaced by cosmopolitan species, with the result that entire ecosystems resembling each other now occur in disparate parts of the country. The interacting mechanisms that result in homogenization are introductions of nonnative species (the key factor homogenizing fish faunas), extirpations of native species, and habitat alterations (e.g., abiotics, stream flow patterns, sedimentation) that facilitate the first two processes. The nonnative species are usually pollutant tolerant and have a wide range of tolerance to a host of abiotic conditions at the extreme end of the spectra (e.g., low dissolved oxygen concentrations, high temperatures). Native endemic species are usually highly specialized, having adapted to a narrow set of abiotic conditions. Miltner et al. (2004) provided a list of 53 fish species native to Ohio rivers and streams that are highly sensitive to either habitat degradation, pollution, or both. They found significant declines in biological integrity detectable when the amount of impervious surface cover exceeded 13.8%. Table 7.5 illustrates both the causes of nonnative species introductions and suggested strategies to reduce the rate of biotic homogenization.

Table 7.5   Causes of Nonnative Species Introductions and Suggested Responses

Causes of Species Introduction	Strategies for Reducing Homogenization
Deliberate transfers (e.g., stocking by sport fishing industry and state agencies)	Stop deliberate transfers and by-product introductions
By-product introductions (e.g., canal building, ballast water discharge, aquaculture operations)	Minimize habitat alterations, e.g., restore natural flow regimes
Human introductions in urban landscape designs (e.g., garden ponds).	Increase public education concerning the magnitude of the problem and ecological ramifications
	Use sterile hybrids
	Remove naturalized populations of nonnative species prior to re-establishing native species

7.5 CARING FOR THE WATER CYCLE

The process of urbanization has led to the concentration of many people into a relatively small area, which has had profound negative effects on the water cycle and its filters. The most immediate effect is the rate at which water is withdrawn compared to nature's capability to replace it — called the recharge rate. Urbanites waste water!! They use far more water than is required to accomplish the task at hand. For example, the introduction of many exotic plants (e.g., lawn grasses) demands extensive irrigation to keep the plants alive. The green golf courses in the middle of the xeric Southwest (e.g., Tucson, Arizona) exemplify how urbanites attempt to manipulate the habitat to produce plant and animal communities that could never survive in that region without heavy water subsidies.

What really captures urban residents' attention regarding the water cycle are those times when a catastrophic event (e.g., hurricanes and tornados) turns the water off altogether. Deprivation seems to make disciples of water quality standards, and they focus their attention on the gravity of degrading the benefits of the water cycle. Since water usually appears to be abundant and seemingly everlasting in urban communities, and catastrophic events are rare, an educational process has to be implemented regarding the benefits of preserving the integrity of the water cycle. Education is always a proactive rather than reactive management approach. Educational programs could focus on:

1. The origin of the community water supply and what processes are involved in providing a continuous supply of potable water to residents;
2. Water consumption rates for various resident activities including business, recreational, and household uses;
3. Opportunities for water conservation at the community, residential, and personal levels;
4. Multiple water use strategies that reduce withdraw rates for community water sources (e.g., use of "gray water");
5. The technologies required to replicate nature's water filters;
6. Adoption of alternative water use life styles such as landscaping with endemic vegetation that is adapted to the climate of the area.

7.6 SOIL

The underlying support of any civilization is a stable agricultural production system. No society can maintain itself for long in a civilized state without an ample and reliable food supply for its citizens, and such a supply can only come from the artificial propagation of various plants and animals (i.e., agriculture). More than 90% of the world food supply is from land-based systems that rely on soil as the fundamental resource. Propagation of plant species is even more fundamental than that of animals, because plants are always at the beginning of the food chain (i.e., the link to sunlight energy). Humans can take solar energy for granted, since it remains essentially constant, and they can do little to manage it in one way or another.

The soil ecosystem has several unique characteristics that distinguish it from ecosystems that exist in air and water. These distinguishing characteristics include:

1. A slow rate of nutrient and energy transfer. Decomposition rates vary from a few months in tropical rain forests to several years in temperate forests.
2. Different textures (e.g., different combinations of sand, silt, and clay) that demand some very unique adaptations on the part of animals that live in the soil. In short, moving through a soil medium is much more difficult than moving through air or water. For example, the earthworm (*Lumbricus terrestris*) is an invertebrate that literally eats its way through topsoil. It is estimated that as much as 15 tons per acre of soil pass through earthworms each year in the course of their feeding (Wright 2004; see Perspective Essay 7.1). Earth-dwelling or fossorial mammals such as pocket gophers (*Thomomys* sp.) have rudimentary eyes, but highly evolved senses of touch and smell, and appendages adapted to rapid digging.
3. A near total reliance on decomposers (bacteria, fungi) to begin the process of nutrient and energy transfer. Literally millions of bacteria can be seen and counted in one gram of soil (Wright 2004).
4. The exclusive use of detritus (dead animal and plant material) rather than green plants as the source of energy and nutrients.
5. An extreme susceptibility to disturbance with very slow recovery times. For example, one of the first impacts of urbanization on soil structure is the removal of topsoil (Horizons O and A, Figure 7.3). The development of topsoil from subsoil or parent material takes hundreds of years, depending on region and climate (Wright 2004).

7.6.1 Soil Structure

The soil ecosystem exists to accomplish two tasks: to provide a substrate to support green plants, and to provide nutrients, water, and oxygen to green plants. As such, soils

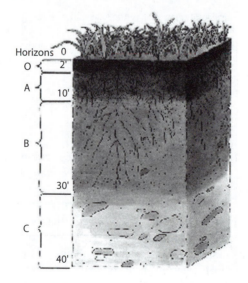

Figure 7.3 **(See color insert following page 276)** A typical soil profile showing four horizons. (USDA photo)

Table 7.6 Description of Soil Profiles

Horizon	Description
O	The top layer or organic horizon consisting of decaying plant material called "humus."
A	The most productive layer of soil commonly referred to as "topsoil." Most plant roots are found here. The A horizon is composed primarily of a mixture of mineral and humus material and is where most soil functions occur.
B	a subsoil layer formed by materials leached or moved from the A horizon including clay, iron, aluminum, and some organic material.
C	The lowest layer of soil composed of disintegrated parent material and other minerals

are complex systems of solid matter, pore spaces filled with water and oxygen, and numerous bacteria, fungi, and other organisms. Soil is a mixture of four basic components:

1. 45% inorganic materials (minerals) including rock, clay, silt, and sand that give structure to the soil. Soil texture refers to the relative proportions of each mineral type in a given soil.
2. 5% organic matter, including living and decomposing organisms and plant parts, supplies nutrients and helps hold moisture in the soil.
3. 25% air that moves through the pore spaces provides oxygen to plant roots.
4. 25% water and dissolved nutrients, important for a number of a plant's life processes, also move through the pores (USDA 2001).

Soil consists of different layers called "horizons" (Figure 7.3). The arrangement and makeup of these horizons influence the amount and type of plant growth that can occur. The thickness of each layer will vary depending on the climate and region and impacts of urbanization. Table 7.6 describes the four general horizons in a soil profile.

In general the impact of urbanization on the soil horizons is to greatly increase the O horizon through over mulching and to lose or greatly diminish the A horizon through construction grading and scraping or erosion. The impacts of urbanization on soil are covered later in this chapter.

7.6.2 Soil Function

Successful and healthy plant growth is dependent upon the soil's capability to allow water, nutrients, and oxygen to infiltrate to the A horizon and hold it so plant roots have access to these essential resources. A soil texture that contains roughly 40% sand, 40% silt, and 20% clay is called "loam" and is considered the optimal mineral composition for soil to facilitate healthy plant growth. The mixture of humus and other organic material from the O horizon might be considered soil sponges that increase soil porosity, enhancing the water-holding, nutrient-holding, and aeration qualities of soil. As earthworms eat their way through the O, A, and sometimes the B horizon, they form open channels that connect the horizons, increase water infiltration and oxygen transfer from the surface, and, as a bonus, increase the humus content in the A horizon. Earthworms have an exceptional ability to displace large amounts of soil and play a major role in topsoil formation (see Perspective Essay 7.1). Some of the same benefits are provided by burrowing mammals such as the pocket gopher.

7.7 IMPACTS OF URBANIZATION ON SOIL STRUCTURE AND FUNCTION

The soils of urban systems are the same as those of the natural systems that exist in the area. So the important thing is to examine what happens to soil during the process of urbanization (Schueler 2000; Davis 2003; DNREC 2004).

7.7.1 Grading and Scraping

Grading and scraping are the most visible effects that urbanization has on the local soil (Figure 7.4). As new development begins, often the first step is to bring in heavy equipment and begin "sculpting" the land to meet the needs of the project. At the low-impact end of things, this may involve moving soil around and mixing up horizons. At the high-impact end, it may involve completely removing the topsoil and depositing it elsewhere. Removing topsoil eliminates the organic matter and its associated microbes that help make the soil productive and destroys the soil community.

7.7.2 Compaction

The most damaging process that happens to urban soil is compaction. Healthy soil is loose and contains many voids or pockets filled with air or water. Compaction

Figure 7.4   Soil grading and scraping. (John M. Davis/Texas Parks and Wildlife Department)

occurs when anything applies sufficient pressure to the soil to compress these air pockets. Foot traffic compacts soil to a degree. Riding lawn mowers compact the soil to a larger degree. Automobiles compact the soil even more. Bulldozers, road graders, and other heavy construction equipment cause so much soil compaction that healthy full-grown trees may die within a year of completing an urban development. Plants rely on air pockets in the soil to provide oxygen to their roots and to the microorganisms that make up the soil community. Without oxygen in the soil, all life forms that depend on soil oxygen soon perish. Furthermore, roots often have a difficult time penetrating compacted soil to obtain nutrients.

When soils are compacted by development impacts, there is decreased water absorption, and increased surface water runoff. Compounds such as fertilizers, pesticides, herbicides, gases, oils, and other pollutants accumulate on these "created" impervious surfaces during dry weather conditions. These pollutants form a concentrated first "flush" to water bodies following a storm event. Soil compaction is so widespread in urbanized landscapes that impervious surfaces are the second highest source of pollutant concentration; only piped sources were higher.

7.7.3 Exposure

Due to construction, urban soil is often left exposed to the elements for extended periods. Wind, rain, and ultraviolet radiation all take their toll on the soil. Wind and rain erode the soil, which removes topsoil, and UV radiation kills soil bacteria that are essential in maintaining nutrient cycling.

7.7.4 Contaminants

Often urban homeowners have problems growing things in certain areas of their yard. A common reason for this is that the contractor who built their home had apparently buried all of the scrap mortar, bricks, lumber, etc. in their yard!

Other homeowners have discovered that one chemical or another was spilled at the site and is the cause of soil problems. Urban homeowners use prodigious quantities of salt-based rather than organic fertilizers (e.g., Milorganite, http://www.milorganite.com/). The use of so much salt-based fertilizer on the property causes salinization or "salty soil," which sucks the water out of plants rather than allowing water to go from the soil into the plant. As mentioned earlier in this chapter, urban soils have a large number of contaminants that ultimately end up in urban streams, rivers, and lakes.

7.8 PHYSICAL PROCESSES THAT INFLUENCE URBAN SOIL

7.8.1 Additions

Anything that is added to the soil, such as organic matter or salts, is an addition. Additions can occur naturally or when amendments are added, such as fertilizer, construction material, and mulches. It is important to consider where a bag of mulch, sold at local garden centers, came from, given the other chemicals and organisms that might be mixed in with the mulch.

7.8.2 Losses

Minerals can be washed out or leached from the soil, and both inorganic and organic matter can be lost through erosion.

7.8.3 Translocations

Soil materials may be transported from one site to another but not lost from the system. Sometimes topsoil is removed from one construction site and sold to other developers or homeowners who want to restore their soil fertility. Soil eroded on a hill can be deposited in a low-lying area below.

7.8.4 Transformation

The chemical and physical properties of the soil can be changed over time. As water moves through the soil and evaporates, it can leave salt accumulations behind. Microorganisms that feed on dead plant and animal matter transform the matter into humus. Heavy irrigation of urban soils can change pH conditions. A pH of 7 is neutral, a pH above 7 is alkaline or basic, and a pH below 7 is acidic, as illustrated below (USDA 2001).

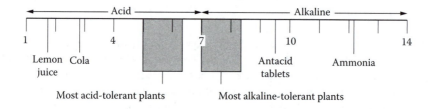

7.9 TAKING BETTER CARE OF URBAN SOIL

In summary, it is important to understand the value of a healthy soil ecosystem in sustaining a high quality of life for all organisms. The process of urbanization has seriously altered or even removed some of the services provided by the soil ecosystem. However, there are ways to restore and sustain the physical and biological properties of soil that benefit humans and all other organisms (USDA 2001). The golden rules of soil conservation are:

- Cover the soil.
- Minimal or zero tillage.
- Mulch for nutrients.
- Maximize biomass production.
- Maximize biodiversity.

Following the golden rules of soil conservation will produce productive soil with:

- A good supply of nutrients and nutrient-holding capacity,
- Good infiltration and water-holding capacity, and ability to resist evaporative water loss,
- Porous structure for aeration,
- Near-neutral pH,
- Low salt content.

There are many sources of free information that provide strategies to restore soil fertility even in the most devastated urban soils. This information can be obtained from the state Extension Service, producers cooperative stores, and Web-based sources under the search terms "urban soils." Homeowners who restore natural soil conditions reap the benefits of watering less, using less fertilizers, healthier plant growth, and more plant and animal diversity.

PERSPECTIVE ESSAY 7.1: DARWIN'S EARTHWORMS

Charles Darwin (Figure 7.5) began and ended his career with "worms." His 1881 treatise on worms: *The Formation of Vegetable Mould through the Action of Worms with Observations on Their Habits,* was published in the 72nd year of his life and

Figure 7.5 Charles Darwin. (Public domain picture)

only 6 months before his death. The earthworm book is perhaps the third best-known among the general public of Darwin's books following the 1872 publication of *The Origin of Species* and *The Descent of Man*.

Darwin spent 36 years observing earthworms, meticulously recording their behavior, measuring the amount of soil moved through their activities, and evaluating their influence on soil fertility (Brown et al. 2003). To most people earthworms are unpleasant, slimy, ugly, blind, deaf, and senseless animals, of little use except for fish-bait, and a general nuisance, particularly because of their unsightly surface castings (Feller et al. 2003). However, the resurgence of interest in organic farming and "biological agriculture" (in which earthworms play a more important role influencing soil fertility) in recent years has brought renewed recognition of Darwin's book and earthworms (Brown et al. 2003).

There has also be renewed interest in raising earthworms, called "vermicomposting," which is a fairly simple procedure, requires no more space that a wash tub, is inexpensive, and is an excellent technique for recycling food waste in the apartment as well as composting yard wastes in the backyard. The resultant "vermicompost" contains 5 to 11 times more nitrogen, phosphorus, and potassium as the surrounding soil and makes an excellent mulch and soil conditioner for potting plants and the home garden.

Night crawler hunting is nearly a lost form of recreation by urban residents. A generous supply of earthworms can be captured at night as they emerge to feed on surface litter and mate. The best conditions for hunting night crawlers is just after a spring rain, at dusk or early darkness, with a flashlight covered with red cellophane. The best places to hunt are the yards of homes in the older areas of town, cemeteries, school yards, or other areas where the soil has not been disturbed by recent construction. The night crawlers will be lying on the surface with a portion of their body still in their hole. They are quick to retreat into their hole when disturbed!! Holding the flashlight with one hand, use two fingers on the other hand to quickly pinch down on the crawler, clamping it gently but firmly between your fingers. The crawler will put up a struggle trying to retreat back into its hole, but a slow and gentle pulling action will cause the crawler to release its grip on the sides of its hole. Pull too hard and only a half of a night crawler will be collected. On a good night, an entire gallon can of night crawlers can be captured using these procedures.

REFERENCES

Adams, L. W., Dove, L. E., and Leedy, D. L. (1984). Public attitudes toward urban wetlands for stormwater control and wildlife enhancement, *Wildlife Society Bulletin*, 12:299–303.

APWA (1981). Urban Stormwater Management, Special Report No. 49, American Public Works Association, Chicago, IL.

Barbour, M. T., Gerritsen, J., Snyder, B. D., and Stribling, J. B. (1999). Rapid Bioassessment Protocols for Use in Streams and Wadeable Rivers: Periphyton, Benthic Macroinvertebrates and Fish, 2nd ed., EPA 841-B-99-002. U.S. Environmental Protection Agency, Office of Water, Washington, DC.

Brown, G. G., Feller, C., Blanchart, E., Deleporte, P., and Chernyanskii, S. S. (2003). With Darwin, earthworms turn intelligent and become human friends, *Pedobiologia*, 47:924–933.

Davis, J. M. (2003). Urban systems, in *Texas Master Naturalist Statewide Curriculum,* 1st ed., Haggerty, M. M. ed., Texas Parks and Wildlife Department, College Station, TX, 753 p.

DNREC (2004). Influence of common landscaping and grading practices in the creation of impervious surfaces, Delaware Department of Natural Resources and Environmental Control, Water Resources Division, *Tributary Times,* 3(1):10–12.

DeVivo, J. C. (1995). Impact of introduced red shiners (*Cyprinella lutrensis*) on stream fishes near Atlanta, Georgia, in *Proceedings 1995 Georgia Water Resources Conference,* Hatcher, K., Ed., University of Georgia, Athens, pp. 95–98.

Feller, C., Brown, G. G., Blanchart, E., Deleporte, P., and Chernyanskii, S. S. (2003). Charles Darwin, earthworms and the natural sciences: various lessons from past to future, *Agriculture, Ecosystems, and Environment,* 99:29–49.

Ferguson, B. K. (1998). *Introduction to Stormwater,* John Wiley and Sons, New York.

Hammer, D. A. (1997). *Creating Freshwater Wetlands,* 2nd ed., CRC Press, Boca Raton, FL.

Holmlund, C. M. and Hammer, M. (1999). Ecosystem services generated by fish populations, *Ecological Economics,* 29:253–268.

Lane, K. F. (1991). Landscape planning and wildlife: methods and motives in *Wildlife Conservation in Metropolitan Environments,* Adams, L. W. and Leedy, D. L., Eds., National Institute for Urban Wildlife, Columbia, MD, pp. 139–142.

Marsh-Matthews, E. and Matthews, W. J. (2000). Geographic, terrestrial and aquatic factors: which most influence the structure of stream fish assemblages in the Midwestern United States? *Ecology of Freshwater Fish,* 9:9–21.

Miltner, R. J., White, D., and Yoder, C. (2004). The biotic integrity of streams in urban and suburbanizing landscapes, *Landscape and Urban Planning,* 69:87–100.

Page, M. and Brooks, B. M. (1991). A *Field Guide to Freshwater Fishes,* Houghton Mifflin, New York.

Paul, M. J. and Meyer, J. L. (2001). Streams in the urban landscape, *Annual Review of Ecology and Systematics,* 32:333–365.

Phipps, R. (2001). Got fish? Already tired of that holiday gift aquarium? Think before you dump and create an even bigger problem, USGS News Release.

Rahel, F. J. (2002). Homogenization of freshwater faunas, *Annual Review of Ecology and Systematics,* 33:291–315.

Riley, A. (1998). *Restoring Streams in Cities: A Guide for Planners, Policymakers, and Citizens,* Island Press, Washington DC.

Schueler, T. (2000). The compaction of urban soils? Technical Note #107. *Watershed Protection Techniques,* 3(2):661–665.

USDA (2001). *Urban soils,* United States Department of Agriculture, Natural Resource Conservations Service, National Soil Survey Center, Lincoln, NE.

Wright, R. T. (2004). *Environmental Science: Toward a Sustainable Future,* Prentice Hall, Upper Saddle, NJ.

SECTION IV

Sociopolitical Issues

CHAPTER 8

Human Dimensions in Urban Wildlife Management

You see, one of the most important things to managing wildlife has nothing to do with wildlife at all. It's managing people!

Dr. Neil A. Waer, Wildlife Biologist

CONTENTS

As Americans have become more urbanized, their relationship with wildlife has changed considerably. Traditional consumptive-use activities, such as hunting and fishing, continue to be popular forms of recreation, albeit for a decreasing percentage of the population. Wildlife watching, on the other hand, has become increasingly popular. Activities such as observing, feeding, and photographing wild animals are now a big business, accounting for expenditures of over $38.4 billion in 2001 alone (U.S. Department of the Interior 2001). During the same period fishing and hunting accounted for $36.6 billion and $20.6 billion, respectively.

Not all interactions between people and wildlife, however, are enjoyable. As human development sprawls into formerly wild habitat, and as wildlife species attempt to adapt to the presence of humans, conflicts between humans and wildlife have increased. Some of the most common forms of conflict include damage to homes and gardens, crops, and livestock, and threats to the health and safety of people and their companion animals. As a result, conflicts may arise between the general public and wildlife managers as well. Several regional studies have indicated that urban residents often have conflicting goals, such as the desire to reduce human–wildlife conflict with particular species while at the same time enhancing wildlife viewing opportunities of the same species (Conover 1997). In addition, the American public has widely diverse attitudes, expectations, and tolerance levels concerning wildlife. Developing management solutions and communication strategies to meet the needs of all stakeholders has proven to be a challenge for both wildlife management agencies and the wildlife professionals they employ.

8.1 THE "PEOPLE FACTOR"

Wildlife professionals have long joked that their job is really about managing people — the importance of the "people factor" was recognized by no less notable a figure than Aldo Leopold in his groundbreaking text on *Game Management* published in 1933. But it wasn't until the 1970s that the wildlife profession as a whole began to give serious consideration to the effect of human activities and perceptions on wildlife management.

Human dimensions (HD) is the study of the human considerations in wildlife management. Stout (1998) found several definitions of human dimensions research including:

1. The application of social-science principles about human behavior for biological and ecological aspects of natural resource management (Gigliotti and Decker 1992).
2. An approach consisting of an explanation of human actions based on social science concepts and methods and the application of these findings in wildlife decision-making, with emphasis on theory-building (Manfredo et al. 1995).
3. A science which integrates multiple disciplines into natural resource management; specifically, HD research is "the scientific investigation of the physical, biological, sociological, psychological, cultural, and economic aspects of natural resource utilization at the individual and community levels," (Ewert 1996:5).

It is evident that the term covers a broad set of research topics, including communication, behavior of both groups and individuals, program assessments, public participation in management planning and implementation, as well as economic, social, aesthetic, and other types of wildlife values.

The key element in HD research is a diversified approach that integrates sociology, psychology, anthropology, economics, ecology, and education in natural resource management. Stout (1998) provided a summary of the types of study themes and management applications addressed in HD research (Table 8.1).

Table 8.1 Examples of Research Themes and Management Applications of Human Dimensions Studies

Research Themes	Management Applications
Attitudes, beliefs, motivations, satisfactions, preferences, and ethics of user Groups	Agency and program needs assessments, design specifications, and evaluation
Modeling of recreational participation, specialization, and involvement	Natural resources education, interpretation, and communication
Quantification of socioeconomic and noneconomic values (cost and benefit)	Public involvement, conflict management, and dispute resolution strategies.
Risk perception and communication	Human impacts on resources, situational analyses, and demand projections
Contributions of wildlife to human quality of life.	Human aspects of species management
Identification of socialization and social worlds.	Market segmentation, stakeholder identification and characterization.
	Integrated resource planning and policy analysis
	Appraisals of natural resource professionals

Sources: Bryan, H., in Ewert, A. W., Ed., Westview Press, Boulder, CO, pp. 145–166, 1996; Decker, D. J. et al., in Ewert, A. W., Ed., Westview Press, Boulder, CO, pp. 29–45, 1996; Ditton. R. B., in Ewert, A. W., Ed., Westview Press, Boulder, CO, pp. 73–90, 1996; and Stout, R. J., Texas A&M University, College Station, 1998.

The first book dedicated to an aspect of human dimensions appeared in 1964, Douglas Gilbert's *Public Relations in Natural Resources Management.* This seminal text was followed by a small flurry of publications from the 1980s through today: *Valuing Wildlife: Economic and Social Perspectives,* by D. J. Decker and G. R. Goff (1987); *American Fish and Wildlife Policy: The Human Dimension,* by W. R. Mangun (1992); *Wildlife and People: The Human Dimensions of Wildlife Ecology,* by G. G. Gray (1993); *Natural Resource Management: The Human Dimension,* by A. W. Ewert (1996); and *Human Dimensions of Wildlife Management in North America,* edited by Decker et al. (2001).

Presently, there are several journals that publish human dimensions research, including: *Fisheries* (American Fisheries Society), *Journal of Human Dimensions of Wildlife* (Taylor & Francis Group), *Journal of Leisure Research* (National Recreation and Park Association), *Leisure Sciences Journal* (Taylor & Francis Group), *Proceedings of the Southeastern Association of Fish and Wildlife Agencies,* the *Wildlife Society Bulletin* (The Wildlife Society), and *Transactions of the North American Wildlife and Natural Resources Conference.* Clearly, interest in human dimensions of wildlife management has grown substantially since Leopold first suggested the concept.

8.2 CONDUCTING HUMAN DIMENSIONS RESEARCH

The complexity of HD studies makes them difficult to conduct, and in many cases this type of research is held to a more rigorous standard than traditional wildlife research. For example, human dimensions researchers can easily find reliable data on the size of the human population and their subset of interest (e.g., birdwatchers, hunters, anglers, hikers). Strict attention must be paid to these numbers in order to draw a sample that will accurately represent the entire population; failure to adhere to the rules of sampling protocol will cause the study to be labeled as "biased" by peer reviewers. Conversely, researchers involved in the study of wildlife species rely on population estimates as a matter of course. The fact that methods used for estimating wildlife numbers can have huge margins of error often is ignored.

Other standards that must be met when undertaking a human dimensions study include: (1) a study sample that is a large enough subset of the target population to reduce sampling error to ±3%; (2) factors that contribute to bias must be anticipated and controlled; and (3) the study must include measurable objectives and research hypotheses. Measurable objectives and hypotheses focus on "who, what, where, and when" questions. For example, a human dimensions study can reasonably expect to measure whether or not gender plays a role in acceptance of a particular management technique, but it cannot reasonably expect to measure why men and women react differently to a proposed method.

Researchers attempting to design a human dimensions study should follow seven basic steps (Salant and Dillman 1994):

1. *Determine the target audience and the best method for reaching them.* Considerations may include:
   - Whether the target audience is broad (e.g., general public) or narrow (e.g., birders);
   - How potential respondents will be identified;
   - How contact will be made (e.g., by telephone, by mail, or in-person interviews);
   - Available resources (e.g., money, people, and time).
2. *Identify control groups for comparison of responses with target group.* For example, if a study involves resident attitudes concerning urban deer management, both neighborhoods with and neighborhoods without deer conflicts should be included in the research design.
3. *Develop survey instrument.* This step takes both time and skill to do well and should use the study objectives and research hypotheses as the blueprint for development.
4. *Determine appropriate statistical measures to use in data analysis.* First establish whether the study is attempting to establish a qualitative or quantitative difference. Keep in mind there are situations where statistical tests are irrelevant. For example, if a survey of all state natural resource agencies is conducted, and responses are received from all 50 agencies, then the study has captured all of the possible respondents. Any difference in response is statistically significant.
5. *Administer the survey instrument.* Response rates need to be high enough (e.g., >50%) to allow for a statistically significant representation of the target audience. The smaller the target audience, the larger the sample needed to provide significant representation.
6. *Follow up with nonrespondents.* This step is necessary to determine if there is some inherent bias in the nonrespondent group. For example, there may be a common factor (e.g., age, gender) within the nonrespondent group that prevents them from participating in the study.
7. *Communicate results.* The final step of any research project is to share the results in peer-reviewed journals, popular literature, the news media, and at professional and public meetings.

It is possible to short-circuit the rigor in HD research by taking shortcuts or adopting a study design that is not grounded in measurable objectives or testable hypotheses. A common misconception by individuals not schooled in reputable HD research methods is that survey instrument development involves little more than coming up with some questions to ask constituents — in other words, "anyone can create an effective questionnaire." Not so! Such simplistic attempts will provide quick answers, in particular, the answers desired, but with little consideration given to sampling protocols, control of bias, nonresponse follow-ups, hypothesis testing, and communicating the data to anyone beyond those who initiate the study. The data reported are usually in the form of frequencies (percent of this or that) without an examination of connections to other factors that might cause a response. The information gathered — often at considerable expense — usually can not stand the rigor of HD peer review. Those who do not have a background in questionnaire development and/or statistical analysis should take advantage of the personnel resources available through universities and even the private sector rather than use the "do-it-yourself" approach.

The National Survey of Fishing, Hunting, and Wildlife-Associated Recreation (referred to hereafter as the FHWAR Survey), conducted by the U.S. Department of the Interior every 5 years since 1955, is one of the oldest and most comprehensive continuing human dimensions surveys available (U.S. Department of the Interior 2001). Wildlife-associated recreation, formerly referred to as "nonconsumptive wildlife-related recreation," includes observing, photographing, and feeding fish and wildlife. The FHWAR Survey's three interest categories (fishing, hunting, and wildlife-associated recreation) are not mutually exclusive — many of the individuals surveyed report enjoying fish and wildlife in a variety of ways.

The purpose of the FHWAR Survey is to gather information on the number of anglers, hunters, and wildlife-watchers in the U.S. — the activities they participate in, how often they participate, and how much they spend on these activities. Survey results are presented on a national and a state-by-state basis. Results include participation rates by various demographic characteristics. Total days of participation, trips, and expenditures also are reported, as are the common types of animals hunted, fished, or watched. The results of the 1991, 1996, and 2001 surveys are available online (http://www.census.gov/prod/www/abs/fishing.html). Because the FHWAR Survey methodology and questions were similar during these three survey periods, estimates and results are comparable.

Human dimensions researchers have used specific data from the FHWAR Survey as a starting point to conduct more detailed analysis of the economic impacts of hunters, anglers, and wildlife-watchers, such as:

- The extent of under-representation by gender, racial, and ethnic groups;
- Changes in participation rates during 5-year periods;
- Changes in wildlife recreation expenditures.

These studies are published as separate reports available online from the U.S. Census Bureau (see URL provided above) or from private, for-profit research groups (e.g., Southwick Associates, http://www.southwickassociates.com).

The general public may not consider the FHWAR Survey to be of great importance, or even be aware of its existence, but the document provides valuable information to communities and businesses on the potential economic impact that can be realized from local wildlife resources. The 2001 Survey, for example, reported $56.2 billion in total expenditures for hunting and fishing and $33.7 billion for wildlife watching.

8.3 THE ROLE OF HUMAN DIMENSIONS IN URBAN WILDLIFE MANAGEMENT

There are many roles for human dimensions in urban wildlife management. For the sake of this discussion, however, four broad categories of assessments will be used:

- Public participation in wildlife-related programs;
- Wildlife values;
- Quality of life issues;
- Conflict resolution.

Each of the categories will be examined in the order listed above.

8.3.1 Public Participation in Wildlife-Associated Recreation

According to data compiled in the 2001 FHWAR Survey, nearly 82 million U.S. residents participated in some type of wildlife-related activities (Table 8.2). Nearly 22 million Americans participate in more than one of the three FHWAR Survey activity classifications.

Birds and mammals were the most popular wildlife observed by the U.S. public (Figure 8.1). The popularity of birds and mammals may have a connection to the charismatic association that people have regarding animals with feathers and fur compared to animals with scales and exoskeletons (aka reptiles and insects).

A look through a current issue of *TV Guide* provides additional evidence of the growing interest in all things wild. Not only are shows about wildlife commonplace, there are now entire networks (e.g., Animal Planet and Discovery Channel) devoted to wildlife and nature programming. The 1982 Conservation Directory of the National Wildlife Federation listed 90 wildlife-related organizations representing about 8.5 million members; by early 2004 the list had expanded to 1219 organizations and 23 different environmental issues that nonprofit organizations may address

Table 8.2 Total U.S. Wildlife-Related Recreation in 2001

Activity	Participants (million)	Expenditures (billion)
Fishing	34.1	$34.1
Hunting	13.0	$20.6
Wildlife watching	66.1	$38.4
Total	82.0[a]	$108.0[b]

[a] Some individuals surveyed participate in more than one activity category.
[b] Total does not include "unspecified" expenditures of $13.8 billion.

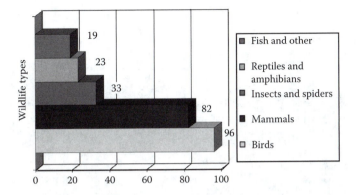

Figure 8.1. Percent of wildlife observers by type of wildlife observed. (Data Source: U.S. Department of the Interior, Fish and Wildlife Services, and United States Department of Commerce, Bureau of the Census, 2001. National survey of fishing, hunting and wildlife-associated recreation.)

(Table 8.3). Nearly three dozen magazines available in the U.S. focus on wildlife, and Americans age 16 or older spent $395 million on these publications in 1996 (Decker et al. 2001).

There are many public wildlife-related programs that use backyard wildlife for appreciation and the betterment of the community. Two examples — bats in bridges and Eagle Days — illustrate how these programs are implemented. Alternatively, there exist some highly questionable public wildlife-related programs that lead to serious negative consequences. Rattlesnake roundups are one example that represents a form of wildlife exploitation that is difficult to justify by any standard. Several studies have examined the HD consequences of this type of program (Thomas and Adams 1993; Adams et al. 1994; Fitzgerald and Painter 2000). A second example is that of constructed habitats (e.g., wildscaping). These programs, while well intentioned, may have unintended ecological consequences. Few HD assessments of constructed habitat programs have been conducted.

Table 8.3 Wildlife Issues Nonprofit Organizations May Address

Agriculture and farming	Pollution (general)
Air Quality and atmosphere	Population
Climate change	Public health
Development and developing	Public lands and greenspace
Ecosystems	Recreation and ecotourism
Energy	Reduce, reuse, and recycling of waste
Ethics and environmental justice	Sprawl and urban planning
Executive, legislative and judicial	Transportation
Finance, banking, and trade	Water habitats and quality
Forest and forestry	Wildlife and species
Land issues	Other
Oceans, coast, and beaches	

Data Source: National Wildlife Federation, http://www.nwf.org/conservationDirectory/search.cfm, 2004.

Figure 8.2 Crowds gather on summer evenings in Austin, Texas, to watch the evening emergence of Mexican free-tailed bats from under the Congress Avenue Bridge. (© Merlin D. Tuttle, Bat Conservation International)

Many urban communities use local wildlife resources to generate revenue for their citizens and cities. The Mexican free-tailed bat (*Tadarida brasiliensis*) colony under the Congress Avenue Bridge in Austin, Texas, is a perfect example and is discussed in more detail in Chapter 11. The number of tourists and the revenue generated from those who visit the bridge to watch the bats emerge during the summer months is enormous (Figure 8.2). The following excerpt from the website of Bat Conservation International illustrates this point:

> As the city came to appreciate its bats, the population under the Congress Avenue Bridge grew to be the largest urban bat colony in North America. With up to 1.5 million bats spiraling into the summer skies, Austin now has one of the most unusual and fascinating tourist attractions anywhere. The *Austin American-Statesman* created the Statesman Bat Observation Center adjacent to the Congress Bridge, giving visitors a dedicated area to view the nightly emergence. It is estimated that more than 100,000 people visit the bridge to witness the bat flight, generating $8 million in tourism revenue annually.
>
> *http://www.batcon.org/*

Another example of marketing the wildlife in your own backyard is the highly successful Eagle Days program (Figure 8.3), started in Missouri in the late 1970s (Witter et al. 1979). Bald eagles (*Haliaeetus leucocephalus*) use the Mississippi, Missouri, and Arkansas Rivers, along with various reservoirs, in their southern migration from Canada. They establish wintering grounds (stopovers) where they feed on fish and rest near urban communities and then move on. Several urban communities (Clarksville, Missouri; Tulsa, Oklahoma; Cassville, Wisconsin; and LaClaire, Iowa) have realized that they can convert a "free" natural phenomenon into a revenue-generating opportunity. Regrettably, there are no hard data on the

Figure 8.3 Eagle Days festivals are held annually in Missouri and other states. (Harold Selby)

amount of additional revenue generated by communities that promote Eagle Day events. It is known that what started out as a few hundred observers has grown into the thousands causing some communities to "gear up" with enhanced goods and services during Eagle Days.

In both of the above cases people — particularly urbanites — are spending time and money to observe wildlife "up close and personal." The wildlife resource is used in a safe and unobtrusive manner that allows each species to go about its business, but under the watchful eyes of curious humans. It is assumed that humans are increasing their awareness, knowledge, and appreciation of wildlife. Whether these transformation conditions translate into an increased stewardship ethic and concern for habitats is unknown. The last statement is rather tragic; the tools used in the human dimensions of wildlife management could be easily used to determine the attitudes, activities, expectations, and knowledge of program participants concerning wildlife and their habitats.

There is a flip side to the story on how humans use the wildlife resources around them; rattlesnake roundups provide an example of exploitation under the guise of recreation and public education (Figure 8.4). Unlike the previous two examples, rattlesnake roundups have been studied extensively in terms of the human dimensions, educational, ecological, and economic ramifications (Thomas and Adams 1993; Adams et al. 1994; Fitzgerald and Painter 2000). Rattlesnake roundups are events in which rattlesnakes (*Crotalus* sp.) are collected from the wild for public display, sale, and consumption. It would be difficult to identify another public event that uses a wildlife resource in this manner. In the words of the Humane Society of the United States (HSUS),

> Rattlesnake roundups are cruel and ecologically damaging events that have taken place in many parts of the U.S. since the late 1920s. Every year, in the states of Texas, Oklahoma, Kansas, New Mexico, Pennsylvania, Alabama, and Georgia, thousands of rattlesnakes are captured and slaughtered or used in competitive events in ways that violate the most basic principles of wildlife management and humane living. Though

Figure 8.4   (See color insert following page 276) A Texas rattlesnake roundup holding pen for captured rattlesnakes. (C. E. Adams)

they are often promoted as fundraising events for local civic causes, rattlesnake roundups primarily benefit their organizers and corporate sponsors.

The Truth behind Rattlesnake Roundups, http://www.hsus.org/ace/12078

In Texas, one of the states in which these events are held, a hunting license is required to pick the snake up off the ground. As long as this requirement is met, there are no additional regulatory statutes that govern the harvest of rattlesnakes. A snake in the bag is considered to be the property of the individual holding the bag — no questions asked (Adams et al. 1994).

Rattlesnake roundups may bring as much as $3 million into a community in one weekend. Civic organizations host the events and use the revenue generated from the sale of snakes to dealers to provide student scholarships and to support other community programs. Roundup spectators number in the thousands. They are local residents, tourists, the media, and members of environmental protest groups. Most are between the ages of 24 and 55, reside in large (50K) cities, and are predominantly white males (Thomas and Adams 1993).

The promoters and supporters of rattlesnake roundups promote these events as educational programs that enlighten the public on the natural history of the animals, biological aspects of rattlesnake existence, and how to treat a rattlesnake bite. However, the "educational" programs also include how to butcher a rattlesnake, remove one's body from a sleeping bag occupied by 20 or more rattlesnakes, or safely walk among hundreds of rattlesnakes. Serious consideration needs to be given to public attitudes toward rattlesnakes, and snakes in general, before and after attending one of these events. At least one study has shown that, when spectators leave the educational events, their knowledge of rattlesnakes is similar to those who did not attended the programs (Fitzgerald and Painter 2000).

Some of the most popular educational efforts in urban areas are programs undertaken by both private organizations and government agencies to encourage homeowners to develop wildlife habitat on their properties. The objectives of many of these programs include: (1) increasing awareness and appreciation of wildlife,

(2) educating homeowners and community leaders regarding habitat creation and restoration, and (3) increasing the total amount of quality habitat or "green space" in urban and suburban areas. The National Wildlife Federation (NWF), a nonprofit organization, for example, sponsors the well-known "Backyard Habitat Program." Since 1973, the NWF has been providing tools and resources for homeowners to create and restore native wildlife habitats in their backyards. As of July 2004, the organization had certified a total of 46,000 sites as backyard wildlife habitats (www.nwf.org). Several groups have similar "wildscaping" programs, and there are many websites on this topic — a Google search using this keyword resulted in nearly 1,400 hits.

Incorporating wildlife habitat into residential areas is becoming increasingly common for homeowners and developers and is touted as a way to reduce some of the impacts of residential development on wildlife populations. A study by Hunter (2002) conducted over a 2-year period in Bexar County, Texas, found bird diversity at the wildscaped neighborhood was significantly greater than in traditional neighborhoods or natural areas. Unfortunately, similar studies are rare. Assessment of whether stated objectives have been met often is a missing component in these types of wildscaping programs. In addition to more studies on the impact of wildscaping on species diversity, many other questions need to be asked: Do the programs actually promote more awareness about and appreciation of urban wildlife? Do they increase homeowners' knowledge of wildlife habitats in and outside of urban areas? Do they increase public sensitivity to conservation issues such as loss of biodiversity, exotic species, and pollution?

There are several possible reasons behind the lack of evaluation for public wildlife education and awareness programs. To begin with, program evaluation is usually an afterthought rather than part of the up-front planning of the program in the form of clearly articulated and measurable program objectives. Often the primary goal is simply to get the program "on the street," after which the number of participants is used to "measure" program success. There may be few, if any, personnel associated with the program who have training and experience in program assessment. Lack of professional interest in investigating the questions listed above may be the result of the low representation of urban biologists at the university level. Another explanation may be the fact that dollars drive the direction of research. Organizations may not be eager to find out that a popular program is having ecological repercussions that are not all positive. And finally, program managers may find it difficult to allocate money to do assessments when other programs are waiting for funding.

8.3.2 Wildlife Values

Economic benefit is one way to value wildlife, but there are many others. The terms *value* and *assigned value* are used interchangeably here to mean the level of worth a person gives to one object relative to another (Brown and Manfredo 1987). Researchers have attempted to categorize and measure both values and attitudes toward wildlife; in fact, neither is directly measurable. Using classification models or values scales as measurement tools, the underlying values of individuals are

Table 8.4   Comparison of Several Wildlife Value Categorizations

King (1947)	Kellert (1980)	Purdy and Decker (1989)	Fulton et al. (1996)
Aesthetic	Aesthetic	Problem-Acceptance	Bequest & Existence
Biological	Dominionistic	Societal-Benefits[a]	Fishing/Anti-fishing
Commercial	Ecologistic	Traditional-Conservation	Hunting/Anti-hunting
Educational	Humanistic		Recreational Experience
Recreational	Moralistic		Residential Experience
Social	Naturalistic		Wildlife Education
	Negativistic		Wildlife Rights
	Neutralistic		Wildlife Use
	Scientistic		
	Utilitarian		

[a]Consists of social-significance and ecological-significance components.

inferred. King (1947) began the process of categorization by defining six over-lapping types of values; several decades later Kellert (1980) described ten attitude categories. Both of these typologies are commonly cited in human dimensions publications. Purdy and Decker (1989) developed the wildlife attitude and values scale (WAVS), patterned after King's classification model, to describe noneconomic values of wildlife. A comparison of various wildlife value categorizations can be found in Table 8.4.

Recently, other authors have developed their own classification systems for wildlife values. Gilbert and Dodds (2001) listed six wildlife values generally recognized by biologists. Conover and Conover (2003) argue there are three wildlife values not yet articulated by biologists that people sense at a subconscious level (Table 8.5).

Because people hold a variety of opinions, attitudes and values concerning natural resources, and because those values are difficult, if not impossible, to measure, wildlife managers face the challenge of trying to make science-based decisions that will also address the needs and expectations of as large a portion of the public as possible. This is not an easy task. Managing wildlife, urban or otherwise, is a relatively simple task compared to managing people.

8.3.3 Quality of Life Issues

Through the ages wildlife has inspired art, music, dance, drama, storytelling, and poetry. Wildlife continues to add to our quality of life. It doesn't seem to matter much whether it's watching visitors to a backyard bird feeder, listening to frogs calling on a summer night, or spotting a white-tailed buck from a hunting blind — millions of North Americans would consider their lives to be less enjoyable without wildlife. In fact, many quality of life indices include wildlife and nature along with good schools, above-average annual income, and cultural opportunities.

The benefits of wildlife and wildlife habitat go well beyond intangibles, as many studies have shown. In Salem, Oregon, urban land adjacent to greenspace was found to be worth approximately $1,200 more per acre than urban land 1,000 feet away, all other things being equal (Nelson 1986). In the neighborhood of the Cox Arboretum in Dayton, Ohio, the proximity of the park and arboretum accounted for an

Table 8.5   Additional Wildlife Value Typologies

Gilbert and Dodds (2001)

Monetary values (aka commercial or economic)	Uses of wildlife that can be exchanged for money or have an economic impact on an individual or society.
Recreational values	The enjoyment people experience from hunting, birding, camping, or wildscaping. There is some overlap between recreational and monetary values when people spend money to pursue these activities.
Scientific/educational values	The roles wildlife species play in advancing human understanding, including the creation and dissemination of knowledge.
Biological, ecological, or ecosystem values	The benefits wildlife species provide by maintaining a functional ecosystem.
Option values	Related to monetary because it is based on the amount a person is willing to pay today to ensure that a resource will be available for his or her use in the future (Krutila 1967; Randall 1991).
Existence Values	The sense of well-being a person experiences from just knowing that wildlife exist in the natural world (Randall 1991).

Conover and Conover (2003)

Historical value	Based on the premise that humans, by their basic nature, are nostalgic and long for the way things used to be, including the presence of certain endemic wildlife species.
Empathetic value	Based on the human ability to empathize with other living beings, such as an animal's fear, distress, or pain.
Character-assessment value	Based on the idea that people who behave ethically or altruistically toward animals are likely to behave similarly toward people, those allow people to ascertain the true character of others.

estimated 5% of the average residential selling price, and in the Whetstone Park area of Columbus, Ohio, the nearby park and river were estimated to account for 7.35% of selling prices (Kimmel 1985). There has been increased interest in developing urban and suburban wildlife habitat as a way to attract real estate buyers. Housing developers have caught on to the fact that nature sells and, as a result, many have begun to promote the incorporation of "green space" in their development plans.

Studies of groups considered "uncommitted" on attitudes about wildlife issues have found that, while these individuals did not seek out wildlife-related activities, they appreciated wildlife in the context of their other activities (Fleishman-Hillard Research 1994). Access to wildlife may even improve human health. Some researchers have proposed that humans have an innate need to associate with other species, termed *biophilia* by Harvard biologist Edward O. Wilson (1984). Emory University Rollins School of Public Health scientist Howard Frumkin has presented evidence for the health benefits of four kinds of contact with the natural environment: contact with nonhuman animals, contact with plants, viewing landscapes, and contact with wilderness (2001). Several studies have linked nonhuman animals with human health. Frumkin pointed to research that concludes pet owners have fewer health problems than non-pet owners. Examples include lower blood pressure, improved survival after heart attacks, and enhanced ability to cope with life stresses.

People can be involved with wildlife near their homes in a variety of ways, from watching wild animals to participating in management by providing food, water, shelter, and space for wildlife (Leedy and Adams 1984; Young 1991). As we discussed earlier in this chapter, *wildscaping* — a form of landscaping that relies heavily on native plants to attract birds, butterflies, and other types of wild-life — and other methods for attracting wild animals have become a fulfilling hobby for many Americans (Figure 8.5). State and federal wildlife management agencies have responded to the increased popularity of residential wildlife watching by creating books, pamphlets, and videos, and developing and promoting "watchable wildlife" programs, often in cooperation with local chambers of commerce and tourism departments.

However, urbanization has isolated humans from the natural world. In general, people who live in urban and suburban areas have a lower overall level of knowledge about wildlife compared to rural residents (Van Druff et al. 1994). Since the majority of human population in North America can now be classified as urban, there is a greater need than ever before to educate the public about wildlife. It's hard to enjoy something you don't understand, or even fear (see Perspective Essay 8.1).

Studies have shown that, while interest in wildlife and the environment may be high, urbanites tend to be unaware of the names of all but the most common species and lack understanding about the way the natural world works (Adams et al. 1987; Penland 1987). One study of high school students (Adams et al. 1987) found that many could not correctly identify common urban wildlife species (e.g., confusing opossums and rats), the relative numbers of selected wild species in Harris County, Texas (e.g., raccoons were considered rare, while cougars were thought to be abundant), the diets of 16 common mammals, and the effect of human habitation on relative abundance of those mammals.

In addition, urbanites have a tendency to misinterpret wildlife behavior. Examples include thinking that a wild animal that does not maintain a safe escape distance "wants to be friends." This lack of understanding can result in a variety of unwanted consequences, including human injury (e.g., tourists mauled by seemingly tame park bears), removal of animals from the wild (e.g., "rescuing" deer fawns), or contributing to exponential proliferation of certain wildlife populations (e.g., feeding Canada geese on city ponds). Lack of public information, education, and awareness of wildlife has been identified as an important wildlife issue (Van Druff et al. 1994) (Figure 8.6).

People from all walks of life want and need information about wildlife. They include adults and school children, elected officials and business owners, local governments, apartment managers, home owners, and members of the media. Their

Figure 8.5   Wildscaping offers the chance to enjoy nature close to home. (John M. Davis/Texas Parks and Wildlife Department)

needs and reasons for obtaining wildlife information are just as diverse, ranging from a desire to increase opportunities to enjoy wildlife on a daily basis, to avoiding property damage, to improving public health (Young 1991). Unfortunately, when questions about living with wildlife arise, Americans often are unsure whom to contact for help (Reiter et al. 1999).

Natural resource agencies have much at stake in shaping public understanding of wildlife management and conservation (Mankin et al. 1999). Agencies have been quite successful identifying and communicating with traditional clienteles (Decker et al. 1987; Witter 1990; Hesselton 1991; Kania and Conover 1991; Gray 1993; Jolma 1994). Some agencies also have attempted to reach out to nonconsumptive users; Texas Parks and Wildlife Department's (TPWD) Nongame and Urban Wildlife Program, for example, has made a major effort to develop programs specifically to appeal to adults excluded by hunting and fishing programs (Adams et al. 1997). On the whole, however, the wildlife profession has had difficulty communicating effectively with the general public (Decker et al. 1987; Jolma 1994).

8.3.4 Conflict Resolution Issues

Urban wildlife conflict resolution issues can be placed into one of two subsets: conflicts between humans and wildlife, and conflicts between humans over wildlife resources. Often, although not always, human–wildlife conflicts are the easier of the two to address. Many how-to books and other readily available reference materials provide detailed instructions for solving common problems that arise between humans and wildlife. Rather than repeat those efforts, this section offers an introduction to some of the more prevalent human–wildlife conflict issues natural resource professionals and the public are likely to face.

Figure 8.6   Urbanites have a tendency to misinterpret wildlife behavior which can lead to a variety of unwanted consequences, including human injury. This man was charged by a small black bear while attempting to take a photo. (National Park Service Historic Photograph Collection, c.1950)

By the same token, hundreds of books devoted exclusively to solving disputes between humans using well-researched methods of negotiation and conflict resolution are easy to find at almost any bookstore. To repeat those efforts here, even in abbreviated form, would not do justice to the subject. However, the topic is one that is vitally important to wildlife managers in general and urban wildlife managers in particular. Anyone involved in this field is strongly advised to consider taking one or more of the many excellent courses available on this subject. Many reference materials on negotiation and conflict resolution are also available. Discussion here will be limited to an overview of the issues that may arise and the ways in which human dimensions can be applied.

Conflicts between humans and wildlife species have increased as urbanization has increased. This is due in part to greater exploitation of urban and suburban habitats by wildlife. Because people enjoy feeling a connection with nature and wildlife, many homeowners actively work to attract wild animals to their property by offering supplemental food (bird, squirrel, and butterfly feeders), water (bird baths and misters), and shelter (bird, bat, and toad houses). State agencies and nongovernmental organizations (NGOs) such as the National Wildlife Federation and others encourage homeowners to wildscape their yards to provide both cover and food for wildlife. In other cases, people "accidentally" welcome wildlife to their homes and yards by leaving garbage cans and pet food and water bowl outside.

Regardless of a homeowner's intent, often the result of providing supplemental food, water, and shelter is a higher density of animals than would normally be supported by an equal amount of natural habitat. Most members of the public remain largely uninformed about the consequences of attracting wildlife. For example, many do not understand that the number of animals will increase as the amount of food offered increases. This concept might sound quite attractive to the novice wildlife aficionado, but the number of animals expecting to be fed can quickly exceed the homeowner's budget or tolerance level — not to mention that of their neighbors! Furthermore, large congregations of animals promotes the transmission of both disease and parasites that can decimate local populations of one or more species — hardly the scenario most wildlife-watchers have in mind when they begin to offer supplemental food.

Research has shown that effective education programs are needed to introduce the public to the realities of wildlife conflict and damage prevention (Adams et al. 1988; Lowery and Siemer 1999). For example, food may be plentiful in the city, but often natural denning sites are not. This can cause raccoons, opossums, skunks, and squirrels to set up housekeeping in attics, decks, and out-buildings. This, in turn, increases the number of wildlife complaints from homeowners.

Not only is the public uneducated concerning human–wildlife conflicts, relatively few are knowledgeable about which agencies are responsible for managing wildlife. A national study on public attitudes toward wildlife damage and policy found that only 19% of survey respondents had ever heard of the U.S. Department of Agriculture's Wildlife Services program, formerly known as Animal Damage Control (Reiter et al. 1999). A study of Texas residents found the majority were unable to identify their state wildlife agency by name (Adams and Thomas 1998). A study in Alabama had similar results (Rossi and Armstrong 1999).

A survey of agency responsiveness to wildlife damage found that most of the organizations were concerned about the issue (Hewitt and Messmer 1997). The majority (75%) of responding state and provincial wildlife agencies indicated they had written wildlife damage management policies. However, many did not actively publicize or articulate their policies. Few agencies evaluate their wildlife damage policies, and when evaluations are done, often they are reactive rather than proactive.

An earlier publication on the responsibilities of various agencies for animal damage management found that agency responsibility hierarchies can be confusing (Berryman 1994). In some states, responsibilities for certain wildlife species are vested with the state agriculture agencies, while other species are the responsibility of the department of natural resources. At the federal level, authority over migratory birds is vested within the U.S. Fish and Wildlife Service (Department of Interior), but depredation control of birds is vested in Wildlife Services (Department of Agriculture). Little wonder the public perception of responsibility for wildlife damage management is unclear at best.

People will attempt to find their own solutions to human–wildlife conflicts when they are unable to get professional help (Case Study 8.1). Not all remedies are the ones wildlife professionals would promote or prefer. Some methods are harmless, and often ineffective, but it is not uncommon for homeowners to try bizarre and even dangerous cures. Stories of damage control efforts gone awry are commonplace. One dramatic example involves pouring gasoline down burrow holes and igniting it to exterminate chipmunks (*Tamias* spp.) (San Julian 1987).

Wildlife professionals tend to regard human–wildlife conflict management in urban and suburban situations as difficult, due to the perceived resistance of the public to a full range of management options (Decker and Loker 1996). Attitudes among urban and suburban dwellers continue to shift away from utilitarian perspectives and toward moralistic and humanistic perspectives (Kellert 1997; Hadidian et al. 1999), meaning they are less likely to accept lethal management techniques. Control of wildlife in urban environments is more technically complex than control in an agricultural situation, even when the same species is involved. Wildlife management agencies face additional impediments when trying to work with the public on conflict issues, including: (1) agency image and credibility problems, (2) conflicts between recommended solutions and the personal values of a diverse constituency, and (3) public animosity toward regulatory agencies (Lowery and Simer 1999).

Given these obstacles, as well as funding limitations, state and federal wildlife agencies often have left resolution of human–wildlife conflicts to individual initiative or to private wildlife control operators (WCOs) (Hadidian et al. 1999). A 1993 national study found that about 63% of licensed trappers had been contacted to remove problem animals (International Association of Fish and Wildlife Agencies 1993). There appears to be rapid growth and privatization of the wildlife control field over the past several decades due to the increase in human–wildlife conflicts (Braband 1995; Barnes 1997).

The role of government in the business of wildlife damage control has been to regulate activities through a licensing or permitting process, and to provide extension or educational services. Legal authority for such regulation is vested in federal and state governments, but in reality, authority often is divided among different agencies, such as natural resources, agriculture, and public health (Barnes 1997). Specifics on the nature, scope, and extent of WCO activities are limited. A 1997 study found that only 36% of states required a license for individuals to practice nuisance wildlife control; only 14 of those states required an annual report of activities (Barnes 1997).

A study of private WCOs found that most have a high school diploma, but little specific training in wildlife damage management or wildlife management in general; more than half of WCOs surveyed had not attended a trapper-education course (Barnes 1995a,b). WCOs come from a diversity of backgrounds, including wildlife biologist, pest control operators, fur trappers, and wildlife rehabilitators (Braband 1995). Barnes proposed a model for nuisance wildlife control licensing containing three key requirements: education, continuing education, and liability insurance (1997). Schmidt discussed the importance of training in wildlife identification and wildlife ecology, state and federal wildlife and pesticide laws, parasites and diseases of concern to wildlife and humans, chemical immobilization and euthanasia, and current and emerging technologies in wildlife damage management (1998). Additional studies are needed to determine the extent and volume of human–wildlife conflict resolution activities and the impact these have on wildlife populations.

The effects of wildlife on human health and safety are a subset of human–wildlife conflict, but one that deserves special attention. Examples of this type of conflict include: zoonotic diseases (e.g., Lyme disease, hantavirus), transportation hazards (e.g., deer-automobile collisions, beaver dams flooding roads), and sanitation problems (e.g., rodents, roosting birds). No national summaries are available on losses in human lives and economic productivity due to these types of human–wildlife conflicts. However, published and unpublished data have been compiled to assess potential cost of wildlife encounters in the U.S. in terms of human illnesses, injuries, and fatalities (Conover et al. 1995).

Table 8.6 Reported Cases of 10 "Notifiable Diseases" for 1999 and 2001 in the U.S. for Which Wildlife Species May Serve as a Vector or Reservoir[a]

| Disease | Cases | | Fatalities[b] |
	1999	2001	1999
Brucellosis	82	136	0
Encephalitis[c]	80	216	3
Hantavirus	31	8	— [d]
Lyme disease	16,273	17,029	7
Plague	9	2	1
Psittacosis	16	25	0
Rabies (human)	0	1	0
Rocky Mtn. spotted fever	579	695	5
Trichinoisis	12	22	0
Tularemia	— [e]	129	— [f]
Total	17,082	18,263	16

[a] Humans, companion animals, and livestock also can serve as the vector or reservoir to the infectious agent. Wildlife are involved in an unknown proportion of these cases.
[b] Fatalities are unavailable for 2001.
[c] Includes California serogroup viral, Eastern equine, St. Louis, and Western equine.
[d] Hatavirus data unavailable for 1999.
[e] Tularemia was not classified as nationally reportable in 1999.
[f] Tularemia data unavailable for 1999.

Source: U.S. Center for Disease Control and Prevention, United States Department of Health and Human Services, Centers for Disease Control, Atlanta, GA, 2001.

There were 18,263 reported cases of the 10 wildlife-related reportable diseases in the U.S. during 2001 according to the U.S. Centers for Disease Control and Prevention (U.S. Centers for Disease Control and Prevention 2001). Lyme disease, discussed in Chapter 3, accounted for over 93% of these cases (Table 8.6). Records for the West Nile strain of encephalitis/meningitis were not kept before 2002. During that year, 4,156 human cases and 284 deaths were reported in the U.S. Humans, companion animals, and domestic livestock can serve as vectors or reservoirs for these diseases, so the proportion of cases attributable to wildlife is unknown.

Human mortality data were lacking for most zoonotic diseases and parasitic infections. For example, histoplasmosis, a respiratory disease caused by inhaling *Histoplasma capsullatum* fungus spores growing in bird or bat feces-enriched soil, is a common disease not included in the CDC's list of reportable diseases (U.S. Centers for Disease Control and Prevention 2001). There are more than 140 diseases or parasitic infections in the U.S. for which nonhuman mammals or birds may serve as a vector or reservoir (Bisseru 1967; Beran 1994).

One of the more common ways in which wildlife impact human health and safety is in the area of ground and air transportation, which were discussed in detail in Chapter 6. Aircraft and avian species often compete for the same airspace. When that happens, collisions may occur, resulting in property damage, injuries, and even fatalities. The problem exists nationwide, but airports in the Eastern and Southeastern U.S. experienced the greatest number of wildlife–aircraft collisions (Wildlife Services 2001). Bird strikes have resulted in the loss of at least 190 lives and 52 aircraft in civil aviation (Thorpe 1996). There have been 283 military aircraft lost and 141 deaths recorded, in the limited number of Western nations from which data are available, between 1959 and 1999 (Richardson and West 2000).

The annual number of collisions between vehicles and deer (*Odocoileus* spp.) has been estimated at >726,000 and increasing (Conover et al. 1995). Approximately 600 collisions between moose (*Alces alces*) and vehicles reportedly occur in Alaska annually (Romin 1994). The incidence of human injury has been reported as 4% nationwide; less than 3% of deer-vehicle collisions resulted in a human fatality (Rue 1987). Approximately 29,000 human injuries and 211 human fatalities occurred as a result of collisions with deer (Conover et al. 1995). Unfortunately, in many states there is no formal mechanism to identify loss of personal property or human life to deer/auto collisions. In fact, more often than not, these accidents may be reported as uncategorized. Therefore, the numbers reported in published literature do not always agree.

Of course, deer and moose are not the only wildlife species involved in vehicle collisions. Unfortunately, few if any records are kept on collisions with other mammalian species, such as black bear (*Ursus americanus*), coyote (*Canis latrans*), and elk (*Cervus elaphus*), much less the number of accidents which were caused by drivers who attempted to avoid smaller wildlife species, such as tree squirrels (*Sciurus* spp., *Tamiasciurus* spp.), raccoons (*Procyon lotor*), and birds.

Despite the positive effects of beaver (*Castor canadensis*) on water quality and the creation of wetland habitat, when these animals begin constructing dams they can create public safety hazards by flooding roadways, plugging culverts, and damaging roadbeds and bridges (Harper 2002). In the northern U.S. 27–40% of total beaver damage complaints involved culvert plugging and road flooding (Payne and Peterson 1986). Flooded and washed-out roads can present serious human safety concerns (Jensen et al. 2001).

When wild animals congregate they can create sanitation problems that pose a threat to human health. Large flocks of starlings, grackles, and red-winged blackbirds may take up residence in urban neighborhoods, and the resulting accumulation of feces under roost trees can be considerable (Figure 8.7). Accumulation of fecal material also causes conflicts between humans and urban waterfowl, particularly Canada geese. While waterfowl are not a health threat to humans, their droppings are causing concerns about water quality control in municipal lakes and ponds. Mammals also can create sanitation concerns; raccoon latrines may increase the potential for transmission of the parasite *Baylisascaris procyonis,* while rodent middens and droppings can be a source for hantavirus infection in humans.

A wealth of information is available to the public from governmental and nongovernmental sources concerning disease transmission and prevention and human safety — if they can figure out where to find it. Many studies have been done which identify the need for education on any number of zoonotic diseases. A study conducted at Cornell University's Human Dimensions Research Unit on human perceptions and behaviors associated with Lyme disease included two information transfer

Figure 8.7   Roosting birds — and their droppings — can become an annoying problem. (John M. Davis/Texas Parks and Wildlife Department)

objectives: identifying communication mechanisms and networks, and developing an information base and guidelines for land and wildlife managers (Siemer et al. 1992). The California Department of Fish and Game developed a program to educate the public about a behavioral approach for managing predators at the urban-wildlife interface which was reported to have reduced the number of complaints received by that agency (Wehtje 1998).

A 2002 study examined the role of wildlife disease monitoring programs to provide early detection of new and emerging diseases, some of which may have serious zoonotic and economic implications (Mörner et al. 2002). The authors suggested that monitoring programs be integrated within national animal health surveillance infrastructures. These programs should have the capacity to respond promptly to the detection of unusual wildlife mortality and institute epizootiological research.

In 1996, researchers at the Centers for Disease Control and Prevention (CDC) in Georgia surveyed 100 small- or exotic-animal veterinarians in California, considered to be a primary source of information on reptile-associated salmonellosis for reptile owners, to assess the veterinarians' knowledge and information transfer practices (Glynn et al. 2001). Veterinarians have a unique opportunity to inform the public about zoonotic disease prevention strategies, immunization recommendations, and risk assessment following potential exposure due to their knowledge of the natural history of vertebrate animals (Herbold 2000). In 2002, the Colorado Wildlife Commission published a policy statement on chronic wasting disease that included a mandate for health experts to provide precautionary guidelines to people hunting in game management units. However, a review of the literature did not reveal any research into the existence of formal information transfer structures or evaluation of the effectiveness of agency efforts to educate the public.

Although conflicts have always been a part of human culture, conflicts over wildlife between members of the general citizenry, and between the public and wildlife agencies, began to increase during the late 1960s and early 1970s. At this time, people other than traditional clientele of wildlife management began to express interest in all types of wildlife, not just game species or "nuisance" wildlife. They wanted an equal voice, along with traditional consumptive users, in how wildlife resources were managed. Hikers, campers, and birders wanted nongame species protected, so they could be enjoyed during nonconsumptive recreation. Environmentalists wanted certain species to be protected to ensure ecosystem health (Decker et al. 2001). By the 1980s, advocates for animal rights had begun to question not only specific wildlife management practices, but whether managing wild animals, especially for the purpose of "harvesting" them, should be done at all.

Conflicts between humans over wildlife usually are the result of different beliefs and attitudes regarding how wildlife resources should or should not be used. In some ways they are quite similar to the kinds of conflicts that arise over other natural resources, including water, air, timber, and minerals.

Stories of conflicts over wildlife can be found almost daily in both local and national newspapers, as well as in other types of media. For example, some New Mexico residents strongly support federal mandates limiting the use of water from the Rio Grande for irrigation in order to protect the endangered silvery minnow, while others, particularly those who depend on irrigated crops for their livelihood, question the fairness of assigning more value to the welfare of fish than the welfare of farmers. Hunters in various western states have pushed wildlife agencies to add a spring hunting season for black bear that, environmentalists argue, will devastate bear populations. In neighborhoods across the country homeowners decide it would be nice to plant a butterfly garden, unaware that they'll soon have an irate neighbor at their door complaining about the damage ravenous caterpillars have caused in a prized vegetable garden.

How can urban wildlife managers hope to solve these types of human–human conflicts? Are solutions even possible? Surely there will be cases in which, despite the best efforts of the manager involved, decisions will need to be made that result in one side "winning" and the other "losing." How can such an outcome be deemed a true solution?

Finding appropriate solutions to urban wildlife management issues requires creativity, skill, and patience. Each situation is unique. In many cases a broad range of conflicting interests and concerns must be addressed, along with the economic, political, and social ramifications of any decision. This can be daunting, in part because techniques have not been fully developed (Carpenter et al. 2000). Classes and books are available that can help managers improve their conflict resolution skills. In addition, Decker and Brown have described some key traits of wildlife managers who have been successful at integrating human dimensions into wildlife management (Decker et al. 2001).

First and foremost, managers need to *be receptive* to many different viewpoints expressed in many different forms — telephone calls, office visits, letters to the editor, editorials, news coverage, demonstrations. This type of information isn't easily interpreted and must be kept in proper perspective. Minority interests can make a large impact using the media, while majority interests may not be as well organized or vocal.

Second, Decker and Brown suggest that managers must *be inquisitive.* Don't expect information to fall into your lap. Think about the many ways in which biological and human dimensions data may be found and gathered: surveys, literature searches, newspaper archives, questioning colleagues, and at public hearings.

Finally, effective managers need to *be diagnostic.* Carefully define the problem and its elements. Develop a plan for approaching the problem. Think about who should be involved in the decision-making process.

Decker et al. (2001) described five general approaches to problem solving using public involvement:

- Expert authority;
- Passive-receptive;
- Inquisitive;
- Transactional;
- Comanagerial.

Expert authority is a top-down approach typical of agencies in the days when they served a narrow constituency with shared values. Today the public is far less tolerant of a paternalistic "we know what's best" attitude. While still in use, there are very few cases in which it can be used effectively.

The *passive-receptive* approach is a slight variation on the top-down paradigm; managers welcome outside input but don't seek it in any systematic way, and the power to make decisions remains with the manager-expert.

Managers using the *inquisitive* approach are proactive in their attempts to gather information. They recognize that unsolicited input can lead to a biased perspective. Decision-making power is still in the hands of the manager, but decisions are made based on a more thorough understanding of the issues.

The *transactional* approach is distinguished from the other approaches described so far by the fact that managers work with interested parties to find acceptable objectives and actions. Depending on the type of decision, this method may allow interested parties to make binding decisions within bounds set by the wildlife agency.

The *comanagerial* approach is, in some ways, the most radical of the five methods described here. Comanagement involves actually sharing decision-making power by giving local communities greater responsibility for solving local wildlife problems. It requires rethinking the role of state wildlife agencies and wildlife professionals, placing more emphasis on providing biological and human dimensions expertise, training community participants, approving community produced management plans, certifying private consultants and community wildlife managers, and monitoring management activities.

Before managers can choose the best approach for solving a particular problem, they must have a thorough understanding of not only the environmental factors, but the human factors as well. The realization that wildlife management is as much about people as it is about natural resources was a driving force behind the creation of a method that could help wildlife professionals address a wide spectrum of societal needs and values — the stakeholder approach — discussed in Chapter 9.

CASE STUDY 8.1: DUCKS AND TRAFFIC

There is a "no pedestrian crossing" sign to discourage visitors from braving the traffic on Jefferson Street to reach the Buena Vista Lagoon duck-feeding area across the road, but ducks and geese can't read. The sight of dead birds along Jefferson Street after being hit by vehicles prompted a Carlsbad, California, resident to take action in hopes of getting motorists to slow down along that stretch of road.

Robert Feliciano's concern over the situation led him to submit a letter to the city asking officials to post "duck crossing" signs near the entrance to the duck-feeding area. In the letter, Feliciano offers to donate $200 to pay for the signs and supply the labor to install them. "I'd be happy to pay for the signs and put them up," Feliciano said. "I just want to get people to slow down."

Feliciano said the situation could result in an accident, since drivers tend to swerve into an adjacent lane to avoid hitting the birds, Feliciano said. "I love animals, so if I see a duck, I'm going to put on my brakes and let them get by. A lot of people are in such a hurry, they get upset and cut into another lane to get around the animal."

A visit to the duck-feeding area emphasized the conflict between nature and urban activity, when a mother and her 11-year-old daughter arrived holding a duck-ling they rescued from the nearby Sears parking lot. On Feliciano's advice, they agreed to take the duckling to an animal shelter. "Obviously, these animals were probably there before the mall was and now the mall has displaced them," said the mother, Mary Lou Jenkinson. Jenkinson said she supported Feliciano's idea.

Finding ducks and geese in the nearby streets is commonplace at this time of the year, said Petey Tade, comanager of the Buena Vista Audubon Nature Center. Tade said the problem was much more prevalent before the feeding area was enclosed by a fence in 1990. She agreed that signs warning motorists are a good idea, especially because of the large number of children visiting the area.

August 21, 2003
Excerpt from North County Times, San Diego, California
Reported by Michael J. Williams, Staff Writer

PERSPECTIVE ESSAY 8.1: URBANITES' FEAR OF THE NATURAL WORLD AROUND THEM

I talk to people all of the time about the compulsion to mow tall grass or remove understory. When I get to the heart of the matter, it always involves some type of fear. I hear statements like, "I mow the tall grass to keep snakes away." There it is.

Fear. To that person, tall grass equals snakes, which equals fear. Though they have never been bitten or harmed by a snake, they remove all tall grasses to avoid them.

I hear the same fear-based arguments pertaining to understory. I hear people voice opposition to maintaining natural habitat or say that they want the brush and leaf litter "cleaned up" because it will attract rats, or snakes, or criminals, or whatever. It's all fear. And much of it is simply not based on fact or even personal experience on the part of the person experiencing the fear.

I try to explain to them that rodents living in and around our homes are almost always an introduced species. The house mouse (*Mus musculus*), roof rat (*Rattus rattus*), and the Norway rat (*Rattus norvegicus*) are the most common offenders. They are introduced species and prefer to associate with human habitation. They are not the ones that prefer to be in the woods and fields. So a homeowner saying that a woodlot or prairie is causing a rat problem in their home is assigning blame in the wrong place.

However, rats may just be the scapegoat serving as a convenient excuse to get rid of habitat when the real reason the person is opposed to the habitat is fear of some sort. Several years ago I began working with a local park to reestablish native understory to the woodland areas as well as encourage prairie grasses in the openings. About 3 years into the restoration project when the understory and grasses were progressing nicely, I got a call from a homeowner living across the street from the park. She objected to the restoration project stating that it was causing a rat problem at her home.

I was intrigued and asked her what type of rats she was seeing and after some discussion she admitted that she hadn't actually seen any rats on her property. I asked her if she had seen any rats crossing the street from the park and entering her neighborhood. She admitted she hadn't. I then gently asked her if she had seen any rats actually in the park. She replied, "Oh I don't go over there." Puzzled, I asked her why not and she replied, "I'm afraid of snakes." Fear had shown up again and it wasn't related to rats. I did my best to ease her fears. I can't say that she became a supporter of the project, but she no longer voiced opposition.

Law enforcement officials often oppose maintaining open space in a natural condition. In my experience, they push for understory to be cleared for visibility purposes. They fear that understory hides criminals. I believe this fear, like many others, is blown out of proportion. I recently attended a meeting with a local municipality to determine how a local creek corridor should be maintained by the parks department. The corridor had sections that were highly maintained and other sections that were left natural. At the meeting were representatives from various community interests. Each community group was allowed to present concerns and suggestions as to how the corridor should be managed. The law enforcement officer representing the police department suggested that the understory along the creek be removed for safety concerns. He indicated that natural areas such as this would encourage violent crime. When asked if there had ever been a violent crime in that city associated with that creek corridor he answered, "No." When asked if there had ever been a violent crime associated with natural areas anywhere in the city, he answered, "No."

You see, that city did not have a problem with understory encouraging crime, but that's the fear-based argument that was used against managing the corridor in a natural fashion. This is not to say that crime never occurs in natural areas. There are examples where it has, but according to the Justice Bureau, violent crimes most often occur within our own homes. Should a city have a problem with violent crimes occurring in a particular natural area, then certainly something should be done about it, but to universally eliminate understory out of fear for potential crime seems overreactive to me.

Some people will inevitably argue that removing understory simply "looks better." They'll say their actions aren't based on fear, but aesthetics. I submit that in this situation we determine what "looks good" based indirectly upon fear. That which eases our fear comforts us, and we are attracted to that which we find comforting. Orderly landscapes imply that we have a sense of control and human dominance. This eases our fears.

Natural landscapes do not imply human dominance and, thus, contain a degree of uncertainty for some people. To those who are fearful, a wooded landscape that looks less fearsome is aesthetically pleasing. However, to people who aren't fearful of understory, a woodland with the shrub layer intact is preferred.

Tradition is yet another reason that people will clear understory. It's the "that's the way it's supposed to be" syndrome. I contend that our traditions come from practices that we become accustomed to seeing. Our society has been clearing understory out of fear for so long, that it has now become tradition.

By John M. Davis

Urban Biologist, Texas Parks and Wildlife Department, Dallas, Texas

REFERENCES

Adams, C. E., Leifester, J. A., and Herron, J. S. C. (1997). Understanding wildlife constituents: birders and waterfowl hunters, *Wildlife Society Bulletin*, 25:653–660.

Adams, C. E., Stone, R. A., and Thomas, J. K. (1988). Conservation education within informational and education divisions of state natural resource agencies, *Wildlife Society Bulletin*, 16:333–338.

Adams, C. E. and Thomas, J. K. (1998). *Statewide Survey of the Texas Public Outdoors: A Vision of the Future*, Texas A&M University, Department of Wildlife and Fisheries Sciences, College Station.

Adams, C. E., Thomas, J. K., Lin, P.-C., and Weiser, B. (1987). Urban high school students' knowledge of wildlife, in *Integrating Man and Nature in the Metropolitan Environment*, Adams, L. W. and Leedy, D. L., Eds., National Institute for Urban Wildlife, Columbia, MD, pp. 83–86.

Adams, C. E., Thomas, J. K., Sternadel, K. J., and Jester, S. L. (1994). Texas rattlesnake roundups: implications of unregulated commercial use of wildlife, *Wildlife Society Bulletin*, 22:324–330.

Barnes, T. G. (1997). State agency oversight of the nuisance wildlife control industry, *Wildlife Society Bulletin*, 25:185–188.

Barnes, T. G. (1995a). Survey of the nuisance wildlife control industry with notes on their attitudes and opinions, *Proceedings of the Great Plains Wildlife Damage Control Conference,* 12:104–108.

Barnes, T. G. (1995b). A survey comparison of pest control and nuisance wildlife control operators in Kentucky, *Proceedings of the Eastern Wildlife Damage Control Conference,* 6:39–48.

Beran, G. W., Ed. (1994). *Handbook of Zoonoses,* Vol. 1. 2nd ed., CRC, Boca Raton, FL.

Berryman, J. H. (1994). Animal damage management: responsibilities of various agencies and the need for coordination and support, *Proceedings of the Eastern Wildlife Damage Control Conference,* 5:12–14.

Bisseru, B. (1967). *Diseases of Man Acquired from His Pets,* Lippincott, Philadelphia, PA.

Braband, L. A. (1995). The role of the nuisance wildlife control practitioner in urban wildlife management and conservation, *Proceedings of the Eastern Wildlife Damage Control Conference,* 6:38.

Brown, T. L. and Manfredo, M. J. (1987). Social values defined, in *Valuing Wildlife: Economic and Social Perspectives,* Decker, D. J. and Goff, G. R., Eds., Westview Press, Boulder, CO, pp. 12–23.

Bryan, H. (1996). The assessment of social impacts, in *Natural Resource Management: the Human Dimension,* Ewert, A. W., Ed., Westview Press, Boulder, CO, pp. 145–166.

Carpenter, L. H., Decker, D. J., and Lipscomb, J. F. (2000). Stakeholder acceptance capacity in wildlife management, *Human Dimensions of Wildlife,* 5(3):5–19.

Conover, M. R. (1997). Wildlife management by metropolitan residents in the United States: practices, perceptions, costs, and values, *Wildlife Society Bulletin,* 25:306–311.

Conover, M. R. and Conover, D. O. (2003). Unrecognized values of wildlife and the consequences of ignoring them, *Wildlife Society Bulletin,* 31:843–48.

Conover, M. R., Pitt, W. C., Kessler, K. K., DuBow, T. J., and Sanborn, W. A. (1995). Review of human injuries, illnesses, and economic losses caused by wildlife in the United States, *Wildlife Society Bulletin,* 23:407–414.

Decker, D. J., Brown, T. L., Driver, B. L., and Brawn, P. J. (1987). Theoretical developments in assessing social values of wildlife: toward a comprehensive understanding of wildlife recreation involvement, in *Valuing Wildlife: Economic and Social Perspectives,* Decker, D. J. and Goff, G. R., Eds., The Wildlife Society Annual Meeting Symposia on Wildlife Damage, October 1–5, pp. 76–95.

Decker, D. J., Brown, T. L., and Knuth, B. A. (1996). Human dimensions research: its importance in natural resource management, in *Natural Resource Management: the Human Dimension,* Ewert, A. W., Ed., Westview Press, Boulder, CO, pp. 29–45.

Decker, D. J., Brown, T. L., and Siemer, W. F., Eds. (2001). *Human Dimensions of Wildlife Management in North America,* The Wildlife Society, Bethesda, MD.

Decker, D. J. and Goff, G. R. (1987). *Valuing Wildlife: Economic and Social Perspectives,* Westview Press, Boulder, CO.

Decker, D. J. and Loker, C. A. (1996). Human dimensions insights for successful wildlife damage management in suburban environments, *The Wildlife Society Annual Meeting Symposia on Wildlife Damage,* October 1–5.

Ditton, R. B. (1996). Human dimensions in fisheries, in *Natural Resource Management: the Human Dimension,* Ewert, A. W., Ed., Westview Press, Boulder, CO, pp. 73–90.

Ewert, A. W. (1996). *Natural Resource Management: The Human Dimension,* Westview Press, Boulder, CO.

Fitzgerald, L. A. and Painter, C. W. (2000). Rattlesnake commercialization: long-term trends, issues, and implications for conservation, *Wildlife Society Bulletin,* 28:235–253.

Fleishman-Hillard Research. (1994). *Attitudes of the Uncommitted Public toward Wildlife Management,* Fleishman-Hillard, St. Louis, MO.

Frumkin, H. (2001). Beyond toxicity: human health and the natural environment, *American Journal of Preventive Medicine,* 20:234–240.

Fulton, D. C., Manfredo, M. J., and Lipscomb, J. (1996). Wildlife value orientations: a conceptual and measurement approach, *Human Dimensions of Wildlife* 1(2):24–47.

Gilbert, D. L. (1964). *Public Relations in Natural Resources Management,* Burgess Publishing Company, MN.

Gilbert, F. F. and Dodds, D. G. (2001). *The Philosophy and Practice of Wildlife Management,* 3rd ed., Krieger Publishing Co., Malabar, FL.

Gigliotti, L. M. and Decker, D. J. (1992). Human dimensions in wildlife management education: pre-service opportunities and in-service needs, *Wildlife Society Bulletin,* 20:1–14.

Glynn, M. K., Mermin, J. H., Durso, L. M., Angulo, F. J., and Reilly, K. F. (2001). Knowledge and practices of California veterinarians concerning the human health threat of reptile-associated salmonellosis (1996). *Journal of Herpetological Medicine and Surgery,* 11(2):9–13.

Gray, G. G. (1993). *Wildlife and People: The Human Dimensions of Wildlife Ecology,* University of Illinois Press, Urbana, IL.

Hadidian, J., Childs, M. R., Schmidt, R. H., Simon, L. J., and Church, A. (1999). Nuisance-wildlife control practices, policies, and procedures in the United States, in *Wildlife, Land, and People: Priorities for the 21st Century,* Field, R., Warren, R. J., Okarma, H. K., and Sieverd, P. R., Eds., Proceedings of the 2nd International Wildlife Management Congress, The Wildlife Society, Bethesda, MD, pp. 165–168.

Harper, S. (2002). VDOT trying kinder method of beaver control, *The Virginian-Pilot,* 22 June 2002.

Herbold, J. R. (2000). What you and your clients need to know about zoonotic diseases: rabies, Lyme disease, and salmonellosis, *Proceedings of the North American Veterinary Conference,* 14:833–834.

Hesselton, W. T. (1991). How governmental wildlife agencies should respond to local governments that pass anti-hunting legislation, *Wildlife Society Bulletin,* 19:222–223.

Hewitt, D. G. and Messmer, T. A. (1997). Responsiveness of agencies and organizations to wildlife damage: policy process implications, *Wildlife Society Bulletin,* 25:418–423.

Hunter, A. L. (2002). Comparison of avian communities within traditional and wildscaped residential neighborhoods in San Antonio, Texas, thesis, Southwest Texas State University, San Marcos.

International Association of Fish and Wildlife Agencies Fur Resources Committee (1993). Ownership and use of traps by trappers in the United States in 1992. Fur Resources Committee of the International Association of Fish and Wildlife Agencies and The Gallup Organization, Washington, DC.

Jensen, P. G., Curtis, P. D., Lehnert, M. E., and Hamelin, D. L. (2001). Habitat and structural factors influencing beaver interference with highway culverts, *Wildlife Society Bulletin,* 29:654–664.

Jolma, D. J. (1994). *Attitudes toward the Outdoors: an Annotated Bibliography of U.S. Surveys and Poll Research Concerning the Environment, Wildlife, and Recreation,* McFarland, Jefferson, NC.

Kania, G. S. and Conover, M. R. (1991). Another opinion on how governmental agencies should respond to local ordinances that limit the right to hunt, *Wildlife Society Bulletin,* 19:222–223.

Kellert, S. R. (1997). *The Value of Life: Biological Diversity and Human Society,* Island Press, Washington, DC, USA. 263pp.

Kellert, S. R. (1980). Contemporary values of wildlife in American society, in *Wildlife Values,* Shaw, W. W. and Zube, E. H., Eds., U.S. Forest Service, Rocky Mountain Experiment Station, Fort Collins, CO, pp. 31–60.

Kimmel, M. M. (1985). Parks and property values: an empirical study in Dayton and Columbus, Ohio, thesis, Miami University, Oxford, OH.

King, R. T. (1947). The future of wildlife in forest land use, *Transactions of the North American Wildlife and Natural Resources Conference,* 12:454–67.

Krutilla, J. V. (1967). Conservation reconsidered. *American Economics Review,* 57:778–786.

Leedy, D. L. and Adams, L. W. (1984). *A Guide to Urban Wildlife Management,* National Institute for Urban Wildlife, Columbia, MD.

Leopold, A. (1933). *Game Management,* Charles Scribner's Sons, Reprinted in 1986 by University of Wisconsin Press, Madison, WI.

Lowery, M. D. and Siemer, W. F. (1999). Resource agencies as effective sources of information on wildlife damage prevention and control: overcoming the obstacles, *Abstracts of The Wildlife Society Annual Conference,* 6:141–142.

Manfredo, M. J., Vaske, J. J., and Decker, D. J. (1995). Human dimensions of wildlife management: basic concepts, in *Wildlife and Recreationists: Coexistence through Management and Research,* Knight, R. L. and Gutzwiller, K. J., Eds., Island Press, Washington, DC, pp. 17–31.

Mangun, W. R. (1992). *American Fish and Wildlife Policy: The Human Dimension,* Southern Illinois University Press, Carbondale, IL.

Mankin, P. C., Warner, R. E., and Anderson, W. L. (1999). Wildlife and the Illinois public: a benchmark study of attitudes and perceptions, *Wildlife Society Bulletin,* 27:465–472.

Mörner, T., Obendorf, D. L., Artois, M., and Woodford, M. H. (2002). Surveillance and monitoring of wildlife diseases, *O I E Revue Scientifique et Technique,* 21:67–76.

National Wildlife Federation (2004), http://www.nwf.org/conservationDirectory/search.cfm.

Nelson, A. C. (1986). Using land markets to evaluate urban containment programs, *APA Journal,* Spring:156–171.

Payne, N. F. and Peterson, R. P. (1986). Trends in complaints of beaver damage in Wisconsin, *Wildlife Society Bulletin,* 14:303–307.

Penland, S. (1987). Attitudes of urban residents toward avian species and species' attributes, in *Integrating Man and Nature in the Metropolitan Environment,* Adams, L. W. and Leedy, D. L., Eds., National Institute for Urban Wildlife, Columbia, MD, pp. 77–82.

Purdy, K. G. and Decker, D. J. (1989). Applying wildlife values information in management: the wildlife attitudes and values scale, *Wildlife Society Bulletin,* 17:494–500.

Randall, A. (1991). The value of biodiversity. *Ambio,* 20:64–68.

Reiter, D. K., Burnson, M. W., and Schmidt, R. H. (1999). Public attitudes toward wildlife damage management and policy, *Wildlife Society Bulletin,* 27:746–758.

Richardson, W. J. and West, T. (2000). Serious bird strike accidents to military aircraft: updated list and summary, *Proceedings of International Bird Strike Committee,* 25:67–98.

Romin, L. (1994). Factors associated with mule deer highway mortality at Jordanell Reservoir, Utah, thesis, Utah State University, Logan.

Rossi, A. N. and Armstrong, J. B. (1999). *Alabamian's Wildlife-Related Activities, Wildlife Management Perceptions, and Hunting Behavior,* Alabama Game and Fish Division.

Rue, L. L., III. (1987). *The Deer of North America,* Crown Publishers, New York.

Salant, P. and Dillman, D. A. (1994). *How to Conduct Your Own Survey,* John Wiley and Sons, New York.

San Julian, G. J. (1987). The future of wildlife damage control in an urban environment, *Proceedings of the Eastern Wildlife Damage Control Conference,* 1:229–233.

Schmidt, R. H. (1998). Required knowledge, *Wildlife Control Technology,* 5:6–7.

Siemer, W. F., Knuth, B. A., Decker, D. J., and Alden, V. A. L. (1992). *Human Perceptions and Behaviors Associated with Lyme Disease: Implications for Land and Wildlife Management,* HDRU Series 92–8. Department of Natural Resources, Cornell University, Ithaca, NY.

Stout, R. J. (1998). Human dimensions of interactions between white-tailed deer and urban dwellers in Missouri, Ph.D. dissertation, Texas A&M University, College Station.

Thomas, J. K. and Adams, C. E. (1993). The social organization of rattlesnake roundups in rural communities, *Sociological Spectrum,* 13:433–449.

Thorpe, J. (1996). Fatalities and destroyed civil aircraft due to bird strikes 1912–1995, *Proceedings of International Bird Strike Committee,* 23:17–31.

U.S. Centers for Disease Control and Prevention. (2001). Morbidity and mortality weekly report, summary of notifiable diseases, United States, 2001. United States Department of Health and Human Services, Centers for Disease Control, Atlanta, GA.

U.S. Department of the Interior, Fish and Wildlife Services, and United States Department of Commerce, Bureau of the Census. (2001). National survey of fishing, hunting and wildlife-associated recreation, United States Government Printing Office, Washington, DC.

Van Druff, L. W., Bolen, E. G., and San Julian, G. S. (1994). Management of urban wildlife, in *Research and Management Techniques for Wildlife and Habitats,* Bookhout, T. A. Ed., The Wildlife Society, Bethesda, MD, pp. 507–530.

Wehtje, M. E. (1998). Defensible space: a behavioral approach for managing predators at the urban-wildlife interface, *Vertebrate Pest Conference Proceedings,* 18:290–292.

Wildlife Services. (2001). Wildlife Services assistance at airports, United States Department of Agriculture, Animal and Plant Health Inspection Service, Washington, DC.

Wilson, E. O. (1984). *Biophilia: The Human Bond with Other Species,* Harvard University Press, Cambridge, MA.

Witter, D. J. (1990). Wildlife management and public sentiment, in *Management of Dynamic Ecosystems,* Sweeney, J. M., Ed., North Central Section, The Wildlife Society, West Lafayette, IN, pp. 162–172.

Witter, D. J., Tylka, D. L., Werner, J. E. (1981). Values of urban wildlife in Missouri, *Transactions of the North American Wildlife and Natural Resources Conference,* 46:424–431.

Witter, D. J., Wilson, J. D., and Maupin, G. T. (1979). "Eagle Days" in Missouri: characteristics and enjoyment ratings of participants, *Wildlife Society Bulletin,* 8:64–65.

Young, C. (1991). Fostering residential participation in urban wildlife management: communication strategies and research needs, in *Wildlife Conservation in Metropolitan Environments,* Adams, L. W. and Leedy, D. L., Eds., National Institute for Urban Wildlife, Columbia, MD, pp. 203–209.

CHAPTER 9

The Stakeholder Approach and Urban Wildlife Management

The art of communication is the language of leadership.

James Humes

CONTENTS

Because wildlife in North America belong to the public rather than to individuals, they are a shared resource, a commons. The government, through state and federal agencies, manages the resource for "the people" in the public interest. However, management and decision-making regarding wildlife can be challenging, as is the case with any common property. The same wildlife resource can be valued for very different reasons. People have diverse and even conflicting wildlife values (see Chapter 8). While the uses and values of wildlife vary between individuals, cultures, and regions, the underlying human dimensions problem is consistent: how to balance sustainable use and distribution of resources while recognizing the needs and desires of many different interests.

The latter part of the twentieth century saw a change in the focus of wildlife managers. Beginning in the mid 1980s, concepts about who benefits from wildlife management and who should be considered in management decision-making began to broaden. Whereas, traditionally, agencies focused on landowners and consumptive users, thought of as clients, during the 1990s a stakeholder approach gained increasing support.

9.1 THE POLICY LIFE CYCLE AND URBAN WILDLIFE MANAGEMENT

According to Wright (2004), environmental policy is developed in a sociopolitical context, usually in response to a problem. Policy development can occur at every level of government. Local policy may focus on zoning for conservation and agriculture. Federal policy generally addresses broader problems, particularly those that transcend political boundaries, such as air pollution.

The development of public policy tends to take a predictable course, sometimes referred to as the policy life cycle (Figure 9.1). Typically, this life cycle has four stages: *recognition, formulation, implementation,* and *control.* Each stage can be thought to have a certain amount of political importance or "weight."

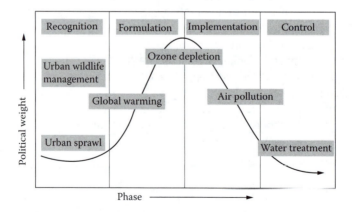

Figure 9.1   Urban wildlife management and other ecological issues and their place in the policy life cycle. (R.T. Wright, Pearson Prentice Hall, Upper Saddle, NJ, 2004)

The *recognition stage* begins with the first perception of a problem, often as a result of scientific research. The political weight of the problem at this point is small. Once the media starts to spread the word the public becomes involved, and often many dissenting viewpoints on what to do, if anything, are expressed. If the problem garners enough publicity, it will eventually receive some attention from at least one level of government.

As the problem gains political weight it enters the *formulation stage*. At this point the public begins to look to legislators and policymakers to do something; debate over exactly what to do occurs in the corridors of power. Political battles ensue over how much regulation is needed, who will be affected, and how much taking action will cost. Media coverage is high. Politicians begin to hear about the issue from their constituents and lobbyists. During this stage, policy makers consider the "Three E's" of environmental policy: *Effectiveness* (will the policy accomplish what it intends to do?), *Efficiency* (will the objectives be accomplished for the lowest possible cost?), and *Equity* (will the financial burden of implementing a solution be fairly distributed among all the stakeholders?).

Once decisions have been made about how to solve the problem, the policy has reached the *implementation stage*. By now, public concern and political weight are starting to decline. Emphasis shifts to development of specific regulations and their enforcement. Industry adjusts to the new regulations. Eventually, greater attention may be given to efficiency and equity as various stakeholders gain experience with the policy.

Finally, we reach the *control stage*. Years may have passed since the recognition stage began. Regulations may become more streamlined and are considered to be a fact of everyday life, although many policies are vulnerable to political shifts. Policymakers focus on keeping the problem under control, and over time the public may forget the problem ever existed. Or the policy may create new problems, in which case the whole process begins again!

We have placed urban wildlife management policy formulation in the recognition stage because, while it still carries little political weight, it does garner some media attention (particularly human–wildlife conflicts) and has been documented as a public service need through scientific research.

Some type of dramatic event will have to take place to move urban wildlife management into the formulation stage, for example, heightened evidence of human losses (health, life, property) connected to urban wildlife. We discussed in Chapter 8 how many animals in urban areas are the vectors of a host of communicable diseases that can be transmitted to humans. There is probably nothing that would capture the attention of the public, politicians, and the media more than a serious outbreak of plague, rabies, or Lyme disease in a densely populated urban community.

9.2 WHAT IS A STAKEHOLDER?

A *stakeholder* is any individual who has an interest in wildlife management; that is, citizens who will be affected by, or will affect, a management decision or action (Susskind and Cruikshank 1989; Crowfoot and Wondolleck 1991; Decker et al. 1996). A person's stake may be economic, recreational, cultural, social, or related to human health and safety. Typically, stakeholders include individuals and groups that have legal standing, political influence, sufficient moral claims connected to the situation, or who have the power to block implementation of a decision (Susskind and Cruikshank 1989). A partial list of potential stakeholders includes birders and other wildlife watchers, hunters and trappers, hikers and campers, farmers and ranchers, homeowners and landlords, recreation-related businesses (e.g., equipment retailers, outfitters, tour guides, restaurants, and motels), local governments (city, county, tribal, state), nongovernmental organizations, and state and federal management agencies.

9.3 THE CHANGING FACE OF WILDLIFE STAKEHOLDERS

Based on a model developed by Aldo Leopold (1933), wildlife agencies have long used a quasi-agricultural approach to management, in which wildlife are managed to allow an annual sustained yield or harvest of specific species of interest — primarily game animals (Decker et al. 2001). The focus of these management efforts is primarily hunters, trappers, and landowners. As discussed in Chapter 1 consumptive users pay for agency services through license fees (Figure 9.2) and special taxes imposed on equipment such as firearms and ammunition. While millions of acres of public lands are managed by various governmental entities, much of the land base that provided habitat for wildlife is controlled by private land owners. Management activities have been designed in large part to serve and build a constituency of consumptive users and to return revenues to the agencies (Dunlop 1991).

These groups fit the traditional definition of a constituency: a group of people (constituents) who authorize or support the efforts of others (such as wildlife professionals) to act on their behalf (Decker et al. 2001). The traditional wildlife constituency was limited to a few specific, relatively homogenous groups, and the system functioned well for those interest groups. Managers understood the values

Figure 9.2   **(See color insert following page 276)** Since 1934, sales of Federal Duck stamps have raised more than $700 million used to acquire more than 5.2 million acres of habitat for the National Wildlife Refuge System. (USFWS)

of their clientele and, in many cases, shared those values. Often wildlife managers were members of one or more of the constituent groups they served.

But during the late 1960s and early 1970s, the concept of wildlife management clients was changing. Not all who paid for management received services, while not all with an interest in management paid for those services. The concept of "user" wasn't completely accurate either, as nonconsumptive recreational activities were becoming more popular. As the public became more interested in the environment, and more distrustful of government, individuals who had never been considered part of the traditional support base for wildlife management began to demand a seat at the table. The public wanted more than to have their opinions heard by agencies; they wanted to play an active role in management decisions.

Historically, American wildlife stakeholders were overwhelmingly Caucasian males. The face of wildlife stakeholders is radically different today, including both genders, diverse races, cultures, and socioeconomic groups, all with wildlife values that are every bit as varied. These individuals have a much wider variety of attitudes and expectations regarding wildlife management than their traditional counterparts. This makes the job of meeting those demands far more difficult than in the past.

9.4 A GUIDE TO MAJOR STAKEHOLDERS

The first step in successfully implementing a stakeholder approach is to evaluate and identify who is substantially affected by the decision or action in question. Stakeholders can be grouped into four major categories: governmental entities (public sector), nongovernmental organizations (NGOs, also known as the private sector), academic institutions, and the public. Keep in mind that an individual stakeholder can belong to more than one of these categories. For example, a U.S. Fish and Wildlife Service employee is also a member of the public, and may have ties to an NGO, particularly some kind of advocacy group. Moreover, the specifics concerning who has a stake in a management decision will vary from case to case.

9.5 GOVERNMENTAL ENTITIES

There are at least five different levels of government to consider whenever a wildlife management decision is being evaluated: federal, state, tribal, county, and municipal. Jurisdictions may overlap — for example, state, tribal, and county governments, or municipal and several different county governments. At each level, one may find several different departments with varying degrees of interest in the issue at hand. To complicate matters further, in some cases a variety of foreign government agencies may be affected as well, particularly when the issue involves migratory species of fish, birds, and/or mammals.

This section will focus on the governmental departments and agencies at each level that are most likely to become involved in wildlife management issues, their jurisdiction, mandate, and/or focus. Discussion of how these levels of government overlap and the resulting challenges for wildlife managers can be found in Chapter 10.

9.5.1 Federal Government

Some federal agencies have direct wildlife management responsibilities, while others are simply affected by wildlife management decisions. Although it's unlikely that all of the departments and bureaus discussed below will need to be involved in any one management decision, we'll take a look at those that should at least be considered in any stakeholder identification effort.

9.5.1.1 U.S. Department of the Interior (DOI)

Four of the Department's eight bureaus have some degree of responsibility for managing the nation's wildlife resources: U.S. Fish and Wildlife Service (USFWS), the Bureau of Land Management (BLM), National Park Service (NPS), and the U.S. Geological Survey (USGS). The DOI manages one out of every five acres of land in the U.S. Over 450 million people visit 388 units of the national park system, 540 wildlife refuges, and vast areas of multiple-use lands (U.S. Department of the Interior 2004).

- *U.S. Fish and Wildlife Service (USFWS):* USFWS is the primary national wildlife resource agency. The agency's mission is "working with others, to conserve, protect and enhance fish, wildlife, and plants and their habitats for the continuing benefit of the American people." The Service traces its origins back to 1871 when Congress created the U.S. Commission on Fish and Fisheries (Department of Commerce) and the Division of Economic Ornithology and Mammology (Depart-ment of Agriculture). In 1939 responsibility for wildlife and fisheries resources was moved to the Department of the Interior and the Fish and Wildlife Service. Among its key functions, the Service enforces federal wildlife laws, protects endangered species, manages migratory birds, conserves and restores wildlife habitat including wetlands, and helps foreign governments with their international conservation efforts. USFWS also oversees the Federal Aid in Wildlife Restoration Act (Pittman-Robertson) and the Federal Aid in Sport Fish Restoration Act (Dingell-Johnson) programs that distribute hundreds of millions of dollars in excise taxes on fishing and hunting equipment to state fish and wildlife agencies. Additionally, since the vast majority of fish and wildlife habitat is on nonfederal lands, the Service is involved in a number of partnership activities that promote voluntary habitat development on private lands, including Partners in Flight and Partners for Fish and Wildlife (U.S. Fish and Wildlife Service 2004).
- *Bureau of Land Management (BLM):* In 1946, the Grazing Service was merged with the General Land Office to form the Bureau of Land Management. When the BLM was initially created, there were over 2,000 unrelated and often conflict-ing laws for managing the public lands. The BLM had no unified legislative mandate until Congress enacted the Federal Land Policy and Management Act of 1976 (FLPMA). It is the mission of the BLM to "sustain the health, diversity and productivity of the public lands for the use and enjoyment of present and future generations." The Bureau is responsible for managing 262 million acres of land — about 1/8 of the land in the U.S. — most of which are located in the western part of the country. BLM manages a wide variety of resources and uses, including wild horse and burro populations, fish and wildlife habitat, and wilderness areas (U.S. Bureau of Land Management 2004).

Table 9.1   National Park Service Biological Resource Management Programs

Ecosystem restoration	Threatened and endangered species
Invasive species	Wildlife health
Integrated pest management	Wildlife management
Migratory birds	

- *National Park Service (NPS):* The National Park Service was created in 1916 as a new federal bureau responsible for protecting the 40 national parks and monuments then in existence and those yet to be established. The mission of the NPS is "…to promote and regulate the use of the…national parks…and to provide for the enjoyment of the same in such manner and by such means as will leave them unimpaired for the enjoyment of future generations." The national park system has grown to comprise 384 areas covering more than 83 million acres in 49 states, the District of Columbia, American Samoa, Guam, Puerto Rico, Saipan, and the Virgin Islands. In many areas, NPS units represent the last vestiges of once vast undisturbed ecosystems. The national parks serve as outdoor laboratories for the study of physical, biological, and cultural systems and their components. Many NPS management programs focus on wildlife issues (National Park Service 2004; Table 9.1).
- *U.S. Geological Survey (USGS):* The USGS was established on March 3, 1879 to study the geological structure and economic resources — including biological resources — of the public domain. The Survey operates sixteen Biological Resource Centers (BRCs), listed in Table 9.2. Although each BRC plays an important role in wildlife management, several are of particular interest.

  The Center for Biological Informatics (CBI) develops, identifies, and provides access to tools that facilitate collection and use of biological information and data. CBI also cooperates with others to improve access to existing information and data. The term "biological informatics" refers to the development and use of computer, statistical, and other tools in the collection, organization, dissemination, and use of information to solve problems in the life sciences (U.S. Center for Biological Informatics 2004).

  - The National Wildlife Health Center, established in 1975, is a biomedical laboratory dedicated to assessing the impact of disease on wildlife and identifying the role of various pathogens in contributing to wildlife losses. Located in Madison, Wisconsin, the Center monitors and assesses the impact of disease on wildlife populations, defines ecological relationships leading to the occurrence of disease, transfers technology for disease prevention and control, and provides guidance, training, and on-site assistance for reducing wildlife losses

Table 9.2   U.S. Geological Survey Biological Resource Centers

Alaska Science Center	Northern Prairie Wildlife Research Center
Center for Biological Informatics	National Wildlife Health Center
Columbia Environmental Research Center	National Wetlands Research Center
Florida and Caribbean Science Center	Pacific Island Environmental Research Center
Fort Collins Science Center	Patuxent Wildlife Research Center
Forest and Rangeland Ecosystem Science Center	Upper Midwest Environmental Sciences Center
Great Lakes Science Center	Western Ecological Research Center
Leetown Science Center	Western Fisheries Research Center

when outbreaks occur. Center staff also provide expertise regarding animal welfare regulations and their application to wildlife. Technical assistance regarding animal welfare matters is often provided to wildlife biologists and others. Preparation of videotapes, publications, consultations, and training are activities commonly carried out by the Center in the animal welfare arena (National Wildlife Health Center 2004).

- The Patuxent Wildlife Research Center was established in 1936 as the nation's first wildlife experiment station. The Center has been a leading international research institute for wildlife and applied environmental research, for transmitting research findings to those responsible for managing our nation's natural resources, and for providing technical assistance in implementing research findings so as to improve natural resource management. Patuxent's scientists have been responsible for many advances in natural resource conservation, especially in such areas as migratory birds, wildlife population analysis, waterfowl harvest, habitat management, wetlands, coastal zone and flood plain management, contaminants, endangered species, urban wildlife, ecosystem management, and management of national parks and national wildlife refuges. The Center develops and manages national inventory and monitoring programs and is responsible for the North American Bird Banding Program and leadership of other national bird monitoring programs. The Center's scientific and technical assistance publications, wildlife databases, and electronic media are used nationally and worldwide in managing biological resources (Patuxent Wildlife Research Center 2004).

9.5.1.2 Department of Agriculture (USDA)

President Abraham Lincoln founded the USDA in 1862. In Lincoln's day, 48% of Americans were farmers, and these individuals needed good seeds and information to grow their crops. Even though 80% of Americans currently live in areas classified as urban, gardening is one of the most popular hobbies in the U.S. Still, it would be stretching the truth to say America is still an agricultural society.

Several agencies within USDA have some level of responsibility for managing wildlife, including the Forest Service (FS), the Natural Resource Conservation Service (NRCS), and Wildlife Services (part of the Animal and Plant Health Inspection Service). In addition, the Cooperative State Research, Education, and Extension Service (CSREES) is responsible for facilitating some wildlife research and disseminating information on wildlife to the public (U.S. Department of Agriculture 2004).

- *Forest Service (FS):* Established in 1905, the Forest Service manages public lands in national forests and grasslands encompassing 191 million acres of land — an area equivalent in size to the state of Texas. Two Service programs address wildlife management issues: the National Wildlife Program and the National Wildlife Ecology Unit. The National Wildlife Program covers terrestrial animal species not considered threatened, endangered, or sensitive (TES) and assists field biologists in meeting the Service's wildlife and habitat goals. Initiatives include "Eyes on Wildlife," "Get Wild!" "Making Tracks," and "Taking Wing." The Wildlife Ecology Unit (WEU) facilitates the transfer of technology and technical information from research to field biologist. WEU staff serve on technical teams addressing terrestrial watershed management issues, including TES species (U.S. Forest Service 2004).

- *Natural Resource Conservation Service (NRCS):* The agency that would eventually become NRCS was started in 1933 as the Soil Erosion Service. Today the NRCS mission is to provide "leadership in a partnership effort to help people conserve, maintain, and improve our natural resources and environment." Of the agency's 20 different conservation programs, at least five specifically target wildlife and wildlife habitat: Conservation Reserve Program, Conservation Technical Assistance Program, Grazing Lands Conservation Initiative, Wetlands Reserve Program, and Wildlife Habitat Incentives Program (National Resource Conservation Service 2004).
- *Wildlife Services (WS):* Wildlife Services, formerly known as Animal Damage Control, is part of the USDA's Animal and Plant Health Inspection Service (APHIS). WS was granted statutory authority for wildlife damage management by the Animal Damage Control Act of 1931. Its mission is "to provide federal leadership in managing conflicts between humans and wildlife." In spite of a traditional focus on rural and agricultural interests, requests for assistance with urban wildlife issues are becoming more common — so much so that WS now employs urban wildlife biologists. Research is conducted to assess the need for wildlife damage control and to develop ways in which to address those needs, including nonlethal methods (Wildlife Services 2004).

9.5.1.3 National Oceanic and Atmospheric Administration (NOAA)

The ancestor agencies of the National Oceanic and Atmospheric Administration include the U.S. Coast Survey established in 1807, the U.S. Weather Bureau established in 1870, and the U.S. Commission of Fish and Fisheries established in 1871. NOAA protects coasts, bays, estuaries, and the Great Lakes and is involved in helping to mitigate damage from oil and hazardous material spills. It operates 13 marine sanctuaries around the U.S. and oversees other Marine Protected Areas, such as national seashores, national wildlife refuges, national estuarine research reserves, and other critical habitat areas.

The National Marine Fisheries Service (NOAA Fisheries) is dedicated to protecting and preserving living marine resources through scientific research, fisheries management, enforcement, and habitat conservation. In addition to marine fish species, the program protects marine mammals, including whales, dolphins, porpoises, seals, and sea lions, and six endangered species of sea turtles.

Other federal departments and agencies that may need to be considered when attempting to identify stakeholders include those listed in Table 9.3 (National Oceanic and Atmospheric Administration 2004).

Table 9.3 Potential Federal Government Stakeholders

Department of Commerce	Federal Highway Administration
Department of Health and Human Services	Federal Railroad Administration
Department of Housing and Urban Development	National Highway Traffic Safety Administration
Department of Transportation	U.S. Army Corps of Engineers
Environmental Protection Agency	U.S. Customs Service
Federal Aviation Administration	

9.5.2 Tribal Governments

There is a unique relationship between the U.S. and the 562 federally recognized tribal governments (Bureau of Indian Affairs 2004). These tribes are considered sovereign governments, a fact that hundreds of treaties, federal laws, and court cases have repeatedly affirmed. Sovereignty is an internationally recognized concept, with treaties acting to formalize nation-to-nation relationships. However, while the U.S. government recognizes tribal governments as sovereign nations, the U.S. Congress is recognized by the courts as having the right to limit the sovereign powers of tribes. State governance is generally not permitted within reservations; only Congress has plenary (overriding) power over Indian Affairs (American Indian Policy Center 2004).

The legal position of Alaska Natives is unique among Native Americans. While the same legal rules apply to Alaskan Natives as to native peoples elsewhere in the U.S., the historical development of the relationship between Alaskan Natives and nonnatives was considerably different. The primary results were that the majority of Alaskan Natives have never lived within reservations. Until recent decades many Alaska Natives, many in remote villages, have been affected very little by nonnative culture. The application of federal Indian law to Alaska Natives is still in its formative stages, with no clear answers to many basic legal questions (Mertz 1991).

American Indians in the lower 48 states have jurisdiction over 45 million acres of reserved lands and an additional 10 million in individual allotments (Figure 9.2). There are another 40 million acres of traditional native lands in Alaska. Tribal governments generally place a high priority on preserving these lands and their natural resources, including many vulnerable wildlife species, for future generations (U.S. Fish and Wildlife Service 2004). Many tribal communities allocate individual use rights to land and particular resources without allowing users to sell the land or the resource rights. The Iroquois and Hopi, for instance, recognize use rights but prohibit sales by individuals. The hunting territories of the Cree are individually allocated and passed down from father to son, but sales are not permitted (Anderson 1995).

Tribal governments may have their own departments of wildlife or natural resource management (e.g., Nez Perce, Navajo, and Southern Utes). Some federal agencies have created American Indian Liaison Offices specifically to work with tribal governments. The relationship between tribal and state governments can be a bit tricky, and friction can arise between these entities. However, many tribes and states are discovering ways to set aside jurisdictional debate in favor of cooperative government-to-government relationships that respect the autonomy of both sides. Tribal governments, state governments, and local governments are finding innovative ways to work together to carry out their governmental functions. New intergovernmental institutions have been developed in many states, and state–tribal cooperative agreements on a broad range of issues are becoming commonplace (Johnson et al. 2000). Tribes are now being recognized as prominent fisheries and wildlife managers and, as such, expect full participation in national fisheries and wildlife initiatives.

Table 9.4   Potential State and Territorial Government Stakeholders

Agriculture	Housing/community development
Animal health	Parks
Commerce/economic development	Pest control board
Environmental protection	Public safety
Forestry	Public utilities/public service commissions
Health	Tourism
Historic preservation	Transportation

9.5.3 State and Territorial Governments

Every U.S. state and territorial government has some kind of department to oversee wildlife management. These departments go by many names — Game and Fish, Conservation, Environment — but for simplicity's sake they are usually referred to as state departments of natural resources (DNRs). Keep in mind, as when dealing with the federal government, other state government departments may be affected by wildlife management issues and decisions. Table 9.4 provides a list of generic state departments to think about during any attempt to identify stakeholders.

9.5.4 County and Municipal Governments

Sometimes referred to as local governments, this is where the levels and overlap of jurisdiction can become incredibly complex. City and county governments often have departments that correlate and, to some degree, overlap with state and federal government jurisdiction and each other; for example, health, animal control, parks, tourism, and transportation departments are common at most levels of government. Additionally, there may also be separate townships and other types of incorporated areas within municipal boundaries, each with its own government and departments.

Of course, wild animals recognize none of the governmental boundaries we have discussed in this section, which is one reason the process of identifying governmental stakeholders can be a time-consuming and tricky business. Urban wildlife managers should familiarize themselves with the governmental structures within their own area of responsibility. In addition, managers should work proactively to identify the names and phone numbers of relevant department heads *before* a management action or decision needs to be made — and keep the list current.

9.6 NONGOVERNMENTAL ORGANIZATIONS (NGOS)

The public sector works with and depends on the nongovernmental private sector, and together they address most of the society's needs. Nongovernmental organizations (NGOs) include three types of businesses: for-profit, not-for-profit, and nonprofit. A "business" is a commercial or industrial enterprise and the people (paid or volunteer) who constitute it. All three types may become stakeholders in urban wildlife management issues and should be considered during the identification process.

For-profit businesses are the most familiar; these are the manufacturers, distributors, wholesalers, retailers, service providers, and any other organization that engages in an activity primarily to generate income and profit for either individuals or the incorporated entity. A variety of for-profit businesses may be directly or indirectly affected by wildlife management decisions, including: manufacturers that use natural resource products; importers that ship wildlife and/or wildlife parts overseas as food, pelts and skins, pets, and folk medicine; wholesalers and retailers offering products for consumptive and nonconsumptive wildlife-related recreation; hotels, motels, restaurants, travel agents, guides, and others tourism-related businesses; and other service industries such as wildlife control operators, taxidermists, and public education providers.

Many people don't think of a *nonprofit* company as a business, but it does fit the definition provided above. There are two primary differences between for-profit and nonprofit businesses: (1) nonprofits are tax-exempt organizations that serve the public interest, and (2) no individual or group is supposed to benefit financially from the activities of a nonprofit organization. In general, the purpose of a nonprofit business must be charitable, educational, scientific, religious, or literary.

Nonprofits are created for diverse reasons: to provide a service not addressed by some level of government; to advocate for a particular interest group or industry; to support governmental and academic activities; to conserve, preserve, or restore natural and cultural resources; to educate, inform or publicize; to support research; or to promote a hobby or activity.

The National Wildlife Federation Conservation Directory lists 23 different environmental issues that nonprofit organizations may address (Table 9.5). Consideration of these issues and their related organizations is a good place to start during the process of identifying stakeholders.

A *not-for-profit business* is an organization established for charitable, humanitarian, or educational purposes that is exempt from some taxes and in which individuals do not accrue profits or losses. While some would argue there is very little difference between nonprofit and not-for-profit organizations, there are subtle distinctions recognized by nonprofit, legal, academic, and other communities. However, these differences have little effect when it comes to identifying and working with stakeholders.

Table 9.5   Environmental Issues Included in the National Wildlife Federation Conservation Directory

Agriculture and farming	Pollution (general)
Air quality and atmosphere	Population
Climate change	Public health
Development and developing countries	Public lands and greenspace
Ecosystems	Recreation and ecotourism
Energy	Reduce, reuse and recycling of waste
Ethics and environmental justice	Sprawl and urban planning
Executive, legislative, and judicial reform	Transportation
Finance, banking and trade	Water habitats and quality
Forest and forestry	Wildlife and species
Land issues	Other
Oceans, coasts and beaches	

9.7 ACADEMIC INSTITUTIONS

University faculty, researchers, and cooperative extension personnel may play several different roles in wildlife management decision-making. Academic personnel may be able to provide insight into public opinion through the development of surveys and focus groups. They may work in cooperation with both the public and private sectors on a variety of wildlife-related projects. Academics may also become stakeholders in their own right, as when agency actions and decisions have the potential to impact ongoing academic research projects. The best way to identify potential academic consultants and stakeholders is to do an extensive literature search on the topic under consideration and then contact those individuals who have demonstrated their involvement in the form of publications.

9.8 THE PUBLIC

Before we can discuss the public as stakeholders we must first define what is meant by the term. According to the WorldNet Dictionary, *the public* is "a body of people sharing some common interest" or "people in general considered as a whole" (World-Net Dictionary 2004). Yet this definition is likely to give the false impression that the public is some homogeneous group that acts and reacts as a whole. Nothing could be further from the truth.

The public is the least homogenous, least organized, and least well-defined of all the stakeholder groups we've discussed so far. Members of the public may, in addition, be associated with the public sector, the private sector, academic institutions, or some combination of these. For example, a wildlife biologist who works for the U.S. Fish and Wildlife Service (public sector) may be a member of both The Wildlife Society and the Nature Conservancy (private sector) in addition to being a member of the interested public.

The values expressed by the public — both as individuals and as a group — are much more diverse than those of an organization. Both public and private sector organizations develop specific sets of goals and values, often expressed in general terms within the group's mission statement. When someone becomes a member or a volunteer of an NGO it's usually because they are in agreement with the values and goals of that organization. If either the organization or the member/volunteer has a change of heart, they will simply part company. The result is that the membership tends to be largely consistent in its values and goals.

Employees of both public and private sector organizations may, as individuals, have values that diverge from, or are even in opposition to, those of the organization for which they work. While they are at work, however, they usually are expected to represent the values of the organization rather than their personal opinions, thus creating a consistent organizational persona.

Luckily for wildlife managers, not every single member of the public will have a stake in every possible decision that needs to be made. Often it can be helpful to divide stakeholders into three types of publics: a nonpublic, a latent-aware public, and an active public (Grunig 1983). The *nonpublic* is the subset that is least aware

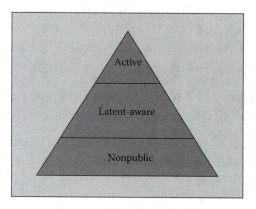

Figure 9.3   Pyramid structure of the number of people in each subset of the public (Decker et al. 2001).

of (and probably least interested in) the issue at hand. The *latent-aware public* is the subset that is aware of the issue and may even be predisposed to take action, but do not for any number of reasons. The *active public* is the crucial subset — the one that has mobilized around a particular issue.

For any issue, the number of people in each subset of the public will form a kind of pyramid (Figure 9.3). It is common for a large portion of the public to not recognize they have a stake in a wildlife issue. Most people will be minimally aware of the issue, if at all, and only a small subset will be motivated enough to actually participate. Keep in mind that a group that is small in proportion to the entire public may still be large in absolute terms.

Identifying active public stakeholders is not an easy task, particularly those individuals who are not also members of the government, NGO, or academic categories. Communication and research can help to reveal stakeholder groups and individuals. Mass media also can be used to solicit participation as well as to educate the public at large about a particular issue.

CASE STUDY 9.1: STAKEHOLDERS DISAGREE ON BEST APPROACH FOR MANAGING FALLOW DEER

They are easily spotted from the road here, lounging in fields and munching grass with little fear of predators. Introduced for hunting six decades ago, fallow and axis deer are popular with tourists eager to see wildlife at Point Reyes National Seashore (Figure 9.4). But park rangers see them as an invasive species that threaten native deer and elk, devour excessive amounts of vegetation, hurt agriculture, and possibly spread disease.

"Invasive species" — organisms introduced into an environment by humans deliberately or by accident — are a problem nationwide. From kudzu in the South to zebra mussels in the Great Lakes, hearty, fast-multiplying invaders can overwhelm an ecosystem, displacing native species and hurting local economies.

Figure 9.4   Fallow deer are an introduced species at Point Reyes National Seashore. (NPS)

Now, Point Reyes officials want to eliminate more than 1,000 nonnative deer — using shotguns and contraception — from the 71,000-acre national park about 40 miles north of San Francisco.

The park's draft plan calls for eradicating the two exotic species by 2020 by sterilizing about one-quarter of them and shooting the rest. After the 60-day public comment period ends on April 8, the park is expected to issue a final plan late this year or early next year, and start eliminating the deer in 2006.

But, as might be expected in liberal-minded Northern California, the idea of killing such attractive creatures has prompted intense debate among wildlife biologists, conservationists, dairy ranchers, and animal activists.

Park scientists and some environmentalists say the invaders must be eliminated to protect the native ecosystem. "Protecting native species and biological diversity should be the park's prime priority," said Gordon Bennett, chairman of the Sierra Club's Marin County group. "These exotic species are basically agents of genocide. They displace and occupy the habitat of native species."

But many nearby residents and animal rights advocates question whether the deer really threaten the environment, and argue that the animals shouldn't be killed just because they aren't natives. "I think it is being both cruel and insensitive to say just because they're not native we should kill them," said Elliot Katz, president of In Defense of Animals, based in Marin County.

Dozens of residents voiced their opposition to shooting the deer at a recent meeting at Point Reyes, where park officials presented their preferred plan, one of five options being considered. "I don't agree with killing the deer at all. I think it should not be an option," said Ilka Hartmann, who lives in the coastal community of Bolinas. "These are beautiful, majestic animals that were brought here against their will to be hunted."

Fallow deer, originally from Asia Minor and the Mediterranean, come in a range of different colors, from white to brown; males grow large antlers. Axis deer, which come from India, are spotted. About three dozen fallow and axis deer, purchased by a landowner from the San Francisco Zoo, were brought to Point Reyes to be hunted in the 1940s, before the region became a national park in 1962.

Park rangers had been allowed to use rifles to cull both herds to keep each population steady at about 350 individuals. The culling stopped about 10 years ago because of public discomfort with shooting the deer, and both populations have grown steadily with no natural predators to keep them in check. The fallow population of 850 is rising 11% per year, while the axis population of 250 is increasing 20% annually. The nonnative deer compete for food and habitat with the park's two native species: black-tailed deer and tule elk. Known for their aggressive behavior, fallow deer have been seen chasing away tule elk, which are three times larger. The nonnatives can also carry Johne's disease, a virus that usually doesn't sicken the deer but can be deadly for tule elk.

Farmers complain that the deer dig up their beets and carrots, and ranchers say they harass cows, eat hay and livestock feed, damage fences, and devour vegetation in their pastures. Under the National Park Service proposal, expected to cost about $4.5 million, meat from the killed animals would be donated to charities, and an experimental form of contraception could be tested.

Eliminating the deer through contraception alone wouldn't work, because it's virtually impossible to sterilize every animal, experts said. It's also expensive — about $3,000 per animal, compared with $300 to shoot them.

"This is a tough decision because it's very emotional," said park Superintendent Don Neubacher. "This is a very complicated issue. In the end I don't think there will be a solution that pleases everybody."

March 29, 2005
Excerpted from an Associated Press article published at Merced Sun-Star.com

by Terence Chea

REFERENCES

American Indian Policy Center. (2004). Website. http://www.airpi.org/pubs/indinsov.html.

Anderson, T. L. (1995). *Sovereign Nations or Reservations? An Economic History of American Indians,* Pacific Research Institute for Public Policy, San Francisco, CA.

Bureau of Indian Affairs. (2004). Website, http://www.doi.gov/bureau-indian-affairs.html.

Crowfoot, J. E., and Wondolleck, J. M., Eds. (1991). *Environmental Disputes: Community Involvement in Conflict Resolution,* Island Press, Washington, DC.

Decker, D. J., Brown, T. L., and Siemer, W. F., Eds. (2001). *Human Dimensions of Wildlife Management in North America.* The Wildlife Society, Bethesda, MD.

Decker, D. J., Krueger, C. C., Baer, R. A., Jr., Knuth, B. A., and Richmond, M. E. (1996). From clients to stakeholders: a philosophical shift for fish and wildlife management, *Human Dimensions of Wildlife,* 1(1):70–82.

Dunlop, T.R. (1991). *Saving America's wildlife: Ecology and the American Mind, 1850–1990.* Princeton University Press, Princeton, NJ.

Grunig, J.E. (1983). *Communication Behaviors and Attitudes of Environmental Publics: Two Studies,* Association for Education in Journalism and Mass Communication, Columbia, SC.

Johnson, S., Kaufmann, J., Dossett, J., and Hicks, S. (2000). *Government to Government: Understanding State and Tribal Governments.* National Conference of State Legislatures, Washington, DC.

Leopold, A. (1933). *Game Management.* University of Wisconsin Press, Madison.

Mertz, D.K. (1991). A primer on Alaska Native sovereignty. Alaska.Net website, http://www.alaska.net/~dkmertz/natlaw.htm.

National Oceanic and Atmospheric Administration. (2004). Website, http://www.noaa.gov.

National Park Service. (2004). Website, http://www.nps.gov.

National Resource Conservation Service. (2004). Website, http://www.nrcs.usda.gov/.

National Wildlife Health Center. (2004). Website, http://www.nwhc.usgs.gov/about_nwhc/index.html.

Patuxent Wildlife Research Center. (2004). Website, http://www.pwrc.usgs.gov/mission3.htm.

Susskind, L., and Cruikshank, J. (1989). *Breaking the Impasse: Consensual Approaches to Resolving Public Disputes.* Basic Books, New York.

U.S. Bureau of Land Management. (2004). Website, http://www.blm.gov/nhp/facts/index.htm.

U.S. Center for Biological Informatics. (2004). Website, http://biology.usgs.gov/cbi/about/.

U.S. Department of Agriculture. (2004). Website, http://www.usda.gov.

U.S. Department of the Interior. (2004). Website, http://www.doi.gov.

U.S. Fish and Wildlife Service. (2004). Website, http://faq.fws.gov/fwsfaq.html.

U.S. Fish and Wildlife Service. (2004). Website, http://endangered.fws.gov/tribal/index.html.

U.S. Forest Service. (2004). Website, http://www.fs.fed.us/biology/wildlife/index.html.

Wildlife Services. (2004). Website, http://www.aphis.usda.gov/ws/.

Wright, R.T. (2004). *Environmental Science: Toward a Sustainable Future,* Pearson Prentice Hall, Upper Saddle, NJ.

WorldNet Dictionary. (2004). http://www.hyperdictionary.com/dictionary/public.

CHAPTER 10

Legal Aspects of Urban Wildlife Management

The problems of pre-empted, overlapping, and concurrent jurisdiction over American wildlife become even more complex when we consider multiple uses and users.

Susan J. Buck, political scientist

CONTENTS

Laws govern just about every aspect of our lives, and urban wildlife is no exception. Many people do not realize that there are hundreds, if not thousands, of federal, state, county, and municipal laws, as well as subdivision covenants and deed restrictions related to wildlife and wildlife management. As the human population grows and encroaches on natural wildlife habitat, more wild animals exploit urban landscapes. Human society attempts to address the concerns associated with urban wildlife through legislation and statutes, and the public is expected to abide by these laws and regulations. In order to comply with the legal constraints on wildlife management, it helps to understand which wildlife laws and regulations are in place, when and why they were created, and how they impact the lives of both humans and wild animals.

As was discussed in Chapter 9, there are at least five different levels of government to consider whenever a wildlife management decision is being evaluated: federal, state, tribal, county, and municipal. In each management case, various jurisdictions may overlap (Figure 10.1). It's the layered characteristic of wildlife law that often creates challenges for wildlife managers. Simply establishing which entity at which governmental level has jurisdiction can be a major undertaking. An important first step in this process is an understanding of the major laws likely to come into play in most urban wildlife management efforts.

10.1 FEDERAL LAWS

There are numerous federal laws that govern wildlife, but there are four of particular importance in the management of urban wildlife. They are, in order of passage: the Lacey Act, the Migratory Bird Treaty Act, the Animal Damage Control Act, and the Endangered Species Act.

10.1.1 Lacey Act of 1900

The Lacey Act of 1900 was the first federal law to address wildlife protection nationwide. This act was founded to help curtail the *interstate* (between states)

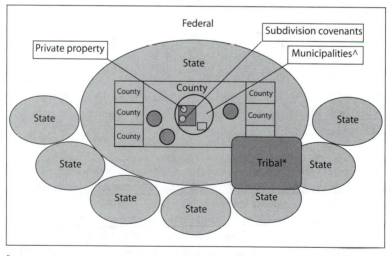

* Sovereign tribal lands may overlap state, county, and municipal boundaries.

∧ Municipalities may contain townships or other incorporated areas within their boundaries.

Figure 10.1 Government jurisdictional overlap.

trafficking of illegally taken wildlife. With the passenger pigeon hunted to extinction and other bird and mammal species in declining numbers, legislation was needed to protect native wildlife. In its original form, the Act focused on problems such as the introduction of harmful exotic avian species, the growing scarcity of many native birds, international trade in bird feathers, and interstate commerce in illegally killed and transported wildlife. Its original language authorized limits on avian importation, initiated programs aimed at protecting native bird species, and targeted illegal wildlife trade by strengthening state laws and requiring accurate labeling of wildlife shipments (Anderson 1995).

As a result of revisions over the past 100+ years, the Lacey Act has been expanded to include rare plant species and fish. As it reads today, the Act prohibits interstate and foreign commerce — import, export, transport, sale, receipt, acquisition, or purchase — of fish, wildlife, or plants that were taken, possessed, transported, or sold in violation of federal, tribal, state, or foreign laws. The Act is enforced by requiring accurate labeling of wildlife shipments and criminalizing most types of trafficking in fish, wildlife, and plants (Figure 10.2). Violators face fines (up to $250,000 for individuals and $500,000 for organizations), forfeiture of wildlife and equipment, and criminal penalties (up to five years in prison).

U.S. agencies that investigate violations of federal wildlife laws include the Fish and Wildlife Service, the Park Service, the National Marine Fisheries Service, the Bureau of Land Management, the Forest Service, and the Animal and Plant Health Inspection Service. State wildlife officers also investigate violations of state wildlife laws, which sometimes develop into federal cases. Partly because of the increasing number and complexity of such cases, the Wildlife and Marine Resources Section (WMRS) was created within the Department of Justice's Environment and Natural Resources Division in 1979. Prosecutors from the WMRS consult with federal investigators and assistant U.S. attorneys who are involved in criminal wildlife investigations or litigation, and sometimes take the lead prosecutorial role in complex or unusual cases (Anderson 1995).

Figure 10.2 **(See color insert following page 276)** U.S. Customs agents shown with some of the items confiscated under the Lacey Act. (Steve Hillebrand/USFWS)

The Lacey Act occupies a central position within the legal framework of the U.S. for three reasons: (1) it applies to a wider array of wildlife, fish, and plants than does any other single wildlife protection law; (2) it provides for a longer potential term of incarceration than do most other wildlife laws containing felony provisions; and (3) the scope of the acts it prohibits is broader than most other wildlife laws (Anderson 1995).

In 1988 Congress made several changes to the Act. It added a subsection explicitly defining a sale of wildlife to include the provision or purchase of guiding or outfitting services for the illegal acquisition of wildlife. Another change involved falsification of documents related to shipments of fish, wildlife, or plants. Prior to 1988, the section outlining this offense had applied only to shipments that were actually imported, exported, or transported; in contrast, the 1988 version applied to documents related to wildlife, fish, or plants intended for import, export, or transport. Other amendments authorized federal wildlife enforcement officers to make warrantless arrests for any federal offense committed in their presence and for any felony based upon reasonable belief that a felony has been or is being committed. The 1988 language also authorized wildlife officers to make warrantless searches and seizures (Anderson 1995).

The Lacey Act is administered by the Departments of the Interior, Commerce, and Agriculture through their respective agencies. These include the U.S. Fish and Wildlife Service, National Marine Fisheries Service, and Animal and Plant Health Inspection Service.

10.1.2 Migratory Bird Treaty Act of 1918

Following on the heels of the Lacey Act, the framers of the Migratory Bird Treaty Act (MBTA) were determined to put an end to the commercial trade in birds and their feathers that had wreaked havoc on the populations of many native bird species by the early twentieth century. The MBTA represented one of the first major federal legislative attempts to protect a particular type of wildlife. This Act was originally created as a treaty between the U.S. and Great Britain (for Canada). Since the first treaty, the U.S. has entered into treaties with Japan, Mexico, and the former Soviet Union (Musgrave et al. 1998).

The original treaty between the U.S. and Great Britain was created because of a concern over the decreasing population of migratory birds in general and waterfowl (geese, ducks, coots, etc.) in particular. At the time, these birds were being hunted to the point of extinction, as was the case of the passenger pigeon (*Ectopistes migratorius*). Also, during this same time period, several exotic species of birds were nearly hunted to extinction as a result of the fashion craze of using exotic bird feathers as adornments for women's hats and clothing (Figure 10.3). However, the MBTA created tension between federal and state management authority that continues to shape federal involvement in the regulation of wildlife (Musgrave et al. 1998).

As the law reads today, it is illegal, except if permitted by regulation (e.g., legal hunting), to take, capture, hunt or kill; attempt to take, capture or kill; possess, offer to sell, purchase, barter, deliver, or ship, export, import, or transport any migratory bird, part, nest, egg, or product. Essentially, what this means is that almost all birds, eggs, and nests are protected by law. The MBTA and a complete listing of birds

Figure 10.3 The popularity of women's hats decorated with wild bird feathers and body parts threatened to push many species to extinction. (National Archives and Record Administration/USFWS)

protected under this Act can be found at the U.S. Fish and Wildlife Service website (http://www.fws.gov).

The MBTA authorizes certain Department of Interior employees to: arrest a person violating the Act in the employee's presence or view without a warrant; execute a warrant or other process issued by an officer or court; and seize and hold all guns, traps, nets, vessels, vehicles, and other equipment used in pursuing, hunting, taking, trapping, ensnaring, capturing, or killing migratory birds (see Case Study 10.1).

10.1.3 Animal Damage Control Act of 1931

In contrast to statutes protecting wildlife, the Animal Damage Control Act was passed in 1931 in part to assist with eradication of wildlife that threatened livestock grazing and agriculture on Western federal and private lands. Federal involvement in predator control actually dates back to the late 1800s and was originally authorized by Congress in 1915. Originally, management of animal damage control was vested in the Department of the Interior, but was transferred to the Department of Agriculture (USDA) in 1985 (Musgrave et al. 1998).

The Act gives the Secretary of Agriculture broad authority to conduct investigations, experiments, and tests to determine the best methods of eradication, suppression, and control of a variety of wildlife species. One purpose of these activities is to protect livestock and other domestic animals through the suppression of rabies and tularemia in wild populations. The Secretary may cooperate with states, individuals, agencies, and organizations to carry out the Act (Musgrave et al. 1998).

Wildlife Services, an agency within USDA, has been given the mission of providing "federal leadership in managing conflicts between humans and wildlife" (Figure 10.4).

Figure 10.4   One activity of Wildlife Services is to conduct experiments to determine the best methods of wildlife control. (APHIS)

The agency's traditional focus was on rural and agricultural interests, but it has become increasingly involved in urban wildlife issues. The primary role of government in overseeing private wildlife control operations has been to regulate activities through a licensing or permitting process — at least in some states — and to provide extension and/or educational services. Legal authority for such regulation is vested in federal and state governments, but often is divided among different agencies, including natural resources, agriculture, and public health (Hadidian et al. 1999).

10.1.4 Endangered Species Act of 1973

The Endangered Species Act has its origins in the 1960s, with the Endangered Species Preservation Act (ESPA, 1966). This Act required a listing of threatened and endangered species and also required that monies from the Land and Conservation Fund be used to purchase suitable habitat for these species. However, the ESPA did not provide for regulations against the killing or trading of threatened and endangered species.

In 1969 The Endangered Species Conservation Act (ESCA) provided additional protection for threatened and endangered species (Figure 10.5). ESCA created two lists of animals, one for foreign species and one for native species, prohibited foreign species from being brought into the U.S., and made the purchase or sale of species on these lists illegal.

These two Acts evolved into the single Endangered Species Act (ESA) of 1973, which replaced the earlier, weaker legislation. The ESA has been considered "the most comprehensive legislation for the preservation of endangered species ever enacted by any nation" (Musgrave et al. 1998). ESA not only requires federal consultation before major federal action impacting threatened or endangered species is undertaken, but it outlaws the taking of such species and provides for acquisition of habitat to protect threatened and endangered (T/E) species. Federal support also

Figure 10.5   (See color insert) Black-footed ferrets are one of the species protected under The Endangered Species Conservation Act. (Tami S. Black/USFWS)

is provided to states that enter into cooperative agreements for conservation of listed species (Musgrave et al. 1998).

The ESA provides broad protection for species of fish, wildlife, and plants listed as threatened or endangered in the U.S. or elsewhere. The ESA requires the Secretary of the Interior to list species as T/E because of a variety of factors, including habitat destruction, overutilization, disease or predation, inadequacy of regulatory mechanisms, or other natural or man-made factors. In the case of marine plants, fish, or wildlife, the Secretary of Commerce determines whether the Secretary of the Interior will list a species or change the status of a species from T/E (see Chapter 9 for additional information on the federal departments involved in management of wildlife). Listings determinations must be made solely on the basis of the best scientific and commercial data available. Any person may file a civil action to compel the Secretary to apply emergency listing procedures or protective measures against the taking of a resident T/E species within a state, or alleging a failure to determine a species as T/E or no longer so, if the determination is not discretionary with the Secretary (Musgrave et al. 1998).

Violations of the ESA may result in criminal penalties of up to $50,000 or imprisonment for one year, or both, and civil penalties of up to $25,000 per violation. District courts have jurisdiction to enforce the Act's provisions and regulations, or to order the Secretary to perform an act or duty.

For more information on the Endangered Species Act and a complete listing of threatened and endangered plants and animals, readers can visit the U.S. Fish and Wildlife Service website (http://www.fws.gov).

10.2 STATE LAWS

Generally speaking, the federal government is responsible for migratory birds and federally threatened and endangered species, and the states are responsible for all other native wildlife. Each state has its own set of laws and regulations that govern wildlife within its borders. Many states have similar laws regarding furbearers, nuisance wildlife, state-listed threatened and endangered species, and game animals; however, variation may exist in how each state chooses to implement these laws, as long as it remains in compliance with federal law.

In some states wildlife management agencies share responsibility for establishing and enforcing regulations pertaining to wildlife with other governmental entities. These include state agricultural, health, and animal health departments or agencies (see Chapter 9).

Texas offers an appropriate example. A law was passed in 1996, as a result of the canine rabies outbreak in the southern part of the state, granting authority to the Texas State Health Department to declare statewide rabies quarantines on raccoons, coyotes, and species of foxes indigenous to North America. In the case of such a quarantine, the Texas State Health Department may forbid individuals from transporting these animals to, from, or within the state.

10.3 COUNTY AND CITY LAWS

Local wildlife laws created by county and municipal government also must be considered when dealing with urban wildlife. Some city and county laws may not directly address wildlife, but the way in which a wildlife problem is dealt with may be affected by the law. For instance, in order to use pyrotechnics to disperse birds from an urban roosting site, one must comply with the laws and ordinances pertaining to the use of fireworks and discharging a firearm within the city limits. Permission from the police department or city leaders may also be required as a result of the pyrotechnics' loud noise violating local noise level laws.

When investigating which laws may be relevant in any particular urban wildlife management situation, don't stop at the state and federal levels. Local governments may not enact laws that negate state and federal laws, but they may enact laws with more rigorous standards. For example, states establish legal hunting seasons for mourning doves (and many other wildlife species), but even during these periods it is illegal to hunt doves within the boundaries of a subdivision or city if ordinances have been passed prohibiting the discharge of a firearm.

10.4 LOCAL ORDINANCES

Some municipalities and subdivisions inadvertently become de facto wildlife sanctuaries as a result of ordinances, public recreation lands, lakes and ponds, and greenways created by developers. Because of the relative abundance of food and protection from natural predators, these areas may be particularly attractive to a number of wildlife species (see Chapter 5). Conversely, many local governments and even neighborhood associations generate regulations which result in prevention of human–wildlife conflicts. For example, some neighborhood covenants may have regulations against feeding deer and other wildlife. This not only discourages deer from eating landscaping plants, but it may also decrease the possibility of a deer–vehicle collision (Davis 2003).

Tree preservation ordinances provide an example of regulations that can benefit wildlife. The forest understory is an important layer of the forest ecosystem for many species. Unfortunately, in many urban and suburban areas, the rich diversity of native tree species is replaced by relatively few species of exotic trees. In order for this trend to change significantly, city planners, developers, homeowner's associations, and private landowners need to understand the benefits of landscaping with native species, which includes reduced maintenance costs (e.g., fertilizers, pesticides). Well-written tree ordinances address the protection of native trees, their root zones, and understory and ensure that replacement procedures increase the diversity of native species.

Weed ordinances often have the opposite effect of tree ordinances. The concept of "weeds" is the product of an agrarian society. Weeds — any unwanted plants — became the bane of farmers because they compete with crops. Early urban weed laws were enacted by communities in the 1940s to protect the public from neglectful landowners (Figure 10.6). One downside of weed laws is that they protect and encourage proliferation of exotic mono-turf (e.g., zoysia grass). A Chicago weed law is a typical example; this regulation flatly outlaws "any weeds in excess of an average height of 10 inches" (USEPA 2004).

As a result of weed ordinances, those charged with enforcement often penalize homeowners who have natural landscapes. "Overgrown" yards may still be commonly considered to attract rats and mosquitoes or present a fire hazard. This idea was soundly refuted in the case of "The City of New Berlin vs. Hagar." New Berlin, Wisconsin, sued Donald Hagar for violating its weed law by practicing natural landscaping on a meadow consisting of several acres. After hearing expert witnesses refute the city's claims that Hagar's property was a health hazard, the court struck down New Berlin's weed ordinance (see Sidebar 10.1).

Ecologically sensitive weed ordinances do exist. Madison, Wisconsin was the first major city to recognize the legitimacy of natural landscapes by enacting an ordinance validating them (Rappaport 1992). Many cities have since enacted ordinances modeled after the Madison ordinance. This type of ordinance should be written so as to

Figure 10.6 Weed ordinances are enacted to protect public health and property values. (L. Causey)

encourage natural landscaping while protecting society from noxious, invasive exotic plants. Well-written weed ordinances do not contain phrases such as "weeds" and "grass." With the possible exceptions of poison ivy and poison oak, other native plans should not be included on an official "weed" list. The term "grass" should be defined as either "turfgrass" or by listing species names, in order to allow maximum turf height to be regulated while exempting ornamental grasses.

There are two versions of modified weed laws: (1) the setback ordinance; and (2) the natural landscape exception ordinance. Setback ordinances generally require an area measured from either the front or the perimeter of the lot, in which the vegetation may not exceed a certain height (e.g., 10 or 12 inches), exclusive of trees and bushes. The vegetation behind the setback and within the yard is unregulated. Setback distances depend on the type of community and size of the typical lot (Rappaport 1992). Setback laws have several advantages and represent a workable compromise between the sometimes diverse interests of the village, natural landscapers, and neighbors. Setback ordinances are also easy to understand and enforce. A setback solves the practical problems caused by large plants and grasses lopping over into neighbor yards or across sidewalks. Neighbor complaints are generally satisfied by such compromise, and living in a community makes compromise essential (Rappaport and Horn 1998).

The second type of modified weed laws are those that limit the blanket weed ordinances with broadly worded exceptions for environmentally beneficial landscapes (Rappaport and Horn 1998). These exceptions include:

1. Native plantings: the use of native plant species for aesthetic and/or wildlife reasons;
2. Wildlife plantings: the use of native and/or introduced plant species to attract and aid wildlife;
3. Erosion control: to offset and control any soil loss problems occurring and/or predicted;
4. Soil fertility building: the enrichment and eventual stabilization of soil fertility through the use of various plant species;
5. Governmental programs: plantings carried out under federal, state, or local programs that require the unimpaired growth of plants during a majority or all of the growing season;
6. Educational programs: plantings designated for educational purposes;
7. Cultivation: any plant species or group of plant species native or introduced and grown for consumption, pleasure, or business reasons;
8. Biological control: the planting of a particular species or groups of species that control or replace a noxious or troublesome species;
9. Parks and open space: any and all public parks and open space lands maintained by federal, state, or local agencies including private conservation organizations;
10. Wooded Areas: all areas that are predominantly woods.

These types of modified weed laws are easy to understand and adequately balance the interests of natural landscapers and neighbors. Additionally, exceptions can be added or deleted from this list to tailor the weed ordinance to the needs of the village and the bioregion in which the village is located.

A fourth and most recent concept in weed ordinances is simple promotion. These ordinances do not "regulate" vegetation. Rather, they promote (or require) natural

Figure 10.7 Volunteers use native plants to modify the banks of a creek in Texas. (John M. Davis/Texas Parks and Wildlife Department)

landscaping. According to Rappaport and Horn (1998), Long Grove, Illinois, has no law regulating vegetation height. Their ordinance actually requires developers to include 100-foot scenic easements between homes and major streets in their subdivisions. The easements are planted with native plants, wildflowers, and grasses (Figure 10.7).

10.4.1 Who's in Charge Here?

By now, it should be apparent that there are many laws and regulations that must be considered when dealing with urban wildlife issues. A search of city ordinances and deed restrictions can assist on the local level. State wildlife management agencies, such as the state wildlife and fisheries agency, state health agency, and state animal health commission, often can assist with general questions. The U.S. Fish and Wildlife Service is a good contact source for information on federal laws. Problems regarding wildlife causing damage or conflicting with human interest can be addressed by Animal Plant Health Inspection Service-Wildlife Services, as well as by state wildlife damage control departments and even certain nongovernmental organizations.

10.5 PROTECTING THE HEALTH AND SAFETY OF ALL

Wildlife laws are in place not only for the protection of the animals, but also for the protection of human health and safety. In the U.S., wildlife are a commons, owned by all and managed by government agencies. But private individuals own and manage the land and habitat on which wildlife depend (Conover 2002). Any species of wildlife can come into conflict with people. Often the problem is caused, however inadvertently, by homeowners and businesses. Dumpsters, trash cans, and pet food are a virtual "Welcome" mat for wildlife. The North Carolina Division of Wildlife Management has created a guide with wildlife damage control principles that urban residents should know (see Sidebar 10.2). Other agencies have developed similar guidelines.

A popular and seemingly humane solution many people choose for solving conflicts with wildlife is to use a live-trap to capture the offending animal, transport it elsewhere, and then release it. However, there are a number of problems associated with relocation of wildlife. First, in many states an individual cannot legally remove an animal and relocate it without permission from the state wildlife agency and/or permission from the owner of the land where they intend to release the animal, regardless of whether it is private, city, or government-owned property. Accessing someone's property without first gaining permission may be considered trespassing. Second, while many believe that relocating an animal is humane, studies indicate this is probably not true. One study found that approximately 75% of relocated raccoons die within the first year after release (Craven 1998). Numerous factors, such as stress, unsuitable habitat, disorientation which leads to not being able to locate food, water and shelter, and the territorial behavior of animals already living in the area contribute to this mortality rate.

Third, relocating wildlife increases the threat of spreading zoonotic diseases (see Chapter 8) such as rabies, leptospirosis, and plague. When a wild animal is relocated, there is always a chance that it may be carrying a disease or parasite that is then introduced into a previously unaffected population of animals. A fine example of how the relocation of wildlife can spread disease is evident in the raccoon rabies outbreak on the east coast of the U.S. Raccoon rabies was first reported in Florida in the 1950s and spread by raccoon-to-raccoon transmission from Florida into Georgia. Between 1977 and 1981, raccoons from Florida were relocated to Virginia to restock raccoon populations; in these shipments were confirmed cases of rabies. Since 1977, when the first report of raccoon rabies was confirmed, 45,000 raccoons have died of the disease along with about 4,200 cats and 3,000 dogs (Centers for Disease Control 2004).

For the reasons cited above, the best solution to most human–wildlife conflicts is a thorough understanding of the underlying factors contributing to the problem, including both human and wildlife behavior. Often conflicts can be solved or avoided completely by making relatively simple changes. Additionally, this type of approach usually provides a more long-term solution than trapping and relocating, which simply provides an opportunity for a new animal to move in to the now vacant territory that proved so attractive in the first place.

SIDEBAR 10.1: NEW BERLIN V. HAGAR

The Hagar decision marks a significant watershed in the natural landscaping movement. It is the first, and best, judicial recognition of the practice and the irrational assumptions that underlie the use of weed laws to prosecute natural landscapers. The City of New Berlin, Wisconsin, elected not to appeal Judge Gramling's decision, and as a result, the opinion is unpublished.

In April 1976, New Berlin sued Donald Hagar for violating its weed law by practicing natural landscaping and cultivating a several-acre meadow. Hager, a wildlife biologist, fought back. He brought in experts to refute the city's claims that his landscape was a health hazard. The testimony was convincing.

Forest Stearns of the University of Wisconsin demonstrated that the Norway rat does not inhabit or find food in a natural landscape. U.S. Forest Service fire expert, David Seaberg, testified that Dr. Hagar's prairie did not create a fire hazard. Philip Whitford, a botanist from the University of Wisconsin, testified that a prairie fire, unlike a forest fire, does not create large and persistent embers that can be carried by the wind. David Kopitzke, a Milwaukee County Public Museum botanist, established that wildflowers and natural landscapes do not create a pollen problem. Exotic plants, like Kentucky Bluegrass, and trees, like oaks, create more allergenic pollen than a native prairie.

After three months of deliberation, on April 21, 1976, Judge William Gramling issued his decision. The Court found nothing in the testimony to justify the fire and pollen hazard claims that the city cited to support the weed ordinance. He found that natural landscaping did not negatively affect neighbor property values. The court struck down the ordinance as violative of the Equal Protection Clause, because the factual underpinning for the law was too thin to be rational. Following his victory, Mr. Hagar continued his natural landscaping, and New Berlin has not bothered any natural landscapers since.

The John Marshall Law Review
Volume 26, Summer 1993, Number 4

SIDEBAR 10.2: URBAN WILDLIFE DAMAGE PRINCIPLES

1. Identify the wildlife species involved (e.g., deer, skunks, bats, squirrels).
2. Be certain the problem is severe and persistent enough to warrant action.
3. Consider alternatives in the following order:
   - Remove food sources that attract wildlife.
   - Remove overhanging tree limbs or other means of access to buildings.
   - Establish protective structures or barriers to prevent wildlife from entering or damaging property.
   - Humanely remove wildlife from buildings and grounds (check first to see if a wildlife depredation permit is needed to trap, transport, or kill wildlife).
   - Permanently repair buildings to prevent reentry.
4. Monitor buildings and grounds periodically for recurring problems, taking appropriate, immediate action to control and prevent damage.

Reprinted from the North Carolina Wildlife Damage Control Agent Program, 2000
North Carolina Division of Wildlife Management.

CASE STUDY 10.1: OPERATION REMOVE EXCREMENT

Before dawn on July 23, 1998, the city of Carrollton, Texas, bulldozed a rookery for egrets, herons, and other avian species near a municipal park, saying it had received numerous complaints about odor and noise (Figure 10.8). City officials said

Figure 10.8 Adult cattle egret (*Bubulcus ibis*). (Lee Karney/USFWS)

they had monitored the rookery and related problems since fall of 1997, consulted experts, and considered many options before deciding to clear the trees. What the city failed to do was apply for and receive a permit to destroy the breeding grounds from the U.S. Fish and Wildlife Service, as required by the federal Migratory Bird Act of 1918.

Residents at the site said they were outraged by the action, dubbed "Operation Remove Excrement" by city staffers. Jack Laivins said he saw the trees being cleared from about a 1-acre site as he drove to work at 4:30 am. "They had big lights on. The thing that struck me was that there were thousands of birds circling overhead in the light. At first, I thought it was smoke. But it was the birds."

Irene Simpson, a 20-year Carrollton resident, said, "This is the worst thing I've seen. It makes me sick."

Laivins, Simpson, and other passersby helped wildlife rescue volunteers find injured baby birds trapped in large piles of brush. Many birds had neck or spine injuries, and one had its left leg nearly severed.

Carrollton Mayor Milburn Graveley said later that day he was not aware of anyone in Carrollton who opposed the city's action. "This is, purely and simply a health issue," Mr. Graveley said, adding he believed the city did not act too hastily and needed no permission from state or federal authorities.

A day later, facing a federal investigation and growing tension between neighbors, the city officials who ordered the rookery bulldozed said they regretted the decision and acting without consulting the law. "I'm very disappointed to learn at this point that there are federal regulations that we did not consult," City Manager

Gary Jackson said. "My expectation is that we would have known what was permissible. We would have done things differently."

If officials had simply waited a few weeks until the nesting season ended and the egrets and other birds had left for the winter, the city could have acted safely and legally without ruffling the feathers of either the avian or human residents. And, ironically, Dallas County health officials said that razing the site, authorized on the basis of concern for human health and safety, may have temporarily increased residents' risk of contracting an illness. In stirring up accumulated bird droppings at the site, the bulldozers may have spread fungus spores that can cause histoplasmosis, Dr. Karine Lancaster, medical director of the Dallas County Health and Human Services Department, explained.

The decision to bulldoze the rookery cost the community dearly, both financially and in terms of public relations. City Council members agreed to pay $126,800 to the Rogers Wildlife Rehabilitation Center in Mesquite to rehabilitate more than 300 wounded and orphaned birds. Following a criminal investigation, the City Council agreed to pay an additional $70,000 to the U.S. Fish and Wildlife Service for violating one of the nation's oldest conservation statutes. The violation is a misdemeanor and could have led to criminal indictments, but no city staff members were prosecuted.

By the following year the birds had come back home. "I think there are more birds now than last year," said Bob Lanier, chairman of a wildlife and environmental advisory committee that Carrollton created after the rookery was destroyed. The egrets now live in a wooded area just a few dozen yards from the old rookery. Residents who live across the street from the wildlife sanctuary said they no longer mind the egrets. Ana Machado, who lives near the park, said, "I love them. They're my friends."

"The only problem I have with them is that they had to cancel the fireworks," said Belinda Burns. The City Council scrapped Carrollton's Fourth of July fireworks display this year after fireworks testing spooked some egrets.

Mr. Lanier said the city — its residents, as well as its officials — had made progress from an environmental perspective. "I really believe that Carrollton will make no environmental or wildlife move without checking with the right people."

Excerpted from articles published in the Dallas Morning News
July 24, 1998 through July 11, 1999

REFERENCES

Anderson, R. S. (1995). The Lacey Act: America's premier weapon in the fight against unlawful wildlife trafficking, *Public Land Law Review,* 16:27.

Centers for Disease Control, Department of Health and Human Services. (2004). Website http://www.cdc.gov/ncidod/dvrd/rabies. Accessed February 17, 2004.

Conover, M. R. (2002). *Resolving Human-Wildlife Conflicts: The Science of Wildlife Damage Management,* Lewis Publishers, Boca Raton, FL.

Craven, S. R., Barnes, T. G., and Kania, G. (1998). Toward a professional position on the translocation of problem wildlife, *Wildlife Society Bulletin,* 26(1):171–177.

Davis, J. M. (2003). Urban Systems, in *Texas Master Naturalist Statewide Curriculum,* 1st
    ed., Haggerty, M. M., Ed., Texas Parks & Wildlife Department, College Station, TX.
Hadidian, J., Childs, M. R., Schmidt, R. H., Simon, L. J., and Church, A. (1999). Nuisance-
    wildlife control practices, policies, and procedures in the United States, in *Wildlife,
    Land, and People: Priorities for the 21st Century, Proceedings of the 2nd Interna-
    tional Wildlife Management Congress,* Field, R., Warren, R. J., Okarma, H. K., and
    Sieverd, P. R., Eds., The Wildlife Society, Bethesda, MD, pp. 165–168.
Musgrave, R. S., Flynn-O'Brien, J. A., Lambert, P. A., Smith, A. A., and Marinakis, Y. D.
    (1998). *Federal Wildlife Laws Handbook with Related Laws,* Government Institutes,
    Rockville, MD.
Rappaport, B. (1992). Weed laws. A historical review and recommendations, *Natural Areas
    Journal,* 12(4):216–217.
Rappaport, B. and Horn, B. (1998). Weeding out bad vegetation control ordinances, *Resto-
    ration and Management Notes,* 16(1):51–58.
USEPA. (2004). Green Landscaping with Native Plants, United States Environmental Protec-
    tion Agency website, http://www.epa.gov/glnpo/greenacres/weedlaws/.

SECTION V

Special Management Considerations

CHAPTER **11**

The Ecology and Management Considerations of Selected Species

CONTENTS

The previous chapters in this book have demonstrated the complex factors operating in urban areas with respect to wildlife issues. Circumstances that arise as a result of wildlife populations inhabiting urban/suburban areas will likely contain nuances specific to the particular geographic area and wildlife species involved. However, it is helpful to analyze each situation with the following conceptual model: Ecological and sociological factors work in concert to create an urban wildlife management dilemma. The ecology and behavior of the species explain the presence and abundance of the population (why the species is here and in what numbers). Sociological factors such as economics, politics, and culture set the framework for how humans respond to that wildlife presence and abundance.

This chapter is not just about direct *conflict* between humans and wildlife species, although that is the most common theme. We have selected four categories of management scenarios that commonly play out in the urban/suburban ecosystem. These include conflicts between predators and humans, management of endangered species, species that roost in large colonies, and feral species. Examples of each scenario will be presented so that the reader can appreciate the unique challenges inherent in each. We collected information from several sources including scientific journal articles, newspaper and other media reports, interviews, government documents, and personal experience. In addition, we invited colleagues to contribute case studies on Florida key deer (*Odocoileus virginianus clavium*) and feral hogs.

11.1 PREDATORS

Many people may be surprised that predators can become a management concern in urban areas. The urban landscape is home to many species of predators, both small and large. Several midsized carnivore species such as raccoons (*Procyon lotor*), skunks (*Mephitis mephitis*), and foxes (*Vulpes vulpes*) commonly maintain healthy populations within urban areas. In fact, densities of midsized predators in urban areas are often higher than in rural areas (Rosatte et al. 1991). These carnivores can exploit urban areas with such success because they are generalists with respect to diet, require smaller areas for hunting/foraging than larger carnivores, and are less sensitive to habitat changes that result from human disturbance.

Larger, more specialized predators sometimes include urban areas in their home ranges but do not rely on these areas exclusively for their biological needs. They may pass through the urban ecosystem when dispersing to new hunting or breeding grounds. For example, juvenile cougars (*Puma concolor*) were found to use habitat corridors in an urbanizing area of southern California to disperse (Beier 1995). Bobcats (*Lynx rufus*) have been shown to include urban areas such as parks in their nocturnal movements. More than 10% of the home ranges of male bobcats in southern California were in urban areas (Riley et al. 2003).

Humans and predators have long had an adversarial relationship outside of the urban ecosystem. Typically, humans have persecuted predators to prevent attacks on humans, livestock, game species, and/or endangered species. Spread of diseases from carnivores to humans has been a concern and has been used as justification for their lethal control (Sillero-Zubiri and Laurenson 2001).

When large predators move into the suburbs/urban centers, the most common concern expressed by most humans is the risk to human health and safety. Direct contact with predators and disease (e.g., rabies) transmission via predators are both understandable concerns for urban residents. However, attacks on humans are relatively rare. For example, approximately 52 human injuries by large predators, such as bears (*Ursus* sp.), alligators (*Alligator mississippiensis*), coyotes (*Canis latrans*), and cougars, occur each year in the United States (Conover 2002). Few of these injuries occur in urban areas.

When predators inhabit urban areas, residents also express concern for the safety of their pets. Humans are emotionally and financially heavily invested in the welfare of their domestic pets. Consider that over 37 million households in the U.S. own at least one dog. Watching Fido become the first dinner course for an alligator or coyote is more than the typical urban citizen is prepared to experience.

Following is a review of information available for two predator species commonly found in urban and suburban habitats. The coyote and black bear (*Ursus americanus*) were chosen as case studies because they can pose a real risks to humans and their pets and have been a source of conflict in several U.S. and Canadian cities.

11.1.1 Coyotes (*Canis latrans*)

No other wildlife species in North America best represents the adversarial relationship between predators and humans than the coyote. Mark Twain best summarized this struggle:

The cayote is a living, breathing allegory of Want. He is always hungry. He is always poor, out of luck and friendless. The meanest creatures despise him and even the flea would desert him for a velocipede.

Mark Twain, *Roughing It*

The coyote (Figure 11.1) has been the bane of farmers and ranchers since the European settlement of central and western North America. According to Conover (2002), U.S. ranchers lose 71.5 million dollars annually as a result of predation on livestock, the majority of which are taken by coyotes. Consequently, coyotes continue to be hunted by ranchers and the federal government. The U.S. Department of Agriculture Wildlife Services killed approximately 90,000 coyotes in 2001 to prevent predation of livestock. Twenty-four percent of farmers (40% in Great Plains states) experience damage from coyotes (Conover 1998). Historically, the conflict between humans and coyotes has been a rural one. However, western cities such as Los Angeles, CA have been dealing with coyote populations for several years. For

Figure 11.1   (See color insert following page 276) Coyote, *Canis latrans.* (Sara Ash)

example, city officials of Glendale, CA have operated a coyote control program for over 20 years (Baker and Timm 1998).

Coyotes have been expanding their ranges into the eastern half of North America (Gompper 2002) and into its urban centers, as well. For example, in 1999, a coyote was captured in New York City's Central Park (Martin 1999), and that same year, another coyote was captured after being discovered hiding under a taxi on Michigan Avenue in Chicago (Kendall 1999). While eastern cities are not experiencing coyote problems to the extent that western cities are, these isolated captures of coyotes may foreshadow future management challenges in the east. We know that coyote populations in the eastern U.S. are rapidly increasing. The coyote population in Mississippi increased 7.5-fold during a 15-year period (Lovell et al. 1998). It would be naïve to think that coyotes would not move into eastern cities with as much tenacity as they did in the west.

As stated before, coyotes are found throughout the continental U.S. and Canada (Bekoff 1999). They inhabit a diversity of habitats and can change their behavior and ecological requirements depending on habitat. They have expanded their range to include eastern Canada and the U.S. in the past 50 to 75 years. This range expansion is considered to be a result of both natural colonization (Gompper 2002) and intentional and accidental introduction by humans (Hill et al. 1987).

Any species that inhabits such a diversity of habitats from the desert to urban America must be able to exploit a diversity of food items as well. The coyote is no exception. Bekoff (1977) suggested that providing a list of common foods of the coyote is impossible. The coyote is described as an opportunistic predator and scavenger, a characteristic that makes it an ideal candidate for life in the city. Rural diet studies revealed that about 90% of their diet is comprised of mammal flesh (Bekoff 1977). Common prey included deer, elk, sheep, rabbits, and rodents. These results show that coyotes are capable of taking large prey items, an important point to remember when considering the potential risk to urban children.

Specifically, coyotes do well in urban ecosystems, in part, because of their ability to scavenge. Urban coyotes eat some garbage, pet foods, and occasionally prey on domestic pets such as cats and small dog breeds. The importance of garbage and domestic pets in *maintaining* coyote populations in urban areas is unknown. Must coyotes prey on pets in order to survive in the urban landscape? Would better garbage management result in fewer coyotes in urban areas? Definitive answers to these questions are not yet known. However, as demonstrated in earlier chapters, urban habitats hold large numbers of generalist species such as mice and rats, thereby providing a food source to coyotes. So it is reasonable to assume that, even if we were able to remove refuse and pets as a food source, coyotes would still be able to survive in the city.

Another aspect of their success in developed areas is their selection of more "natural" areas for denning and resting sites. Some studies (Riley et al. 2003, Grinder and Krausman 2001) showed that coyotes use forests or habitats with sufficient cover in close proximity to residential or urban parks. Additionally, urban coyotes may alter their activity patterns, possibly to avoid contact with humans (McClennen et al. 2001; Tigas et al. 2002; Riley et al. 2003). However, there have been significant exceptions to this behavior. In some cities, coyotes will hunt during the day in areas of high human activity (Baker and Timm 1998).

Several major urban areas have experienced coyote "problems," namely attacks on pets, children, and adults. Baker and Timm (1998) reported that, between 1988 and 1997, coyotes attacked 53 people and harassed more than 32 people in southern California. The authors summarized the signs of coyote behavior that precede attacks on humans as the following (in order of appearance):

1. Preying on pets at night;
2. More coyote activity in yards and streets at night;
3. More coyote activity in yards and streets during day;
4. Preying on pets during the day;
5. Attacking pets on leashes, chasing joggers and cyclists;
6. Presence of coyotes in children's play areas and parks during the day.

Why do some urban coyote populations avoid direct contact with humans while other populations (i.e., southern California cities) harass and prey on humans? Baker and Timm (1998) suggested that coyote attacks are precipitated not by hunger alone but also by a lack of aggression from humans. Without trapping and shooting control efforts, coyotes lose their fear of humans. Couple this loss of fear with hunger and the need to feed pups in the late summer, and coyote attacks consequently result. Trapping and shooting not only removes offending individuals but also modifies the behavior of remaining individuals (Baker and Timm 1998).

Because of the potential of human injury, the existence of coyote populations in urban areas can become a volatile issue for political leaders. Recently, controversy over coyote control came to a head in Vancouver, British Columbia. Coyotes are believed to have colonized the Vancouver area in the late 1980s (Lampa 2001). Some of the coyotes in a local park were active during the day and commonly approached humans. Park visitors often exacerbated this scenario by intentionally feeding the animals (The Urban Coyote 1999).

Seven documented cases of coyote attacks on humans have occurred in the Vancouver area since the late 1980s (Stanley Park Ecological Society website). Local media reports described coyote attacks on pets. Consequently, the issue of coyote control created ping-pong-like action in the local newspaper editorial section and at city council meetings with heated debate as to the best course of action.

Who appointed us God to decide which animals live and which animals we can just eliminate at our discretion? Humans are expendable — two world wars proved that — but wildlife is not. Name omitted, 2000

I strongly urge for official action to take place on this immediately, as the safety and welfare of our children are at risk, at this moment. Otherwise, do you want to see a group of angry parents and grandparents taking this matter into their own hands? (Name omitted, 2000)

In April 2000, a plan to deal with human–coyote conflicts in Vancouver was developed by two government agencies, the Vancouver Park Board and the Ministry of the Environment, Lands and Parks, and one private environmental organization, the Stanley Park Ecology Society (Administrative Report, General Manager Board of Parks and Recreation, September 13, 2001). According to organizers of the plan, "The strategy recognizes both the short term need to deal with problem wildlife and the longer term program of assisting residents of the City to understand the issues and find the solutions to co-existence with the coyote."

This progressive philosophy resulted, in part, from the research findings of Kristin Lampa (formerly Webber), who conducted a survey of Vancouver residents regarding their knowledge and attitudes about urban coyotes. Lampa (Webber 1997) found that while the majority of survey respondents did not have negative attitudes regarding coyotes, they did have specific and genuine concerns regarding the presence of coyotes in Vancouver. Thirty percent believed that coyotes would attack children, while 82% thought domestic pets were in danger of attack by coyotes. Even with these concerns, however, survey respondents preferred using education as the primary management tool in dealing with conflict between coyotes and residents.

The city organized a three-part plan to manage the conflicts between Vancouver residents and coyotes. First, a 24-hour hotline was available for residents to report encounters with aggressive coyotes. Second, the Ministry of Air, Land and Water Protection staff was responsible for removing aggressive individual coyotes. Lastly, the Stanley Park Ecology Society developed an environmental education program called, "Coexisting with Coyotes." The goal of this education program was to reduce conflicts between humans, their pets, and coyotes by providing information in a range of media outlets including online, telephone hotlines, print, video, and school and community presentations by naturalists.

Organizers of the educational program believed that human behavior precipitated more conflict than did the existence of the coyotes. Consequently, they focused their educational information on how to modify human behaviors to reduce the probability of human–coyote conflicts. For example, they explained why it is inadvisable to intentionally feed coyotes or to leave garbage or pet foods unattended. Other types of information described how to react if a coyote approaches a resident. The common ingredient to all suggestions was to act aggressively (Stanley Park Ecology Society).

The Ministry of Environment, Lands, and Parks, the Vancouver Park Board, Vancouver Foundation, and the Vancouver School Board financially supported the first year of the program. In 2001, a request was made for expansion of the program at a cost of $33,000. The Council and Parks Board approved the request to fund expanded operation of the hotline ($8,000), the salary of a Wildlife Ranger to respond to reports of coyote aggression ($8,000), and more education materials by Stanley Park Ecology Society ($17,000).

Frankly, it is too soon to know if the steps taken by city officials will ameliorate the conflict between coyotes and humans in Vancouver. However, we believe that the key elements needed to reduce conflicts between humans and wildlife have been included in the Vancouver plan. First, there was an attempt to understand the human dimension of the conflict. To quote Kristin Lampa, "Since it is neither desirable or practical to eliminate the wildlife, it is probably more feasible to change human behaviour than the behaviour of wildlife" (Lampa 2001). Second, the city did not dismiss the fears of residents regarding the potential risk of a coyote attack. Money was appropriated to fund an officer who dispatches any aggressive coyotes. While the educational program tried to inform residents of the low probability of attack, it educated the residents about how to react if a coyote approaches them. Knowledge in any situation alleviates fear. Finally, this management plan was culturally sensitive and in line with the philosophy of the majority of residents. We are not suggesting that wildlife populations should be managed based on the court of public opinion, but to disregard cultural and sociological issues is to plan for failure.

11.1.2 Black Bear (*Ursus americanus*)

Similar to coyotes, the black bear (Figure 11.2) has had a tenuous relationship with humans in North American rural landscapes since European settlement. On one hand, bears have been revered as iconic figures in several cultures. Teddy bears, Winnie the Pooh, and Paddington Bear all represent comfort and friendship to millions of children and adults. Native Americans appreciated bears for their abilities to hibernate for months (Wilson and Ruff 1999). However, bears are also perceived as dangerous by many people. Additionally, black bears cause economic damage to

Figure 11.2 Black bear, *Ursus americanus*. (Mike Bender/U.S. Fish and Wildlife Service)

forestry operations, apiaries, and livestock, although estimates of the extent of damage are unknown (Conover 2001).

Black bears are widely distributed throughout Canada, the Pacific Northwest, along the Rocky Mountains south into Mexico, and within forested landscapes in the eastern U.S. (Wilson and Ruff 1999). Much like coyotes, they inhabit a diversity of habitats including urban developments. Densities of black bears vary according to habitat type and food distribution. For example, black bear density at an urban–wildland interface was 120 bears/100 km^2, and in the wildland area, density was 3.2 bears/100 km^2 (Beckmann and Berger 2003b).

Also like the coyote, the black bear is a generalist forager able to consume various types of food. Most studies indicate that black bears eat mostly vegetation, fruit, and nuts (Wilson and Ruff 1999), with animal matter constituting a small percentage of their diets (Utah Division of Wildlife Resources 2000). Because of their diet plasticity, black bears are able to exploit urban food sources. With respect to *diet,* black bears are probably more suited to urban areas than are coyotes, because they rely less heavily on animal matter.

While research on urban black bears is extremely limited, two very valuable studies (Beckman and Berger 2003a, 2003b) have contributed greatly to our understanding of how urban sources of food affect both the ecology and behavior of this species. Beckman and Berger (2003a) found, in comparison to bears living in wild areas, urban bears that ate garbage were active for fewer hours each day, more likely to be active at night, entered hibernation later in the year, and consequently, stayed in dens for a shorter duration. Additionally, the researchers (2003b) found that black bears in urban areas were approximately 30% heavier, had significantly smaller home ranges (90% and 70% reduction for males and females, respectively), and gave birth to triple the number of offspring compared to their wildland counterparts. Interestingly, the mortality rate of dispersing cubs in urban areas was 83%, while wildland cubs exhibited 100% survival during dispersal. All urban cub deaths could be attributed to anthropogenic sources such as vehicles.

In addition to differences between the two groups at the individual level, population characteristics such as density and sex ratios also differed. As stated previously, the urban bear density was 120 bears/100 km^2; wildland bear density was 3.2 bears/100 km^2. The urban bear population had a 6.8:1 male to female ratio, while the wildland population exhibited a 1.6:1 male to female ratio (Beckmann and Berger 2003b). Based on their analysis, the authors made an interesting argument regarding the dispersion of bears across the urban and wildland landscapes. When they compared the current population density and distribution (urban vs. wildland) to data from a previous study that was conducted 10 years earlier from the same region, they found that the population size had not changed appreciably. However, no urban bear population existed 10 years earlier and the wildland population density was less than the historic wildland density (20 to 41 bears/100 km^2). The authors argued that, based on these findings, the establishment of an urban bear population and the increases in nuisance bear complaints were more likely a result of a *redistribution* of the bear population across the landscape instead of an actual population increase.

The authors conjectured that urban sources of anthropogenic food were basically luring bears out of the wildland areas and into urban areas. This conclusion has interesting management implications for urban wildlife. When human–wildlife interactions increase, most lay people and some scientists assume that this occurs because the offending wildlife population is increasing or experiencing a population "explosion." While this assumption may be true in certain circumstances, it is important to understand that interspecific interactions (such as human–bear interactions) are also dependent upon behavior of each species and the dispersion of resources for each species across the habitat.

For example, one interesting human behavior is the deposition of extra resources (garbage) at landfills. This rich resource (food for the bears) is highly clumped in both time and space. Consequently, bears will abandon typical territorial behavior and feed in large groups at these landfills (Rogers 1989). The two species' behaviors in this circumstance can increase the probability of contact between them. Abiotic factors such as climate can also affect the likelihood of human–bear interactions. Zack et al. (2003) studied the effect of the El Niño-Southern Oscillation (ENSO) on the probability of human–bear conflicts in New Mexico between 1982 and 2001. The La Niña phase of ENSO is characterized by dry winters and springs, while the El Niño phase of ENSO is characterized by wet winters and springs. The authors hypothesized that, during dry periods caused by La Niña, bears must travel longer distances in search of food. They predicted that the increased movements would make it more likely that bears would come into contact with humans. The results of their analysis showed that high rates of human–bear encounters were 4.7 times more likely to occur in La Niña years (dry phase) than in El Niño years (wet phase), thereby supporting their hypothesis.

Whatever the driving factor, several communities across North America are experiencing increasing numbers of human–bear conflicts. Often solutions to these conflicts are slow-coming and fraught with controversy. As with the coyote example, changes in human behavior, not bear behavior, have the most valuable impact but are typically very difficult to produce. Peine (2001) compared bear management strategies used by four different urbanizing communities including the following: Juneau, Alaska; Mammoth Lakes, California; West Yellowstone, Montana; and Gatlinburg, Tennessee.

In his analysis, Peine (2001) showed that nuisance bears from each community were typically "unnatural-food-conditioned" bears. Sometimes this conditioning of bears to unnatural foods was exploited for monetary gain. For example, "rumours" have circulated within Gatlinburg for many years that some business owners encourage or participate in feeding bears to attract tourists. Hunters found the best opportunity to kill a bear in the county was in and around Gatlinburg, partially because of the bears' dependence on garbage. Consequently, there was some general resistance to implementing measures to reduce bear activity in the city. However, other factors in addition to the unnatural food sources contributed to recent nuisance bear activity in Gatlinburg. For example, the bear population in Great Smoky Mountains National Park, adjacent to Gatlinburg, grew rapidly in the 1990s because of a combination of abundant natural foods and a reduction in poaching practices. However, in 1997, a late frost and a summer drought precipitated the movement of park bears into Gatlinburg in search of food.

All four communities described in the article developed somewhat similar solutions for their bear problems, although Peine (2001) described in great detail the complexity of the decision-making process and how the factors driving the process for each community were very different. The common theme among the solutions was the creation of ordinances about garbage storage and collection, although the level of enforcement of these ordinances varied across communities.

We should expect more human–bear conflicts in the future for various reasons, including the ones discussed above. First, some populations of bears may be increasing in density. Second, some bear populations may be redistributing themselves according to available resources such as garbage in suburban and urban areas. Third, as we have stated previously, the expansion of human habitation into wild or rural areas is showing no signs of deceleration. So in simple terms, bears and humans are moving toward the urban fringe (albeit in opposite directions) and it is inevitable that their paths will cross.

11.2 ENDANGERED SPECIES

As described in previous chapters, the process of urbanization often results in the simplification of habitats. Consequently, the numbers of wildlife species decreases (lower biodiversity). Species that remain or are able to colonize urban habitats are usually generalists capable of exploiting the urban environment and adapting to human presence. These species are typically very widely distributed, and their population numbers are very large. However, it should be noted that sometimes urban and suburban environments provide habitats for threatened or endangered populations of wildlife species.

Endangered species are found in urban or suburban habitats as a result of two mechanisms. First, endangered species are sometimes introduced into urban areas. The peregrine falcon (*Falco peregrinus anatum*) was mentioned in Chapter 6. Second and more importantly, endangered species are found in urban/suburban areas as a result of rapid human development. As development encroaches into the species range, the populations become concentrated in smaller, more fragmented areas. These fragments of remaining habitat become surrounded by development. Essentially, the original habitat becomes islands within the developed matrix, and the species become isolated inhabitants of those islands. The populations must survive on small parcels of remaining habitat and, therefore, become more vulnerable to extinction. Sometimes the species can partially adapt to the urban areas and use them for part of their biological needs.

For example, one subpopulation of endangered bighorn sheep (*Ovis canadensis*) commonly utilizes urban sources of food and water (USFWS 2000; Rubin et al. 2002). Bighorn sheep in southern California were officially listed as endangered in 1998 because of habitat loss and fragmentation, resulting in part from urbanization. Urbanization not only has reduced and fragmented the sheep's habitat, but also has eliminated access routes between the small subpopulations of sheep (USFWS 2000). Other examples of endangered species that live in urban areas include the Pacific pocket mouse (*Perognathus longimemloris pacificus*) and the Perdido Key beach mouse (*Peromyscus polionotus trissyllepsis*).

Figure 11.3   (See color insert) San Joaquin kit fox, *Vulpes macrotis mutica*. (B. Peterson/U.S. Fish and Wildlife Service)

The goal of the Endangered Species Act of 1973, as mentioned in Chapter 10, is to provide protection for ecosystems upon which endangered species depend for survival and reproduction (Meffe and Carroll 1998). Reaching this goal in an urban/suburban area can be problematic for a number of reasons. First, ownership of important habitat for the species is often split into multiple parties. Managing these fragmented parcels of land for a common goal is difficult at best. Second, natural areas that border existing urban locations have high commercial value and are usually sought after by various interests for development. If endangered species inhabit these areas or need them for survival, economic gains by private developers and cities are thwarted.

Following is a review of information available for two endangered species that can be found in urban and suburban habitats. The San Joaquin kit fox (*Vulpes macrotis mutica*) was chosen as a case study because they are just so darned cute (Figure 11.3), which seems to be a contrite reason for including them, but their physical characteristics make them very appealing to the urban public. According to San Joaquin kit fox experts, the management of the urban fox population has caused very little conflict with humans. We selected key deer because one of our colleagues has professional experience in the management of this species and accepted an invitation to write about his experiences. Second, this species is unique in that, while it is highly endangered, it maintains a locally abundant population in a developed area in south Florida.

11.2.1 San Joaquin Kit Fox (*Vulpes macrotis mutica*)

The San Joaquin kit fox is a highly endangered subspecies of kit fox that is currently distributed in the southern San Joaquin Valley and Salinas Valley in southern California (Cypher and Spencer 1998). Kit foxes weigh between 1.5 and 3.0 kg (3.3

to 6.6 lbs) (Cypher personal communication), less than the average weight of a healthy pet cat. Kit foxes prey primarily on small mammals such as mice and ground squirrels. Other prey items include rabbits, hares, birds, eggs, and insects (USFWS 1998).

Like other canid species, the kit fox has been able to adapt to urban areas. Urban kit foxes use habitats such as golf courses, city parks, undeveloped lands, open spaces, school grounds, canal, railroad and powerline right-of-ways, and drainage basins (Cypher personal communication). Studies of urban kit foxes show that they eat some sources of anthropogenic food (Cypher and Warrick 1993), although the importance of these foods in maintaining the population is not understood. As with the coyote, we do not know if kit foxes *must* have these food sources to survive the urban landscape. Cypher and Frost (1999) found that juvenile urban kit foxes weighed more than their rural counterparts. The researchers hypothesized that the consistency of their urban food sources likely contributed to this difference.

Kit foxes use underground den sites as shelter from environmental conditions, to escape from predators, and for reproduction (USFWS 1998). Urban dens sites include culverts, abandoned pipelines, and banks in fenced drainage basins (Speigel et al. in USFWS 1998; Cypher personal communication). According to the USFWS Recovery Plan (1998), urban kit foxes exhibit different behavioral characteristics from their rural counterparts. Many of the urban foxes are reported to be relatively tame. Much like urban coyotes, they are active during the day and scavenge from human food sources such as garbage.

As a result of encroaching development, the San Joaquin kit fox population is highly fragmented, with subpopulations scattered throughout the San Joaquin Valley. In addition to subpopulations in natural lands, kit foxes have been detected in and close to towns such as Tulare, Visalia, Porterville, Maricopa, Taft, and McKittrick. Additionally, a relatively large population (approximately 200) occurs in the metropolitan Bakersfield area (USFWS 1998).

Kit foxes are not generally recognized as a threat to human and/or pet safety, probably because of their petite size (Cypher personal communication). A 5-lb kit fox would have to be psychotic, self-delusional, or rabid to attempt an attack on most pet dogs. According to Cypher, research ecologist for the Endangered Species Recovery Program in Bakersfield, most residents probably are not aware that kit foxes live in Bakersfield. Those residents aware of the foxes are sometimes protective with regards to the ecologist's activities (e.g., setting traps for foxes). So the typical nuisance issues related to other wildlife species have not been relevant to the urban kit fox. However, the Bakersfield area is developing rapidly and consequently, potential exists for conflict between the endangered kit fox and economic interests. Yet, according to Cypher, there has been relatively little conflict because of the Habitat Conservation Plan written by the city of Bakersfield.

The Endangered Species Act (ESA) prohibits the "take" of any federally endangered plant or animal species. This law applies to private landowners and has been a point of contention since its enactment. In an effort to resolve this issue, the ESA was amended in 1982 to authorize "incidental take" by private landowners if they developed and implemented a Habitat Conservation Plan. The objectives of these plans are to address the potential impacts of a given activity (development) and to outline mitigation procedures to offset those impacts on the endangered species (Nelson 1999).

The Metropolitan Bakersfield Habitat Conservation Plan (MBHCP) was written in 1994 by the city of Bakersfield and Kern County in an attempt to help developers and landowners reach compliance with the laws of the ESA. By writing the MBHCP to include the entire metropolitan area, a separate permit application is not required for each proposed activity by each developer or landowner. When a landowner or developer applies for a building permit with the city, they pay a mitigation fee to offset the potential negative impacts, as outline by the MBHCP. Money collected by the city is used to purchase habitat for the endangered species listed within the plan, including the kit fox, Tipton kangaroo rat (*Dipodomys nitratoides nitratoides*), blunt-nosed leopard lizard (*Gambelia silus*), and Bakersfield cactus (*Opuntia treleasei*). The plan allows developers to be compliant with state and federal laws without paying for biological assessments and mitigation for each project.

11.2.2 Florida Key Deer (*Odocoileus virginianus clavium*)

The authors thank Dr. Roel Lopez of the Wildlife and Fisheries Sciences Department at Texas A&M University for this section.

In this section, we will discuss the uniqueness of the endangered Florida Key deer (Figure 11.4), the impact of urban development on the species in the last 50 years, and some of the current management challenges for the population. First, we will provide some background and a historical perspective to the Key deer. Next, we will review some of the changes in habitat due to urban development in the last 50 years and the effect this development has had on the deer population. We will

Figure 11.4   (See color insert) Key deer, *Odocoileus virginianus clavium*. (R. Lopez)

focus on the history and trends of Key deer that reside on two islands, Big Pine (6,296 acres) and No Name (1,139 acres) keys, because these two islands support approximately 75% of the entire deer population (Lopez et al. 2003a) and have been the center of environmental controversy for the most part. Finally, we will conclude with some of the current management challenges for this endangered species due to the urbanization of its habitat.

The endangered Key deer, the smallest subspecies of white-tailed deer in the U.S., are endemic to the Florida Keys on the southern end of peninsular Florida (Hardin et al. 1984). In comparison to other white-tailed deer, Key deer are smaller, with an average shoulder height between 24 and 31 inches, and average weights of 63 lbs (maximum 100 lbs) and 83 lbs (maximum 145 lbs) for females and males, respectively (Lopez 2001). Female Key deer less than 1 year old are not reproductively active; female yearlings and adults average 1 to 2 fawns/year (Hardin 1974). As previously mentioned, approximately 75% of the entire deer population is found on Big Pine and No Name keys (Lopez et al. 2003a). Since 1960, urban development and habitat fragmentation in the Lower Florida Keys have threatened the Key deer population (Lopez 2001; Lopez et al. 2004). In addition to a loss of habitat, an increase in urban development is of particular concern, because highway mortality accounts for the majority of the total deer mortality. Over half of the deer–vehicle collisions occur on U. S. Highway 1, the only highway linking the Keys to the mainland (Lopez et al. 2003b). Approximately 70% of the total Key deer mortality is due to human-related causes (Lopez et al. 2003b).

The Florida Keys are a series of islands located off the southern coast of peninsular Florida. It is believed that Key deer were isolated from the mainland after the melting of the Wisconsin Glacier following the last ice age nearly 4,000 years ago (Maffei et al. 1988; Folk and Klimstra 1991). The Florida Keys stretch 100 miles from end to end, which forms an effective barrier from the mainland to gene flow and has allowed the Key deer to adapt to its unique habitat (Maffei et al. 1988; Folk 1991). For example, freshwater is a limiting factor for the Key deer, and, as a result, Key deer have evolved to drink brackish water (water approximately as salty as sea water, Folk 1991) in its marine environment. The small population and lack of gene flow with mainland white-tailed deer also has promoted morphological and behavioral divergence in the subspecies (Hardin et al. 1984). For example, shorter snout lengths and wider skulls give the "toy deer" appearance to Key deer (Folk 1991). These are a few of the unique characteristics of the Florida Key deer, which is a product of its unique marine island environment.

The former range of the Key deer is hypothesized to have extended from Key Vaca to Key West but never beyond the Lower Florida Keys (Folk 1991). Currently, Key deer occupy 20 to 25 islands from Sugarloaf to the Johnson keys (Lopez 2001). Within this range, deer use all habitat types but show preference for upland areas (i.e., hammocks and pinelands, Lopez et al. 2004). Few reliable estimates of the original population size have been made; however, by the 1930s it was believed that less than 50 Key deer were in existence throughout their range (Dickson 1955). An early cartoon by Jay "Ding" Darling, cartoonist, conservationist, and chief of the U.S. Bureau of Biological Survey (predecessor to U.S. Fish and Wildlife Service), drew national attention to the Key deer's plight and was instrumental in the drafting of early legislation to protect the deer. Early hunting practices used by poachers included burning the vegetation on islands and driving deer into the water where they were either clubbed or shot to death (this was the image depicted in Darling's cartoon that was later published in the *Washington Post*). Increased law enforcement and the establishment of the National Key Deer Refuge in 1957 provided protection for the deer and its habitat. Consequently, the deer population grew to an estimated 300 to 400 animals by 1974 (Klimstra et al. 1974). Currently, the Key deer population is estimated to be 600 to 700 deer throughout their range (Lopez 2001).

Until the 1950s, the Florida Keys had little human population growth (Lopez et al. 2004). This changed when reliable supplies of freshwater and electricity became available in addition to improved transportation along the now contiguous stretch of US 1 (Lopez et al. 2004). Between 1950 and 1980, a dramatic increase in the human population occurred, causing the State of Florida to declare the Keys an "Area of Critical State Concern" in 1975; in that same year, Monroe County adopted a Land Use Plan and policy for habitat preservation and reduced human population growth (Peterson et al. 2002). Despite these regulations, since the mid-1970s the human population on Big Pine and No Name keys (the core of Key deer habitat) has continued to increase and is currently estimated at 5,000 people for these two islands (Lopez 2001; Lopez et al. 2004). As previously mentioned, rapid urbanization is the primary threat to Key deer in the form of highway mortality and other anthropogenic causes of mortality (Lopez et al. 2003b). This rapid urbanization also has been detrimental to other endangered species in the Lower Florida Keys, including the silver rice rat (*Oryzomys palustris natator*), Lower Keys marsh rabbit (*Sylvilagus palustris hefneri*), and the eastern indigo snake (*Drymarchon corais couperi*). It is the Key deer, though, that has drawn the majority of the local and national attention with regards to urban development (Peterson et al. 2002).

Native flora of the Lower Keys is primarily of West Indian origin (Dickson 1955); vegetation types include pinelands, hardwood hammocks, freshwater marshes, buttonwood forests, mangrove forests, and urban areas (Lopez et al. 2004). Vegetation varies by elevation with red (*Rhizophora mangle*), black (*Avicennia germinans*), and white mangroves (*Laguncularia racemosa*), and buttonwood (*Conocarpus erecta*) forests occurring near sea level (maritime zones). As elevation increases inland, maritime zones transition into hardwood (e.g., gumbo limbo [*Bursera simaruba*], Jamaican dogwood [*Piscidia piscipula*]) and pineland (e.g., slash pine [*Pinus elliottii*], saw palmetto [*Serenoa repens*]) upland forests with vegetation intolerant of salt water (Dickson 1955; Folk 1991). Key deer occupy many islands in the Lower Florida Keys; however, islands with freshwater and upland habitats that provide important plants for the Key deer's diet are preferred (Lopez et al. 2004).

Since the arrival of settlers, approximately 1,583 acres (21% of total area) on Big Pine (1,507 acres, 24%) and No Name (76 acres, 7%) keys have been developed. The largest land conversion, clearing for large subdivisions, occurred prior to 1985, while the majority of home construction occurred in later years (Lopez et al. 2004). In the last 50 years, Big Pine and No Name keys have seen a significant amount of urban development. During this same time period Key deer numbers have increased from less than 100 in the 1950s (Frank et al. 2003) to approximately 400 to 500 in 2000 (Lopez et al. 2003a). As we mentioned earlier, previous research suggests that rapid urbanization is the primary threat to Key deer, in the form of highway mortality and other anthropogenic mortality factors (Lopez et al. 2003b). Why then has the relationship between urban development and deer densities been *positive* rather than *negative*? Is urban development really a threat to the Key deer? The answer lies in reviewing where and how development occurred.

Historically, urban development occurred in tidal areas (e.g., mangrove, buttonwood), that was attributed to market demand to build home sites with water access. Since the enactment of federal and state laws such as the Florida Coastal Management Act of 1985 (Title XXVII, Chapter 380, Part II, F.S.), development of these vegetation types has been restricted (Gallagher 1991). The result was an increase in the development of upland areas (e.g., pinelands, hammocks) or the infilling (i.e., building on existing scarified lots) of existing subdivisions with scarified lots. Lopez (2001) hypothesized that islands with high deer densities were those with a substantial upland component, while islands that were mostly tidal (e.g., Summerland, Ramrod) supported fewer deer than similar-sized islands with more upland area (e.g., NNK, Little Pine Key).

Empirical data supports this idea, and the change in the amount of "useable space" (upland area that is readily available to Key deer; Guthery 1997; Lopez et al. 2004) might explain the increase in Key deer numbers in the last 50 years (Lopez et al. 2003b). Early urban development occurred in tidal areas and included the clearing of large tracts of land for subdivisions. Early practices used by developers in subdivision construction included (1) the clearing of native vegetation in primarily tidal areas (remember, people want to be close to water), and (2) the digging of canals to allow "backyard" water access for each of these homes. Thus, most subdivisions in the Florida Keys have a system of "alleys" that serve as direct water access for human residents. The clearing and increase in elevation from canal fill, in essence, resulted in an increase of "upland" areas for Key deer, which provided a variety of additional food resources (i.e., planted ornamentals, pet food, etc.) for the deer!

With changes in federal and state laws in the mid-1980s prohibiting the development of tidal or wetland areas, urban development pressure shifted toward upland habitats. Recent development (post-1985) includes the infilling of subdivisions and the conversion of pinelands and hammocks to urban areas. Infilling during this period actually was greater than the conversion of native upland areas due to the previously mentioned Monroe County Land Use Plan (1975; Peterson et al. 2002), thus, the net loss of upland habitats was in fact negative. Assuming that upland areas (e.g., pinelands, hammocks, and urban areas to a certain degree) is equal Key deer numbers, we would predict the conceptual model between Key deer density and urban development would be bell-shaped (Lopez et al. 2004). If current development trends continue, the amount of useable uplands will likely decrease, which in turn will decrease Key deer numbers and increase secondary impacts such as road mortality and fence entanglement (Lopez et al. 2003b; Lopez et al. 2004).

The story of the Key deer is interesting in that it offers some unique management challenges for wildlife managers. The role of urban development in this story is critical to understanding many of the historic and current management issues, namely (1) deer overabundance/domestication, (2) habitat fragmentation and anthropogenic mortality factors, and (3) social conflict. We will conclude this section by offering you some insight into some of these challenges.

Like other subspecies of white-tailed deer, the response of Key deer to urban development has been favorable (McShea et al. 1997). We would caution the reader in believing that urban development is equivalent, however, in value to pinelands and hammocks. Urban development does not necessarily provide for all the Key deer's life history requirements (e.g., fawning areas) as compared to pinelands and hammocks (Folk 1991). In the last 50 years, Key deer numbers have increased due to the previously mentioned habitat changes. In fact, current deer estimates suggest the population is at or near carrying capacity (Lopez et al. 2003a; Nettles et al. 2002) and experiencing problems with density-dependent diseases, domestication, and habitat damage.

Ironically, due to the endangered status of Key deer, many commonly used management practices that could be used in the control of deer numbers are prohibited both legally and philosophically. Wildlife managers are faced with some difficult questions — Why is the Key deer "endangered" if population numbers are "overabundant"? Can you legally control deer numbers, for example, with the use of a contraceptive that aborts a fetus? Would that abortion be a "take" under the Endangered Species Act?

Obviously these questions do not have simple answers; however, part of the answer stems from the Endangered Species Act itself, which emphasizes population *numbers* rather than other population factors (Lopez et al. 2001). Future Key deer management will require a radical change in the current management paradigm that should include changes in recovery criteria that emphasizes other parameters such as habitat quantity/quality and herd health (Lopez 2001). Another issue with deer overabundance is the "taming" of deer with increasing human–deer interactions. Most deer disease issues are related to the domestication or "ganging" behavior currently observed in Key deer (Folk and Klimstra 1991; Nettles et al. 2002). For example, high, concentrated deer groups increase the likelihood of disease in urban areas. Furthermore, the carrying capacity for Key deer is artificially raised with illegal feeding and watering of deer in urban areas (Folk and Klimstra 1991; Lopez et al. 2004). Attempting to reduce Key deer–human interactions will be a future challenge for wildlife managers.

Approximately 65% of Big Pine and No Name keys are in public ownership. Currently, approximately 1,700 lots are privately owned, undeveloped, vacant lots (Lopez 2001). The average lot size for privately-owned property is one third acre (Lopez 2001). Monroe County has the highest cost of living and the fifth highest per capita income in Florida (Peterson et al. 2002), which means that land value is high due to demand. For example, a scarified 1/4 acre lot in a subdivision is approximately $50,000 to $75,000 (2004 prices).

High land value and small, mixed ownership patterns in the Florida Keys offer yet another suite of management headaches for wildlife managers. First, land acquisition programs are limited due to the high cost of land purchases. The mosaic of property ownership also is problematic in the implementation in many land management practices. For example, the use of prescribed fire in native pinelands is necessary for their continued maintenance; however, the use of prescribed fire is limited in areas that are a "checkerboard" of both public and private lands.

A high demand to build by developers/landowners and restrictions in obtaining building permits invariably has led to conflict among stakeholder groups (Lopez et al. 2001; Peterson et al. 2002). Conversely, environmental groups also have great expectations of wildlife managers to fulfill their mission statement from their respective natural resource agencies. As the debate continues between these two groups, we find that the Key deer is normally at the center of the debate. "We would want to build a school for our children on the island" or "We want to preserve the 'small town' atmosphere of Big Pine and No Name keys [these islands have fewer residents than other islands]" are both legitimate, reasonable requests. As a manager of the Key deer, you have to be prepared to offer acceptable solutions to both of these questions.

Undoubtedly, the coexistence of a unique wildlife species with humans will continue to offer challenges for managers, biologists, and community leaders in the management of this endangered deer population. Conflict resolution will require an attempt to balance the needs of the Key deer and offer regulatory relief to residents of these two islands. In this process, decisions will need to be made based on the current biology of the Key deer in addition to incorporating other social values. Management of an endangered, urban wildlife species requires such an approach in the successful recovery of the species.

11.3 ROOSTING SPECIES

Urban and suburban locations sometimes provide habitat for species that roost in large groups. Roosts are simply locations where animals such as birds and bats sleep. Examples of animals that roost in urban areas include grackles (*Quiscalus quiscula*), starlings (*Sturnus vulgaris*), pigeons (*Columba livia*), crows (*Corvus* sp.) (Figure 11.5), and several species of bats. Roosts may contain several thousand

Figure 11.5 American crows, *Corvus brachyrhynchos*, roosting on sumac. (John M. Davis/Texas Parks and Wildlife Department)

individuals. For example, a crow roost in Fort Cobb, Oklahoma had more than one million crows (Johnson 1994). While the majority of roosts do not support individuals numbering in the millions, even a relatively small number of animals in a single location can create conflict with human residents. Consider the potential fecal wastes produced by a few hundred grackles in a single tree. Can't imagine the potential mess? Let me relate a personal experience to you. As a former student at Texas A&M University in College Station, Texas, I learned quickly that parking you car under the trees in various parking lots was a definite no-no, especially overnight. It takes only about 50 grackles perched in a roost in that same tree to make your car look like it was bombed with thick, foul-smelling white paint. Additional concerns about roosts include transmission of diseases such as toxoplasmosis and the high levels of noise sometimes generated by the animals, and damage to roost trees (Johnson 1994).

Following is a review of information available for two species that have been documented to roost in large groups in urban and suburban habitats. The American crow was chosen as a case study because it is a common widespread species across the U.S. and has begun to frequently roost and nest in urban areas within the past 50 years. We selected the Mexican free-tailed bat (*Tadarida brasiliensis*) as the second case study to highlight the potential for positive association between humans and an urban roosting species.

11.3.1 American Crows (*Corvus brachyrhynchos*)

The crow doth sing as sweetly as the lark
When neither is attended (i.e., when no one is listening.)

William Shakespeare

Many people have opinions about crows as inferred in the above quotation. Crows have been historically and are currently maligned as dirty harbingers of death. Although highly intelligent, crows are one of the least appreciated bird species by most people. Human lack of respect for this bird increases exponentially when crows gather in large roosts. Similar to coyotes, conflicts involving crows previously occurred only in rural areas. Within the past 40 to 50 years, crows have been colonizing urban and suburban habitats. A survey of wildlife damage professionals showed that 110 cities in 28 states have active crow roosts (Gorenzel et al. 2000). As mentioned before, when large numbers of birds roost together in a small area close to human development, conflict often results.

Crows are found throughout most of the continental U.S. and Canada (Stokes and Stokes 1996). They inhabit landscapes that provide open areas for feeding and wooded areas for nesting and roosting (Johnson 1994). Like the coyote, listing the common food items of the crow would be difficult. Approximately one third of the crow's diet consists of invertebrates, small vertebrates, and eggs. Crows also eat a wide variety of plant matter including wild foods such as acorns, and cultivated foods such as corn. Crows also eat carrion and garbage (Johnson 1994).

Crows commonly defend feeding territories during the day. During the fall and winter, they will leave their territories to join large flocks to roost for the night (Stokes and Stokes 1996). Scientists have been trying for years to understand why some bird species such as crows gather in large communal roosts. Two hypotheses suggest that foraging behavior influences selection of communal roost sites. The first of these hypotheses is called the information center hypothesis (Ricklefs and Miller 1999). It proposes that birds will gather together at night to collect information about foraging success from the other birds. Individual crows assess if other members of the roost were successful at finding food that day. In the morning, unsuccessful foragers from the previous day can then follow successful crows to the food source.

The second hypothesis related to foraging is the diurnal activity center hypothesis. It asserts that a roost of birds is merely a fortuitous aggregation of individuals that have strategically placed their roost site to minimize distance to their supplemental and main food sources (Caccamise and Morrison 1986; Ricklefs and Miller 1999). This hypothesis suggests that the foraging site is selected first and then a decision about where to roost is made subsequently. Imagine your workplace as your foraging area and your house as your roosting site. Most people look for a house that is in close proximity to their workplace. You don't usually buy a house and then look for a place to work. This hypothesis has not been tested with crows. However, if true, crows would not be faithful to a particular roosting site. Instead, we should see shifts in the memberships of roosts on a regular basis. Because many birds form communal roosts in the late fall and winter, some scientists have hypothesized that individuals minimize heat loss when roosting in large groups. Finally, it has been suggested that communal roosting decreases the risk of predation (Weatherhead 1984).

We may never know why communal roosting behavior originally evolved in some species, but benefits derived from it *can* be experimentally tested. In reality, crows probably gain several benefits from roosting in a large group. We will have to wait for scientists to test these hypotheses before we know the relative importance of each factor.

Another question that we have no definitive answer to is why crows have moved into suburban and urban areas to roost. Are they leaving the country life for the excitement of the city? Is the North American crow population growing so rapidly that there is no room in the country and surplus crows have to move to the city? Crow experts have hypothesized that both circumstances may be occurring. Crow populations in the west are growing rapidly, and crows that are not ready to breed disperse into urban areas (Marzluff et al. 2001). In the east, McGowan (2001) asserts that the suburban and urban habitats provide new breeding and roosting opportunities for the crows.

Urban and suburban roosts may provide benefits to crows not found in rural habitats. For example, many urban and suburban areas contain large mature trees that crows find especially attractive for roosting. Especially in the Midwest, urban and suburban areas may be the only source of large trees. Another factor that may influence urban roost site selection is the relative warmth of cities. Urban areas are usually several degrees warmer than the surrounding countryside. Gorenzel and Salmon (1995) showed that roost trees selected by urban crows had higher temperatures than randomly selected trees. The researchers also showed that urban crows selected roost sites with street lighting. The lighting may provide extra protection from great-horned owls, the main predator of American crows.

As stated before, large numbers of crows in an urban or suburban area can result in problems for human residents. An example of this conflict is currently taking place in Auburn, New York, home of approximately 29,000 human residents (Auburn city website) and the winter home of between 25,000–75,000 crows (McGowan 2001). Crows began roosting in the town about 15 years ago (McGowan 2001). Residents have complained about the feces and noise produced by the roosts. Business owners complained that the crows were damaging their property with their feces (Fox 2003).

City council members had discussed the crow issue occasionally (City Council Minutes), but no official action had been taken until an Animal Nuisance Advisory Committee was formed in November 2002. The city granted the committee $10,000 to deal with urban nuisance animals including crows and feral cats. The committee has been criticized as largely ineffective because no official attempts have been made to solve the crow problem.

According to media reports (theithacajournal.com January 2003), a private group of business owners and citizens organized a crow-shooting event in the late winter of 2000. The goal of the event was to reduce the numbers of crows roosting in the Auburn area. Hunters were asked to shoot crows in the rural areas immediately surrounding Auburn. The event received little attention until the winter of 2003 when the organizers advertised the hunt as a contest in which the four-man team that killed the most crows would receive a monetary prize. Needless to say, the proverbial crow feces hit the fan.

Animal rights activists from several organizations came to Auburn to protest the crow-shooting tournament (Varley 2003). Members from The Fund for Animals threatened to sue the crow-shoot organizers (Fox and Broach 2003). The protest received national attention from several media outlets including the *New York Times* (February 2, 2003). Approximately 300 crows were killed in the surrounding countryside during the tournament, which accounts for less than one percent of the total population that roosts in Auburn. In the end, the city of Auburn received negative media attention and is no closer to a solution for dealing with the crow roosts.

11.3.2 Mexican Free-Tailed Bat (*Tadarida brasiliensis*)

Of all small wildlife species, bats have one of the most contentious relationships with humans. Bats have been persecuted by man for generations and only recently have begun to garner respect among nonscientists for their unique and substantial role in our ecosystems. Because of their nocturnal nature, many people do not realize that urban and suburban habitats are homes to several species of bats. However, several species of North American bats have been found to both forage and reproduce in developed areas (Barbour and Davis 1974). Usually conflict arises at a localized scale, with an occasional group of bats or a single bat taking up residence under a house shutter or in the attic of an older building. However, because many bat species will roost in very large groups, the potential for major conflict exists.

Bats are one of the most diverse groups of mammals, with over 900 species (Wilson and Ruff 1999). North American bats and specifically bats found in urban and suburban locations are usually insectivores. Some of these bat species will migrate south when their food supply (insects) runs out. However, other species of bats in North America hibernate during the fall and winter months. This behavior allows them to reduce their metabolic needs until food becomes available again.

Some bat species form large communal roosts and/or hibernacula (roost sites for hibernation). For example, the Mexican free-tailed bat is well known for forming large colonies in caves, some of which contain millions of bats (Davis et al. 1962). Scientists do not definitively know why bats form such large colonies, but their hypotheses are similar to the explanations for the formation of large bird roosts. Because of their small size and specialized thermoregulatory requirements, bats are limited to where they can roost and/or hibernate. Some species (e.g., *Myotis grisescens*) need to roost close to foraging areas to reduce the energetic costs associated with flying (Tuttle 1979). As a result, when suitable sites are found, several bats will take up residence. By roosting in close proximity to other individuals, bats also reduce their energetic costs both in and out of hibernation (Herreid 1963). Transfer of information about foraging sites and protection from predators may also be benefits of communal roosting (Altringham 1996).

Many bat species, including the Mexican free-tailed bat, form what are referred to as maternity colonies. These roosts are largely comprised of females and their young (McCracken 1999). By roosting together in warm sites, the young bats are able to direct most of the energy into rapid growth (Altringham 1996).

A well-know maternity colony of Mexican free-tailed bats is found under the Congress Avenue Bridge in Austin, Texas. The bridge is the spring and summer home to approximately 1.5 million females and young. This colony has received incredible attention from scientists and the media (*National Geographic*), which both extol the ecological importance of this colony. However, when the bats first colonized the bridge in the early 1980s, public reaction was not positive. Bats have been historically and are currently feared by many people as disease-ridden vermin. Media reports in the local papers (Stanley, September 1989; *Austin American-Statesman*) played into these fears by focusing on issues such as rabies transmission.

Visit Austin, Texas, now and you are likely to hear only positive comments about the bats. The bridge has become a tourist hotspot, with approximately 100,000 people visiting during summer months to watch the bats emerge from their roost. The direct economic revenue generated by the bat bridge was estimated at $3.2 million (Ryser and Popovici 2000). What brought about this change? How has the city of Austin capitalized on an animal that usually provokes fear and disgust in so many people?

The bat–human conflict in Austin was resolved largely as a result of an expertly organized educational campaign championed by the private organization Bat Conservation International (BCI). BCI moved its headquarters from Minneapolis to Austin in 1986 (Murphy 1990). They immediately began to educate the residents of Austin about the ecological virtues of the bats. For example, the group estimated that on a typical summer night the bats eat between 10,000 and 20,000 pounds of insects, many of which are agricultural pests (BCI website). Mexican free-tailed bats have been reported to fly 50 km to their foraging sites (McCracken 1999), so farmers in the area also benefit from the existence of this urban roost. The group also worked at dispelling myths about bats, especially concerning rabies. Rabies is certainly a potential risk, however, less than 1% of all bats are infected with the disease. Handling a sick or grounded bat is the most common way that humans are exposed to rabies through bats (Greenhall and Frantz 1994).

Of course, it is unfair to compare the conflict about crows in Auburn to the conflicts about bats in Austin. The bats were not roosting and defecating over parked cars in downtown Austin like the crows do in Auburn. The bats also provide an ecological service (consumption of agricultural pests) and make very little noise doing it. In fact, except for the large bat statue and the hordes of people waiting under the bridge on warm summer nights, one could live in Austin without ever knowing the bats exist. Ask any Auburn resident and they would probably tell you that you would need to be blind, deaf, and olfactorily challenged not to notice the crows living there.

11.4 FERAL SPECIES

A feral species is defined as a domestic species that has escaped from direct human control and consequently established free-ranging populations. Examples of feral organisms that are common both in rural and urban areas in the U.S. include cats, dogs, and hogs. These populations are of concern to wildlife managers because of their potential impacts on native wildlife populations. While not all urbanites recognize and empathize with different species of native wildlife, nearly all of them will recognize and have some level of emotional response to dogs, cats, and/or hogs. Consequently, it becomes especially difficult for wildlife managers to identify control measures that will be acceptable to urbanites.

11.4.1 Free-Ranging Domestic Cats *(Felis catus)*

The relationship between North Americans and the domestic cat (*Felis catus*) has been a relatively positive one. The pet cat population in the U.S. was recently estimated at 59 to 65 million (Center for Information Management 1997) with one in three households owning at least one cat. Consequently, where you find humans, you will undoubtedly find cats as well. As a result, large populations of unowned, free-ranging cats have become established in both rural and urban landscapes. Cat rescue groups claim that between 10 and 50 million unowned cats live in the U.S., but accurate population estimates are presently unavailable (Mahlow and Slater 1996; Patronek 1998). The presence of large numbers of cats in urban areas has created conflict with humans because of nuisance behaviors such as caterwauling, fighting, spraying, defecating, and the production of numerous litters of kittens. In addition, some have cited concern about flea infestations and diseases such as ringworm, cat scratch fever, toxoplasmosis, toxocariasis, rabies, and salmonellosis (Ablett 1981; Passanisi and Macdonald 1990). Wildlife professionals also object to the presence of free-ranging cats because of their potential negative impact on native wildlife populations such as small mammals and birds.

The domestic cat has a wide distribution and occupies a diversity of habitats (Kerby and Macdonald 1988). Dependence on humans for food and shelter varies among cat populations. Some cat populations exhibit behaviors similar to wild species of cats. In these populations, individual cats may occupy exclusive territories in "wild" habitats and subsist entirely on prey such as small mammals (Genovesi et al. 1995). Studies (Dickman 1996) linked predation by domestic cats to the decline of several wildlife species in Australia. Many authors (Errington 1936; McMurray and Sperry 1941; Hubbs 1951; Parmalee 1953; Eberhard 1954; Toner 1956; George 1974, 1978; Fitzgerald 1988; Soulé et al. 1988; Coleman and Temple 1996; Lepczyk et al. 2004) in the U.S. have written about the impacts of domestic cat predation on wildlife.

Cats in urban and suburban areas usually live in groups and obtain their food directly from humans or by scavenging (Dards 1978; Natoli 1985; Yumane et al. 1997). Even though cats living in developed areas benefit from feeding by humans, they commonly prey on wildlife (Churcher and Lawton 1987; Clout and Gillies 2003; Woods et al. 2003). What effect this predation has on native wildlife populations has not been experimentally demonstrated. We do not know if cats are killing surplus individuals that would have died anyway or if the cats are limiting and/or regulating the populations of their prey.

American college campuses are ideal locations to find populations of feral cats (Figure 11.6), for two reasons. First, the transient nature of students is believed to result in a high incidence of pet loss and/or abandonment, thereby establishing and maintaining cat populations on campus and in nearby neighborhoods. Some support is given to this assumption by Oppenheimer (1980) who found a positive correlation between cat density and the proportion of renters in Baltimore neighborhoods. Second the typical campus environment meets the habitat requirements for feral cats including ample cover (e.g., basements, maintenance sheds, dense landscaping) and food (e.g., direct feeding, garbage, and prey). Conditions on campuses

Figure 11.6 (See color insert) One of several feral cats, *Felis catus,* commonly seen on Texas A&M University campus near authors' office building. (Sara Ash)

are so ideal for feral cats that control measures are often necessary to reduce the populations.

The traditional method of control in U.S. cities (including college campuses) is eradication. Shelters euthanize approximately 5 to 7 million domestic cats each year in the U.S. (American Humane Society 1993). However, a growing number of people are supporting the use of TTVAR method in which cats are trapped, tested for and vaccinated against infectious diseases, altered (sterilized), and returned to the capture site where volunteers feed and monitor the cats daily (Zaunbrecher and Smith 1993). This approach was first promoted as a method for reducing numbers of unwanted cats in Europe over 20 years ago (Remfry 1996). Volunteer organizations on several U.S. college campuses have implemented TTVAR for the management of their campus cat populations. Stanford University was the first to promote its use of TTVAR (Johnson 1995), but employees of other campuses like the University of Washington had been using the approach for several years prior to Stanford's public endorsement.

Following is a comparison of the TTVAR programs at the Universities of Washington (UW) and Texas (UT) and Texas A&M University (TAMU). For each university, we will identify and describe pertinent stakeholders and their agendas; resolution to conflicting stakeholders' agendas; political, cultural, economic, and ecological ramifications of the TTVAR program; and future considerations.

On each campus, female staff members were responsible for initiating the use of TTVAR. Sharon, UW staff member, started trapping cats on her campus in 1988. Sharon is a self-described cat lover with experience volunteering for animal shelters and providing foster care for abandoned animals. She first got involved with the feral cat issue when she began taking campus cats home with her. She quickly realized her home could not support the entire campus population, which she estimated at over 100 cats. She began secretly feeding and trapping the cats on campus and found veterinarians who would test, vaccinate and sterilize the cats at a reduced rate. Her primary objectives were to improve the cats' quality of life and reduce the population. In 1994, after providing several reluctant interviews with the local media about her activities, she and another staff member formed the organization Friends of Campus Cats (FCC). The university sanctioned their organization and TTVAR activities.

Two UT staff members and a representative from the Austin Society for the Prevention of Cruelty to Animals (SPCA) formed the volunteer organization Campus Cat Coalition (CCC) in 1995 as a reaction to the feral cat eradication attempt at the University of Texas. According to one of the cofounders of CCC, during the 1994 Christmas break, the UT Office of Environmental Health and Safety hired a pest control company to remove the feral cat population from campus. Approximately 14 cats were trapped and euthanized. When the campus community returned after break, many staff members were outraged by the university's actions. Three staff members were asked to form CCC, which would assume responsibility for the campus cats and implement a TTVAR program to control the population. CCC emphasizes the welfare of the individual cats and eventual reduction in the population.

Dawn, a former staff member of College of Veterinary Medicine at TAMU, first became interested in the feral cat issue when she witnessed the euthanasia of kittens that were trapped by TAMU pest control. Believing that destroying healthy kittens was unnecessary and a waste, she searched for the campus policy regarding the issue of cat control, but found no written policy. Dawn submitted a proposal to the College of Veterinary Medicine outlining a plan to use the TTVAR approach for controlling the campus cat population. By using this approach, veterinary students were afforded the opportunity for additional surgeries during their 3rd and 4th year rotations. Although Dawn believed that TTVAR is the most humane and effective method of controlling the numbers of feral cats, her main objective in promoting its use was to create additional educational opportunities within the College of Veterinary Medicine.

Each cat rescue group expected that their respective university administrations would have similar concerns about the use of TTVAR on campus including costs, liability, and public reaction.

According to Sharon, the UW administration's only request was that she not conduct FCC activities on university time and not use her employee Internet account for FCC use. The UT administration's reason for initiating the eradication attempt included concerns about human safety and complaints about offensive odors associated with feeding the cats and the cats' spraying behaviors. However, because the campus community reacted negatively to the eradication, the administration was open to suggestions which would appease the disgruntled staff. The TAMU administration's primary concern was costs of the program. Dawn and her faculty advisor procured external funding for the program, resulting in support and approval from the College of Veterinary Medicine.

Vocal opposition to the TTVAR program was expressed by a UW faculty member of the Zoology department who was concerned about the impact of the cats on local wildlife. Similar wildlife concerns were expressed in a letter to the editor of a local newspaper by two TAMU faculty members of the Wildlife and Fisheries Sciences Department (personal observation). Other opposition included TAMU staff members of the Physical Plant and Athletics Department who were concerned about human health and safety. Cats commonly lived in the football stadium and could sometimes cause "problems." For example, during the opening kickoff at a TAMU-Nebraska football game, a feral kitten ran out onto the field. A staff member was bitten when trying to capture the cat. However, none of the expressions of opposition on the campuses resulted in change in activities by the cat rescue groups, but rather indicated a lack of support of the program by a small number of the campus community.

All three TTVAR programs garnered attention from the media with articles in several newspapers reporting their efforts. The AFCAT program received international attention from the Discovery Channel-Canada in September 1998. The report aired in Canada in early March 1999. Positive media attention reflects well on each university, thus leading to improved public relations.

Positive cultural consequences of the TTVAR programs included the increased morale of staff members who are invested in the cats' welfare. Several UT and TAMU staff members had been feeding campus cats for many years prior to the TTVAR program, resulting in cherished human–animal relationships. Unlike eradications, the TTVAR approach protects theses bonds. Cat feeders are notoriously committed to their animals, and they will care for a colony of cats for several years (personal observation).

Negative cultural consequences of the TTVAR programs could arise if objectives and intentions of the programs are not clearly defined. TTVAR is not designed to be a solution to the cat overpopulation problem, but rather an approach to stabilize and potentially reduce and existing local population of cats. In addition, it may take several years to substantially reduce the numbers of cats. For example, although Sharon (UW) had been using the TTVAR approach for 11 years at the time of our interview, five colonies of more than seven cats each still existed on campus. The campus community must understand that TTVAR practices alone are not enough to reduce the numbers of feral cats on campus. Abandonment and pet loss must be reduced simultaneously to stem emigration out of the pet population. An alarming suggestion has been made by some cat rescue groups (not the groups from the described campuses) to artificially establish colonies with unwanted or feral cats from animal shelters (Patronek 1998).

Zaunbrecher and Smith (1993) showed that long-term costs for eradication exceed those needed to support a TTVAR program. TTVAR activities on the UW and UT campus are funded by private donations and fund-raisers, while the program at TAMU is supported by a private grant. Each university, therefore, incurs no costs. By using campus cats for educational purposes, the TAMU College of Veterinary Medicine saves money on purchases of laboratory animals for vet student surgeries.

All three university cat rescue groups claimed that their programs may minimize the potential negative impacts of cats on wildlife by providing a reliable food source for the cats. They also stated that neutering may reduce roaming, which brings cats into contact with more wildlife. However, no studies have shown that TTVAR programs minimize the impact of cats on wildlife or that adult cats that are sterilized will reduce their roaming behaviors. In fact, Ash (2001) showed that TTVAR-managed cats continued to roam large distances. Across the country, the use of TTVAR management has become a volatile issue between the cat rescue movement and wildlife managers (The Wildlife Society 2001). Let's examine the scientific reasoning behind this opposition.

Studies have shown that cats hunt regardless of hunger state (Adamec 1976; Churcher and Lawton 1987) and that the presence of cats can lead to a change in the local biodiversity (Hawkins 1998). Additionally, it is important to note that cats behave differently according to environmental factors (Liberg and Sandell 1988). Urban cats kill more birds than their rural counterparts (Churcher and Lawton 1987). However, the diet of urban cats is primarily garbage and direct handouts from humans (Fitzgerald 1988). Cats in rural and suburban areas are more likely than urban cats to come into contact with desirable wildlife species such as native small mammals, game birds, and predators. However, as we learned in the chapter on urban habitats, green spaces within highly urbanized locations can support several species of important wildlife. Therefore, it is important to consider the location of cats before deciding to use TTVAR. Both UW and UT are located in highly urbanized areas; therefore, cats are not expected to have a substantial impact on native wildlife. However, Texas A&M University is located in a rural landscape, and cats on the perimeter of campus will come into contact with several species of native wildlife, potentially creating a conflict of interest among cat rescuers and wildlife interest groups.

Human populations and, consequently, cat populations are not limited in range to highly urbanized areas. As demonstrated by earlier chapters in this book, urban and suburban development of agricultural and ecologically important land is expected to continue. With this progression we will see a continued juxtaposition of humans, their domestic animals, and wildlife. Developed areas with cats, dogs, and humans are currently positioned very close to natural habitats. If TTVAR management is used in close proximity to these areas, we can expect to see the following:

1. Native animals such as raccoons, skunks, and coyotes will be attracted to the feeding stations (Ash 2001).
2. Managed cat colonies will prey on native animals including resident species such as small mammals and migratory species such as songbirds.
3. Humans will dump their unwanted cats in these TTVAR-managed areas (Castillo and Clarke 2003).

11.4.2 Wild Hogs (*Sus scrofa*)

The authors would like to thank Rob Denkhaus and Suzanne Tuttle of the Fort Worth Nature Center and Refuge for their contributions to this section.

The wild hog is the only animal that gives birth to 6 young and 8 survive to adulthood.

Billy Higginbotham circa 2001

Rarely can the origin of a wildlife issue be dated and blame placed on an individual or group, but such is not the case with wild hogs (*Sus scrofa*) in the U.S. Not to be confused with the javelina or collared peccary (*Tayassu tajacu*) that are native to the American southwest, wild hogs (Figure 11.7) first came to the continental U.S. with the Coronado and de Soto expeditions in 1540 and 1541, respectively (Towne and

Figure 11.7 Feral hog, *Sus scrofa*. (Texas Cooperative Extension)

Wentworth 1950). Brought along as a reproducing food source, the hogs were driven along the expedition route. Some escaped and became feral, while others were stolen by Native Americans and released to establish wild populations (Towne and Wentworth 1950).

In the more than 450 years to follow, North American wild hog populations have been supplemented through traditional agricultural practices of free-ranging live-stock, where some animals escape recapture, and through intentional releases to establish huntable populations (Mayer and Brisbin 1991). Wild hogs have long been a sought-after game species prized for their meat and the thrill of the hunt. The Wild Boar Conservation Association and various local groups have developed breeding programs to supply European wild boars for stocking purposes (Myers 1993). In Texas, hunters happily paid from $25 to $1,000 (average $169) for the opportunity to participate in a wild hog hunt (Rollins 1993).

By the late 1970s, wild hogs were concentrated in the southeast portion of the U.S. stretching from Florida west to Texas, and into California (Wood and Barrett 1979). Ten years later, hogs had expanded their range to include 19 states, but were still concentrated in the southeastern states and California (Mayer and Brisbin 1991). By the early 1990s, Miller (1993) reported 23 states with resident hog populations and that populations were increasing rapidly. By 2003, Mayer (J. J. Mayer, West-inghouse Savannah River Company, personal communication) reported that 32 states and 4 Canadian provinces had resident wild hog populations. Obviously, wild hog populations have been on the increase, and they are expanding their range. The question then is, how were they doing it?

Gipson et al. (1998), in a survey of wildlife professionals, found that the two most common explanations for the burgeoning hog populations were that wild hogs from established populations have been captured and relocated to new areas to create additional hunting opportunities, and that wild hogs have escaped from "confined" populations created and maintained by hunting clubs. Of course, some wild hogs dispersed from established populations, and moved into new areas without any human assistance.

Wild hogs include free-ranging domestic, i.e., feral, swine, Eurasian wild boar, and hybrids between the two. Only one population of pure Eurasian wild boar is thought to survive in the U.S., and that is found in Corbin's Park, New Hampshire (Mayer and Brisbin 1991). However, many of the releases in and around areas such as Great Smoky Mountains National Park were reportedly of pure wild stock (Tate 1984). Wild hogs from the Great Smoky Mountains were then used to establish populations in California and other areas (Baber and Coblentz 1986; Mayer and Brisbin 1991). The three types (feral, European stock, and hybrids) exhibit slight differences in morphology (Mayer and Brisbin 1991) and behavior (Jones 1959) that cause individuals to have preferences for which type inhabits any given area.

Regardless of the explanations for how and why wild hog populations have increased and which type is found in an area, it is apparent that this exotic species is quickly invading available habitat throughout the continent. And, like other spe-cies, as available habitat gives way to urbanization, hogs find themselves faced with the opportunity to move into urban areas where their needs can be met but where the problems caused by their existence are magnified.

Wild hogs have much in common with their barnyard cousins, since they are the same species, *Sus scrofa*. However, a life spent foraging for food while exposed to the rigors of the weather causes some modifications in the basic pig template. The best adjective to describe wild hogs is "variable," because they can be found in all shapes, sizes, and colors.

Wild hogs tend to be longer legged and leaner than their captive brethren. While hunters often claim to have bagged hogs weighing over 272 kg (600 lbs), and Nowak (1991) reports domestic hogs weighing as much as 450 kg (1,000 lbs), adult males, known as boars, average 59 kg (130 lbs), and average adult females, or sows, weigh 50 kg (110 lbs) (Stevens 1996). Boars and sows can, and do, frequently attain weights of up to 136 kg (300 lbs) and 91 kg (200 lbs), respectively. Maximum shoulder height for wild hogs is around 0.9 m (3 ft) (Stevens 1996).

Wild hogs tend to have longer, thicker coats than their domestic counterparts with the more wild-type, i.e., Eurasian, hogs exhibiting a brown to black coat with a lighter shade on the tips of the bristles that give a grizzled appearance. Those hogs that flaunt their domestic ancestry can be solid black, brown, blond, white, or red, spotted, or belted. A belted hog has a white band across the shoulder and forelegs. A striped pattern is sometimes found in piglets less than 6 months old, although this pattern is never maintained to adulthood (Mayer and Brisbin 1991).

Boars have two pairs of continually growing tusks (canine teeth) that protrude from the sides of the mouth. The upper and lower tusks wear against each other to maintain a high degree of sharpness and can reach several inches in length unless they are broken or worn from use. Tusks are used for defense and to establish dominance during the breeding season. As protection from a rival's tusks, boars develop a thick, tough layer of cartilage and scar tissue (frequently referred to as a shield) across the shoulder area. Development of the shield continues as the boar ages and engages in combat with other boars. Sows have smaller tusks and do not develop a shield.

Wild hogs are omnivorous, meaning that they eat both plants and animals. They will consume virtually anything that they can find, and their preferences change with the seasons as new foods become available (Wood and Roark 1980). Hogs feed by rooting, which involves churning up the soil with their snouts (Figure 11.8). A feeding area can resemble a plowed field with furrows sometimes a foot or more deep, depending on soil type. Surprisingly, given their mode of feeding, Beyer et al. (1994) found that only 2.3% of a wild hog's diet is ingested soil.

Hog diets are primarily herbaceous, with fruits and seeds making up the bulk of the diet (Baber and Coblentz 1987). When available, acorns are an extremely important dietary item (Henry and Conley 1972; Wood and Roark 1980; Baber and Coblentz 1987). Grasses and sedges are important foods in the spring (Wood and Roark 1980) and less important in the fall (Henry and Conley 1972). Roots, particularly those of plants on hydric, or wet, sites are preferred in the summer, while fungi are fed upon throughout the year (Wood and Roark 1980). Woody material rarely shows up in stomach analysis, but researchers (Wahlenberg 1946; Wakely 1954) described extensive damage to longleaf pine seedlings, and Wood and Brenneman (1977) theorized that hogs chew the root, swallow the sap and starches, and

Figure 11.8   Damage by feral hogs to homeowner's lawn. (Linda Tschirhardt)

reject the woody tissue. They reported finding balls of chewed woody tissue in areas where hogs had been rooting among woody plants.

While animal matter does not normally comprise a large percentage of a hog's total diet, both invertebrates and vertebrates are frequently consumed. Wood and Roark (1980) found invertebrates in at least 62% of the hog stomachs that they examined during all seasons of the year. Invertebrates found included grubs, earthworms, centipedes, clams, and mussels. Howe et al. (1981) found animal matter in 94% of the hog stomachs that they examined, yet it accounted for very little of the total volume. Wild hogs are known to be opportunistic predators on ground-nesting birds (Rollins and Carroll 2001), fawns (*Odocoileus* spp.) (Springer 1975; Hellgren 1993), small mammals (Wood and Roark 1980), reptiles and amphibians (Springer 1975; Wood and Roark 1980; Howe et al. 1981), and domestic livestock (Springer 1975) and have been observed scavenging fish carcasses (Baron 1982).

The wild hog's search for preferred foods through the seasons leads it on a tour of available habitats. Hogs can be found virtually anywhere that has permanent water and sufficient cover (Sweeney and Sweeney 1982). Water is used for drinking and thermoregulation. Hogs lack sweat glands and control their body temperature by wallowing in mud holes. The resulting coat of mud also serves to protect the hog from biting insects. The need for water makes swamps, marshes, and hardwood bottomlands very attractive to wild hogs (Wood and Brenneman 1980; Baber and Coblentz 1986; Ilse and Hellgren 1995). As other foods become available, hogs move into upland areas, especially oak forests.

Wild hogs have the highest reproductive potential of any large mammal in North America (Wood and Barrett 1979; Hellgren 1999). Female hogs reach sexual maturity at approximately 10 months, while male maturity occurs between 5 and 7 months (Sweeney et al. 1979). Wild hogs are polygynous, meaning that one male will breed with many females. Breeding can occur at any time of the year, although Taylor et al.

(1998) observed most breeding in midwinter with a secondary peak in the spring. The gestation period is approximately 115 days (Henry 1968).

Litter size in wild hogs is usually less than that observed in domestic pigs (Baber and Coblentz 1986) and more than is reported in pure Eurasian wild boar (Taylor et al. 1998). Wild hog litter size varies widely depending on region, climate, and food availability but generally falls between four and seven young per litter (Sweeney et al. 1979; Baber and Coblentz 1986; Taylor et al. 1998). Although they generally only produce one litter per year, two litters in one year have been reported (Baber and Coblentz 1986; Taylor et al. 1998).

Sows and their piglets generally form loosely associated groups of eight or less, with rarely more than three adults per group (Sweeney and Sweeney 1982). Boars form small bachelor herds or are solitary except when a sow is in estrous. A sow in estrous will entice boars into the area, where battles for dominance and breeding rights may occur.

With the continuing conversion of undeveloped lands into residential areas, wild hogs are being found closer and closer to urban/suburban areas (Brown 1985), where they have increased contact with humans. Wild hogs are capable of inflicting injury upon humans, especially when cornered or when piglets are involved (Lucas 1977; Tinsley 2002). They have also been known to injure pets and even prey upon livestock (Beach 1993). In Texas, wild hogs are a significant predator of domestic goats and sheep and have been known to prey upon calves (Littauer 1993).

The most common source of conflict surrounding wild hogs is the property damage that they cause. In their never-ending search for food, wild hogs often destroy large amounts of row crops. Much of the total damage is from the actual plants being rooted up, but the amount that is lost due to trampling can also be considerable. Even tree seedlings, such as those on pine plantations, are susceptible to hog depredation (Whitehouse 1999). Fences are often used to prevent hogs from gaining access to agricultural crops, but this frequently results in damaged fences along with the lost crops (Beach 1993).

As the lines that divide hog habitat from human habitat blur, hogs are even found rooting up lawns and golf courses in search of earthworms and other foods. In addition to the destruction of the turf, costly sprinkler systems are sometimes destroyed. This type of damage tends to become more common during drought years because suburban lawns and fairways may be the only areas that receive significant amounts of moisture. Of course, this also brings hogs into closer proximity to people, and the potential for human injury increases.

An increasing hog population has also led to an increase in hog–vehicle accidents. When struck, wild hogs, with their low center of gravity and potential for being large, often cause the vehicle to roll over, causing significant mechanical damage and injuries to the vehicle's occupants (Nunley 1999). Even the hogs' rooting and wallowing activities can lead to mechanical damage when farm equipment inadvertently falls into hidden holes (Nunley 1999).

In addition, hogs are potentially significant disease carriers. Some of these diseases can be transmitted to humans, but most only pose a risk to livestock and wildlife. Diseases such as swine brucellosis and pseudorabies are significant threats to the livestock industry. While their potential as a disease vector is oftentimes emphasized by those who are not fans of wild hogs, David Stallknecht of the Southeastern Cooperative Wildlife Disease Study said in his 2004 presentation for the symposium, The Biology, Management and Control of Wild Pigs, held in Augusta, Georgia, diseases "can be used to support management decisions but may not be significant enough to drive them."

Quantifying the total effect that any species has on the environment is difficult at best, and potentially impossible. This is particularly true for a species as generalist in habitat and food preference as the wild hog. It is important to remember that the wild hog is an exotic species in North America and that our native habitats have not evolved along with it. Therefore, in their native lands, they are a natural component that makes positive contributions to those ecosystems. For example, Eurasian wild boar are thought to enhance the growth of pines in poor soils (Andrzejewski and Jezierski 1979), and their control is suspected to have decreased the rate of nutrient cycling and upset the stability of European forests (Grodzinski 1975).

However, outside their homelands, wild hogs often have negative environmental impacts on plant communities, soils, aquatic systems, and native fauna through their movements, habitat use, and food habits. Hog rooting impacts plant communities through direct consumption of vegetation as well as indirect effects such as trampling and uprooting of uneaten plants. These impacts result in a reduction in plant cover on the forest floor (Tate 1984) and provide an opportunity for early succession plants, including exotic and invasive species, to invade the habitat (Springer 1977). While this may be perceived as a positive impact if an area is being managed for maximum floristic diversity, the reverse is true for lands being managed as natural areas. And, the potential exists for wild hogs to adversely impact endangered plant species through their rooting activities.

In California's oak-dominated ecosystems, Sweitzer and Van Vuren (2002) found that wild hog rooting significantly reduced oak regeneration. Reduced seedling survival was linked to hog rooting, and acorn consumption is blamed for reduced acorn survival to potential germination as well as reduced forage available for native wildlife.

The presence of wild hogs in an area also impacts native animal species either through direct predation or through competition for resources. As stated above, wild hogs are opportunistic predators on a variety of wildlife species including at least two threatened species, Jones' middle-toothed snail (*Mesodon jonesianus*) and the Jordan's red-cheeked salamander (*Plethodon jordani*) (Tate 1984). Hogs are reported to prey upon turtle eggs and hatchlings (Hanson and Karstad 1959). Although they do not frequently prey upon them, intensive hog rooting alters the appropriateness of leaf litter habitat for at least two vertebrate species, southern red-backed voles (*Clethrionomys gapperi*) and northern short-tailed shrews (*Blarina brevicauda*) (Singer et al. 1984).

Wild hogs compete with a number of species for available food sources. In Tennessee, hogs are thought to compete with deer (*Odocoileus virginianus*), bear (*Ursus americanus*), gray squirrels (*Sciurus carolinensis*), turkey (*Meleagris gallopavo*), raccoons (*Procyon lotor*), and opossums (*Dideplhis virginiana*) for acorns and other fruits and seeds. Henry (1969) proposed that, while wild hogs were only minor predators on bird nests, they competed with other native nest predators such as raccoons, opossums, and snakes for the available resource. The wild hog's ability to take advantage of an array of foods gives it a competitive advantage when compared to more specialized feeders.

While there have been no studies that have documented hogs having a significantly adverse direct impact on another species, Roemer et al. (2002) reported how the presence of a large population of wild hogs on the California Channel Islands encouraged golden eagles (*Aquila chrysaetos*) to colonize the islands to take advantage of an abundant food supply, i.e., piglets. Eagles soon became the top predator on the island and fed heavily on the resident island fox (*Urocyon littoralis*) population. With the reduction in foxes, island spotted skunk (*Spilogale gracilis amphiala*) populations increased significantly to fill the available niche. So, while the hogs weren't preying upon any of the other three vertebrates in the study, their presence had a far-reaching ecological impact on all three and the entire island system by causing predation to replace competition as the dominant force shaping the island communities. Another study (Long et al. 2001) hypothesizes that the growing hog population in California is partly responsible for the increase in mountain lions in the state and is therefore indirectly leading to an increase in mountain lion–human encounters.

Some of the potentially more devastating environmental impacts that can be attributed to hogs are directed against the abiotic components of the ecosystem. Hog rooting accelerates decomposition and loss of nutrients from the forest floor and upper soil horizons (Singer et al. 1984). Specifically, rooting reduced phosphorus (P), magnesium (Mg), and copper (Cu) levels available for plant uptake. Singer et al. (1984) found that soil erosion did not significantly increase because of hog rooting in their upland forest study site, but attributed this to the natural porosity of the soil and the decrease in soil bulk density caused by the rooting. However, water sampling in rooted watersheds indicated a significant increase in soil water NO_3 levels. Hog rooting and wallowing is suspected of causing siltation and contamination of Appalachian streams (Howe et al. 1981). Tate (1984) reported higher concentrations of fecal coliform bacteria in areas occupied by wild hogs.

Depending on management goals, management of wild hog populations ranges throughout the spectrum from a desire to eradicate the species to trying to introduce the species into new areas. Because of its status as an exotic species and its adverse impacts to the ecosystem, most land managers are strongly entrenched on the side of eradication, although many admit that eradication of such a prolific species may be impossible because of budget constraints and public perceptions. Although eradication has been (or nearly been) achieved in some island habitats (Kessler 2002; Schuyler et al. 2002), controlling wild hog populations at a level where they are not severely impacting native habitats is the best that can be hoped for in areas where the species is firmly entrenched and such factors as immigration and human-aided introduction cannot be controlled.

Wild hog control efforts take many forms but usually involve a combination of techniques. Traps, controlled shooting from the ground or air, public hunts, capture with dogs, and poisoning have all been used. Attention has been paid to biological controls such as introducing hog diseases and using chemosterilants. Fortunately, the idea of introducing disease into the wild hog population has not been attempted because of the potential for the disease to pass into the domestic livestock industry.

Work continues in chemosterilants, but with little reported, practical success (Killian et al. 2003). The ultimate choice of techniques employed depends upon the availability of money and personnel to conduct a control program, the expertise of the personnel, legal limitations, safety concerns, site conditions, and public sentiment.

Trapping is the most common control method and can be extremely successful, especially when used in conjunction with other techniques. Trapping is a labor-intensive and costly endeavor that requires considerable scouting and moving, pre-baiting, and checking traps. Success is dependent upon the expertise of the personnel doing the trapping (E. K. DeLozier, National Park Service, personal communication), but can result in large numbers of wild hogs being removed from the population in a short period of time. Once trapped, hogs can be euthanized quickly and as humanely as possible via a properly placed gunshot (Beaver 2001). Unfortunately, trapping alone will not eradicate wild hogs from an area, because some hogs, enough to breed and replenish the population, will become trap shy and must be taken by another method.

Controlled shooting, either from the ground or the air, is conducted by trained marksmen. Aerial shooting can result in very large numbers of hogs removed from the population if the hogs inhabit an area with little overhead cover. Ground shooting usually involves stalking hogs in their activity areas at night using night vision equipment and high-powered rifles equipped with scopes. Because of the potentially large numbers of hogs that can be taken in a relatively short period of time, controlled shooting is usually more cost effective than trapping. Safety is of the utmost importance, especially on public lands, when controlled shooting is used, and the potential for an accident can eliminate this technique from consideration in some areas.

In areas where public hunting is allowed, public wild hog hunts, often as an adjunct to deer hunts, are sometimes used as a population control measure. In most cases, this has not proven to be an effective control, because hunters do not remove as many individuals as is necessary either through lack of success or lack of interest in hunting hogs. Many areas that have tried to use public hunts to control hog populations have actually seen an increase in hog populations instead of a decrease.

Hunting with dogs usually involves contracting with a local hunter/dog handler, because few areas can afford to train and support a pack of dogs throughout the year. Dogs can be quite effective in locating, bringing to bay, and capturing wild hogs, although it involves great potential for danger for the dogs and their handlers. Once caught, the wild hog is usually dispatched using a firearm or through exsanguination. Hunting with dogs can raise the ire of animal rights/animal welfare groups, causing them to oppose the program.

Poisoning is used to control wild hog populations in Australia (Choquenot et al. 1990) and New Zealand (Thomas and Young 1999). Yellow phosphorus (CSSP), sodium monofluoroacetate (1080), and warfarin have all been used with success, but the potential for nontarget animal poisoning prevents the use of poison in the U.S.

Because of the wild hog's dual status as both a nuisance and a game animal, public perception of control programs is of utmost importance. Some may believe that nonlethal control measures, such as trapping and relocating, are more appropriate and may oppose other, lethal measures. Trapping and relocation have been used in the Great Smoky Mountains National Park with limited impact on the wild hog population. It is also possible that relocated animals simply return to the area from which they were removed or become a nuisance in their new range.

Wild hogs are here to stay in North America. Total eradication, at least with current technology, is not possible. Because of their generalist nature and great adaptability, wild hogs will continue to expand their range on the continent and invade new habitats. As the human population continues to grow, encounters with wild hogs and their damage, whether in the wilderness or in a suburban yard, will increase, and management of this exotic, invasive species will grow in importance.

The Fort Worth Nature Center and Refuge (FWNC&R) is a 1450+ hectare (3600+ acre) urban green space owned by the City of Fort Worth and managed by employees of the City's Parks and Community Services Department as a natural, native landscape. Resource management objectives are to conserve, maintain and/or restore plant and animal communities native to North Central Texas; this includes controlling the introduction and spread of exotic plants and animals with eradication as the ultimate goal.

The park lies wholly within the political boundaries of Fort Worth on the northwest edge of the city and is bisected by a major river, the West Fork of the Trinity. The wetland and riparian zones of the West Fork and its tributaries provide many acres of prime habitat for feral hogs.

In July 1999, during a routine visit to a remote area on the north side of the park, FWNC&R staff found many areas of major soil disturbance consistent with the type of damage caused by feral hog rooting. Wallows began to appear along watercourses, and the soft, sandy soil in upland areas was plowed both randomly and in long, straight furrows. Soon thereafter, the staff made the first visual records of live hogs and started receiving visitor reports of hog sightings along trails in heavier traffic areas of the park. It is unknown whether the hogs moved into the FWNC&R via natural migration along the river or were purposely introduced by neighboring landowners for illegal hunting.

In keeping with the goal of controlling the introduction and spread of exotic animals, staff immediately began planning for removal of the hogs. An extensive literature search was undertaken to compile information on feral hog natural history, population assessment methods, and existing control options. In addition, staff consulted with local representatives of other agencies including the U.S. Department of Agriculture–Wildlife Services Division, the U.S. Fish and Wildlife Service, and Texas Parks and Wildlife Department. Formulation of a control plan based on the information and experience gathered by others dealing with feral hog problems began in fall 2000, and a draft was ready to begin the approval process by spring 2001.

In formulating the control plan, the staff had to first make a decision about whether to remove the hogs via lethal or nonlethal means. A number of options were considered including live trapping and euthanization by staff, exclusion with fencing, live trapping and relocation to another site, live trapping and euthanization by Fort Worth Police, hiring professional trappers, or conducting a regulated public hunt. As each option was weighed and the unique set of challenges faced by the FWNC&R staff considered, most proved unfeasible for implementation. For example, (1) the cost of fencing for total exclosure was (and still is) prohibitive; (2) staff were informed by the local USDA Wildlife Services agent that permission for relocation of live hogs would not be granted; (3) providing access to police officers and professional trappers to remote and/or environmentally sensitive areas of the Refuge would tax the resources of a small, overburdened staff and create a different set of resource protection challenges; (4) funding for professional trappers was nonexistent; and (5) the risk to the public and potential liability of the City was considered too great to allow a public hunt. Live trapping and euthanization by staff was finally settled on as the only viable option under the circumstances.

Staff also made contact with a variety of external stakeholders during the process of developing the control plan to gather opinions and to generate an inclusive spirit of total disclosure. These stakeholders included the Friends of the Fort Worth Nature Center (a 501(c)3 support organization), Fort Worth Audubon Society, Humane Society of the U.S. (HSUS), People for the Ethical Treatment of Animals (PETA), North Texas Herpetological Society, Dallas/Fort Worth Herpetological Society, and neighbors and other nearby landowners. Input from these external stakeholders was then presented to internal stakeholders, departments within the City of Fort Worth including Police, Risk Management, Legal, and Animal Control, who then provided guidance on safety and liability issues.

Once the trap and euthanize option was chosen, the next steps were to decide trap type, euthanasia means, and carcass disposition method. Staff decided to begin the control program with modified box traps equipped with trip wire, spring-loaded, side swing doors. This type of trap is reasonably portable and small enough to be carried into remote areas with a standard off-road vehicle. Use of snares was considered and has not been totally ruled out as an option, particularly for trap-resistant individuals; however capture of nontarget animals and public perception of the method need to be considered.

Firearms were chosen over chemical means of euthanasia for several reasons. The AVMA Panel on Euthanasia recommends gunshot as a humane method for euthanizing hogs in the field (Beaver 2001). Chemicals are slower to act than firearms, leading to increased trap stress and longer, closer staff interaction with the angry hogs. Cost was also a factor, as the chemicals are expensive and require specialized training in their use. Euthanasia drugs are a controlled substance and are difficult to obtain; they require special secure storage; and a veterinarian must monitor their use. Staff was also concerned that nontarget animals that fed on drug euthanized-hog carcasses could be harmed.

The method chosen for carcass disposition led to the greatest amount of controversy during the formulation of the control plan. Staff chose to move the carcasses to discreet disposition sites and let natural nutrient recycling occur rather than attempt to utilize the meat. There were several reasons for this decision — ecologically, the hogs are robbing resources from the native flora and fauna that should be returned to the native ecosystem; from the Risk Management standpoint of safety and liability, hogs carry several diseases that are transmittable to humans; and practically, staff time and availability are very limited. In addition, this option was the most acceptable to animal rights/animal welfare organizations that agreed in principle to the need for the removal of exotic animals to protect a native landscape, but objected to using those animals for the benefit of humans.

As the process of getting buy-in for a feral hog management plan from the City of Fort Worth governing body and FWNC&R stakeholders continued, three negative hog/human encounters occurred, two at the Nature Center and one at a neighboring ranch, that proved pivotal in steering sentiment in favor of implementing the plan. In April 2001, a woman was walking her dog on leash down an old road in a relatively remote area of the park when she surprised a sow with numerous piglets. The sow tried to escape but a fence blocked its chosen route, so the sow turned and attacked the woman's dog. The woman managed to free her pet, and fortunately, the sow turned and ran away with her offspring instead of continuing the attack. The dog sustained severe injuries but did survive.

Shortly after this incident, a photographer who was a long-time supporter and frequent visitor to the Nature Center reported that he was hiking on a remote trail with his camera and tripod when he encountered a lone hog. The hog turned and charged the photographer, who used his tripod to defend himself against the hog. The animal gave up its attack quickly and departed, leaving the man shaken but uninjured.

A few months after the second incident, a neighboring rancher who had already begun removing hogs from his property, called to report that one of his ranch hands had been attacked by a lone hog while on horseback in an area just across the Trinity River from the FWNC&R. His employee was not hurt, but the horse's legs were badly wounded. It is not known whether or not the horse survived or had to be euthanized. The rancher expressed his concern for the safety of visitors to the Nature Center and urged the staff to implement a control program.

These three negative encounters gave weight to the Nature Center staff's recommendation to implement the new management plan; permission to begin was granted soon afterwards. The staff began devising a public relations strategy to introduce the situation and proposed solution to the general public, but before the PR plan could be implemented, the story hit the media via an internal leak within the City of Fort Worth. A Section B front-page story in the Fort Worth *Star-Telegram* led to featured news items on the local FOX, CBS, and ABC stations (Sidebar 11.1). This media exposure led to extensive public response — approximately 100 comments were received, with less than 6% negatives. Most of those who made comments, in addition to supporting the control program, offered their assistance, which was politely declined.

With the go-ahead given, the staff began final preparations for removing hogs from the property. Members of the FWNC&R feral hog control team first obtained appropriate caliber firearms. Team members then began practicing and familiarizing themselves with the weapons and scouting for prime locations of hog activity to place traps. Trapping began on 09 January 2003, and the program started very well, with seven hogs taken from two traps after the first trap night. Trapping efforts slacked off considerably during the late spring, summer, and fall when the animals had access to plenty of wild foods, but resumed in intensity in December. By the close of 2003, 33 animals had been removed from the park including at least 10 boars and 10 sows of reproductive size.

Trapping continued into 2004 and, as of this writing in August of that year, an additional 41 animals have been removed. Anecdotally, the staff has noticed a significant decrease in hog activity since the program started. A series of exclosures

Figure 5.3 Golf courses have the potential to provide habitat needs for several species of wildlife including these deer. (Ken Hammond/ USDA)

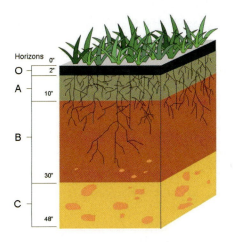

Figure 7.3 A typical soil profile showing four horizons. (USDA photo)

Figure 8.4 A Texas rattlesnake roundup holding pen for captured rattlesnakes. (C.E. Adams)

Figure 9.2 Since 1934, sales of Federal Duck stamps have raised more than $700 million used to acquire more than 5.2 million acres of habitat for the National Wildlife Refuge System. (USFWS.)

Figure 10.2 U.S. Customs agents shown with some of the items confiscated under the Lacey Act. (Steve Hillebrand/USFWS)

Figure 10.5 Black-footed ferrets are one of the species protected under The Endangered Species Conservation Act. (Tami S. Black/USFWS)

Figure 11.1 Coyote, *Canis latrans.* (Sara Ash)

Figure 11.3 San Joaquin kit fox, *Vulpes macrotis mutica.* (B. Peterson/U.S. Fish and Wildlife Service)

Figure 11.4 Key deer, *Odocoileus virginianus clavium*. (R. Lopez)

Figure 11.6 One of several feral cats, *Felis catus,* commonly seen on Texas A&M University campus near authors' office building. (Sara Ash)

have been erected in areas of high hog use for qualitative study of effects on the floral community and to document activity levels. Due to increasing urbanization around the park, which should eventually close most immigration routes, the staff now believes that total eradication is achievable and continues to work toward that goal.

SIDEBAR 11.1: WILD HOGS THREATEN TEXAS NATURE CENTER

Wayne Clark was exploring a remote part of the Fort Worth Nature Center when he saw deep holes in the ground — as if someone had been digging for buried treasure. Over the next few months, he learned that the nature retreat had become infested by a growing population of feral hogs that could put visitors and the environment at risk.

To eliminate the animals' threat, Clark, who supervises the center, and staff members will start a program as soon as next month to track and kill as many feral hogs as they can find.

"People have died from feral hog attacks," Clark said. "It's not extremely common, but the potential is there. You've got a wild animal that weighs up to 400 pounds, has sharp teeth, and isn't afraid to attack." At least one dog has been attacked and nearly killed by wild swine at the center.

At the same time, the feral hogs compete with other animals for available food and have damaged ecosystems by their rooting and wallowing, he said. "We think this program can work," Clark said. "We don't think we can eradicate the problem, but we can control it. We feel we can reduce the hogs' numbers, which reduces damage and risk to visitors."

This has drawn criticism from groups such as People for the Ethical Treatment of Animals, which say the feral hogs shouldn't be killed. "We regret that the Nature Center wants to remove the feral pigs from the area, but to do so using lethal methods is unconscionable when other methods exist," said Stephanie Boyles, a wildlife biologist with the north Virginia-based PETA. "We don't believe the feral pigs should be slaughtered in the name of conservation. They should use methods such as contraception and fencing to keep these animals away from areas of concern."

Others defend the decision to remove the wild swine from the Nature Center by any means necessary. "People must remember and know what the purpose of the Nature Center is," said Bill Meadows, a former city councilman and chairman of the Nature Center Endowment Fund. "The whole point is to preserve and nurture what is natural in this area — the indigenous plants and animals. When a species comes in and serves to significantly alter the mission, there's no question that you have to take a dramatic action. If that means you have to eliminate some of those animals, then that's what you have to do."

There are as many as 300 to 400 feral hogs at the nature center, a population that is rapidly growing, said Richard Zavala, director of the city's parks and community services department.

Over the past few years, city and center officials have researched ways to thin the feral hog population and have consulted with national and state experts at agencies ranging from the U.S. Department of Agriculture to the Texas Parks and Wildlife Department.

Feral hogs can't be transported from one area to another because of the threat of spreading livestock disease such as swine brucellosis and pseudorabies, according to the Texas Animal Health Commission.

"We will trap and euthanize them by shooting them on site," Zavala said. "It will be done in a controlled process, with the traps and euthanasia done by nature center staff in nonpublic areas during nonpublic times. The remains will be left out at the Nature Center to decompose. That's the way we do things at the Nature Center. It's a circle of life."

Feral hogs, which are a mix of domestic hogs and the more aggressive Russian boars, typically are dark and furry and have longer tusks than domestic hogs. They are unprotected, nongame animals. But they are prolific breeders, destroy the landscape, and can be a danger to people and other animals, said John Davis, one of two Texas Parks and Wildlife urban biologists who serve the Fort Worth/Dallas region.

"These hogs can begin breeding by the time they are 6 months old and in a "good" year can bear 20 to 24 piglets. As of 1991, there were an estimated 1 million feral hogs — who have a lifespan of about 8 years — in Texas," Davis said. Trapping and shooting the hogs is a common way to control their population and remove them from areas they are destroying. Leaving their carcasses at the center is an ecologically sensitive way to rid the area of the animals, he said. "This has to be done," he said. "There are a lot of people who aren't going to be comfortable removing the animals, euthanizing a species found on the Nature Center. If it doesn't happen, there will be so much more damage done to the Nature Center and so many other animals will suffer. There's no other answer."

The easiest time to catch the hogs is during the cooler months, when they are more mobile, searching for food such as acorns. That's when Clark and the other nature center employees plan to start laying traps. The traps will be mostly steel boxes or metal fences with doors that allow hogs to enter, to find the bait left inside. After the hogs enter, they will not be able to escape, but others would be able to get inside, Clark said. He hopes to give the City Council a yearly update on the progress of the feral hog management program. "We feel like we're doing the best we can on this," he said.

By Anna M. Tinsley
Star-Telegram

September, 2002

REFERENCES

Ablett, P. (1981). Public reaction to control, in *Proceedings of UFAW Symposium*, Royal Holloway College, University of London, 23–24 September 1980, London, pp. 60–62.

Adamec, R. E. (1976). The interaction of hunger and preying in the domestic cat (*Felis catus*): an adaptive hierarchy? *Behavioral Biology*, 18(2):263–272.

Altringham, J. D. (1996). *Bats: Biology and Behavior*, Oxford University Press, Oxford.

American Humane Association. (1993). Euthanasia stats remain stable, *Shoptalk*, 11:1–2.

Andrzejewski, R. and Jezierski, W. (1979). Management of a wild boar population and its effects on commercial land, *Acta Theriologica,* 23(19):309–339.

Ash, S. J. (2001). Ecological and Sociological Considerations of Using the TTVAR (Trap, Test, Vaccinate, Alter, Return) Method to Control Free-Ranging Domestic Cat (*Felis catus*) Populations, Dissertation, Texas A&M University, College Station.

Baber, D. W. and Coblentz, B. E. (1986). Density, home range, habitat use and reproduction of feral pigs on Santa Catalina Island, *Journal of Mammalogy,* 67:512–525.

Baber, D. W. and Coblentz, B. E. (1987). Diet, nutrition and conception in feral pigs on Santa Catalina Island, *Journal of Wildlife Management,* 51:306–317.

Baker, R. O. and Timm, R. M. (1998). Management of conflicts between urban coyotes and humans in southern California, in *Proceedings 18th Vertebrate Pest Conference,* pp. 200–312.

Barbour, R. W. and Davis, W. H. (1974). *Mammals of Kentucky,* University of Kentucky Press, Lexington.

Baron, J. (1982). Effects of feral hogs on the vegetation of Horn Island, Mississippi, *American Midland Naturalist,* 107:202–205.

Beach, R. (1993). Depredation problems involving feral hogs, in *Feral Swine: A Compendium for Resource Managers,* Hanselka, C. W. and Cadenhead, J. F., Eds., Texas Agricultural Extension Service, Kerrville, pp. 67–75.

Beaver, B. V. (Chair). (2001). 2000 Report of the AVMA panel on Euthanasia, *Journal of the American Veterinary Medical Association,* 218:669–696.

Beckmann, J. P. and Berger, J. (2003a). Rapid ecological and behavioural changes in carnivores: the responses of black bears (*Ursus americanus*) to altered food, *Journal of Zoology London,* 261:207–212.

Beckmann, J. P. and Berger, J. (2003b). Using black bears to test ideal-free distribution models experimentally, *Journal of Mammalogy,* 84(2):594–606.

Beier, P. (1995). Dispersal of juvenile cougars in fragmented habitat, *Journal of Wildlife Management,* 59(2):228–237.

Bekoff, M. (1999). Coyote, *Canis latrans,* in *The Smithsonian Book of North American Mammals,* Wilson, D. E. and Ruff, S., Eds., Smithsonian Institution Press, Washington, DC, pp. 139–141.

Bekoff, M. (1977). *Canis latrans, Mammalian Species,* 79:1–9.

Beyer, W. N., Connor, E. E., and Gerould, S. (1994). Estimates of soil ingestion by wildlife, *Journal of Wildlife Management,* 58:375–382.

Brown, L. N. (1985). Elimination of a small feral swine population in an urbanizing section of central Florida, *Florida Scientist,* 48:120–123.

Caccamise, D. F. and Morrison, D. W. (1986). Avian communal roosting: implications of diurnal activity centers, *American Naturalist,* 128(2):191–198.

Castillo, D. and Clarke, A. L. (2003). Trap/neuter/release methods ineffective in controlling domestic cat "colonies" on public lands, *Natural Areas Journal,* 23(3) 247–253.

Center for Information Management. (1997). *U.S. Pet Ownership and Demographic Sourcebook.* American Veterinary Medical Association, Schaumburg, IL.

Choquenot, D., Kay, B., and Lukins, B. (1990). An evaluation of warfarin for the control of feral pigs, *Journal of Wildlife Management,* 54:353–359.

Churcher, P. B. and Lawton, J. H. (1987). Predation by domestic cats in an English village, *Journal of Zoology,* 212:439–455.

Clout, M. and Gillies, C. (2003). The prey of domestic cats (*Felis catus*) in two suburbs of Auckland City, New Zealand, *Journal of Zoology,* 259(3):309–315.

Coleman, J. S. and Temple, S. A. (1996). On the prowl, *Wisconsin Natural Resources,* 20(6):4–8.

Conover, M. R. (1998). Perceptions of American agricultural producers about wildlife on their farms and ranches, *Wildlife Society Bulletin,* 26(3):597–604.

Conover, M. R. (2001). *Resolving Human–Wildlife Conflicts: the Science of Wildlife Damage Management,* CRC Press, Boca Raton, FL.

Cypher, B. L. and Frost, N. (1999). Condition of San Joaquin kit foxes in urban and exurban habitats, *Journal of Wildlife Management,* 63(3)930–938.

Cypher, B. L. and Spencer, K. A. (1998). Competitive interactions between coyotes and San Joaquin kit foxes, *Journal of Mammalogy,* 79:204–214.

Cypher, B. L. and Warrick, G. D. (1993). Use of human-derived food items by urban kit foxes, *Transactions of the Western Section of the Wildlife Society,* 29:34–37.

Dards, J. L. (1978). Home ranges of feral cats in Portsmouth Dockyard. *Carnivore Genetic Newsletter,* 3:242–255.

Davis, R. B., Herreid, C. F., II and Short, H. L. (1962). Mexican free-tailed bats in Texas, *Ecological Monographs,* 32:311–346.

Dickman, C. R. (1996). *Overview of the Impacts of Feral Cats on Australian Native Fauna.* Institute of Wildlife Research and Biological Science, University of Sydney, Sydney, Australia.

Dickson, J. D., III. (1955). An ecological study of the Key deer, Florida Game and Fresh Water Fish Commission Technical Bulletin 3, Tallahassee, FL.

Eberhard, T. (1954). Food habits of Pennsylvania house cats, *Journal of Wildlife Management,* 18(2):284–286.

Errington, P. L. (1936). Notes on food habits of the southern Wisconsin house cats, *Journal of Mammalogy,* 17(1)64–65.

Fitzgerald, B. M. (1988). Diet of domestic cats and their impact on prey populations, in *The Domestic Cat: The Biology of Its Behavior,* Turner, D. C. and Bateson, P., Eds., Cambridge University Press, New York, pp. 123–150.

Folk, M. L. (1991). Habitat of the Key Deer, Dissertation, Southern Illinois University, Carbondale.

Folk, M. L. and Klimstra, W. D. (1991). Urbanization and domestication of the Key deer (*Odocoileus virginianus clavium*), *Florida Field Naturalist,* 19:1-9.

Fox, C. (2003). City hall fights crows, *The Citizen* online, http://www.auburnpub.com/.

Fox, C. and Broach, L. H. (2003). Crow shoot begins this morning, *The Citizen* online, http://www.auburpub.com/

Frank, P. A., Stieglitz, B. W., Slack, J., and Lopez, R. R. (2003). The Key deer: Back from the brink, *Endangered Species Bulletin,* 28:20–21.

Gallagher, D. (1991). Impact of the built environment on the natural environment, in *Monroe County Environmental Story,* Gato, J., Ed., Monroe County Environmental Education Task Force, Big Pine Key, FL, pp. 226–229.

Genovesi, P., Besa, M., and Toso, S. (1995). Ecology of a feral cat (*Felis catus*) population in an agricultural area of northern Italy, *Wildlife Biology,* 1(4):233–237.

George, W. G. (1974). Domestic cats as predators and factor in winter shortages of raptor prey, *Wilson Bulletin,* 86:384–396.

George, W. G. (1978). Domestic cats as density independent hunters and "surplus killers." *Carnivore Genetic Newsletter,* 3:282–287.

Gipson P. S., Hlavachick, B., and Berger, T. (1998). Range expansion by wild hogs across the central United States, *Wildlife Society Bulletin,* 26(2):279–286.

Gorenzel, W. P., Salmon, T. P., Simmons, G. D., Barkhouse, B. and Quisenberry, M. P. (2000). Urban crow roosts—a nationwide phenomenon? *Proceedings Eastern Wildlife Damage Management Conference,* 9:158–170.

Grodzinski, W. (1975). The role of large herbivore mammals in functioning of forest ecosystems: a general model, *Polish Ecological Studies,* 1:10–15.

Gompper, M. E. (2002). Top carnivores in the suburbs? Ecological and conservation issues raised by colonization of North-Eastern North America by coyotes, *BioScience,* 52(2):185–190.

Gorenzel, W. P. and Salmon, T. P. (1992). Urban crow roosts in California, in *Proceedings 15th Vertebrate Pest Conference,* pp. 97–102.

Gorenzel, W. P. and Salmon, T. P. (1995). Characteristics of American crow urban roosts in California, *Journal of Wildlife Management,* 59(4):638–645.

Greenhall, A. M. and Frantz, S. C. (1994). Bats, in *Prevention and Control of Wildlife Damage,* Hygnstom, S. E., Timm, R. M., and Larson, G. E., Eds., University of Nebraska Cooperative Extension, U.S. Department of Agriculture-Animal Plant Health Inspection Service-Animal Damage Control, pp. D-5–D-24.

Grinder, M. I. and Krausman, P. R. (2001). Home range, habitat use, and nocturnal activity of coyotes in an urban environment, *Journal of Wildlife Management,* 65(4):887–898.

Guthery, F. S. (1997). A philosophy of Habitat Management for Northern Bobwhites, *Journal of Wildlife Management,* 61:291–301.

Hanson, R. P. and Karstad, L. (1959). Feral Swine in the Southeastern United States, *Journal of Wildlife Management,* 23:64–74.

Hardin, J. W. (1974). Behavior, Socio-Biology, and Reproductive Life History of the Florida Key Deer, *Odocoileus virginianus clavium,* Dissertation, Southern Illinois University, Carbondale.

Hardin, J. W., Klimstra, W. D., and Silvy, N. J. (1984). Florida Keys, in *White-Tailed Deer: Ecology and Management,* Halls, L. K., Ed., Stackpole Books, Harrisburg, PA, pp. 381–390.

Hawkins, C. C. (1998). Impact of a Subsidized Exotic Predator on Native Biota: Effect of House Cats (*Felis catus*) on California Birds and Rodents, Dissertation, Texas A&M University, College Station.

Hellgren, E. C. (1993). Biology of feral hogs (*Sus scrofa*) in Texas, in *Feral Swine: A Compendium for Resource Managers,* Hanselka, C. W. and Cadenhead, J. F., Eds., Texas Agricultural Extension Service, Kerrville, pp. 50–58.

Hellgren, E. (1999). Reproduction of feral swine, in *Proceedings of the Feral Swine Symposium,* June 2–3. Texas Animal Health Commission, Fort Worth, pp. 67–68.

Henry, V. G. (1968). Length of estrous cycle and gestation in European wild hogs, *Journal of Wildlife Management,* 32:406–408.

Henry, V. G. (1969). Predation on dummy nests of ground-nesting birds in the southern Appalachians, *Journal of Wildlife Management,* 33:169–172.

Henry, V. G., and Conley, R. H. (1972). Fall foods of the European wild hogs in the southern Appalachians, *Journal of Wildlife Management,* 36:854–860.

Herreid, C. F., II. (1963). Temperature regulation of Mexican free-tailed bats in cave habitats, *Journal of Mammalogy,* 44:560–573.

Howe, T. D., Singer, F. J., and Ackerman, B. B. (1981). Forage relationships of European wild boar invading northern hardwood forests, *Journal of Wildlife Management,* 45:748–754.

Hill, E. P., Sumner, P. W., and Wooding, J. B. (1987). Human influences on range expansion of coyotes in the southeast, *Wildlife Society Bulletin,* 15:521–524.

Hubbs, E. L. (1951). Food habits of feral house cats in the Sacramento Valley, *California Fish and Game,* 37(2):177–189.

Ilse, L. M. and Hellgren, E. C. (1995). Resource partitioning by sympatric populations of collared peccaries and feral hogs in southern Texas, *Journal of Mammalogy,* 76:784–799.

Johnson, K. (1995). National pet alliance report on trap/alter/release programs. *Cat Fanciers Almanac,* July:92–94.

Johnson, R. J. (1994). American Crows, in *Prevention and Control of Wildlife Damage,* Hygnstom, S. E., Timm, R. M., and Larson, G. E., Eds., University of Nebraska Cooperative Extension, U.S. Department of Agriculture-Animal Plant Health Inspection Service-Animal Damage Control, pp. E-33–E-40.

Jones, P. (1959). *The European Wild Boar in North Carolina,* North Carolina Wildlife Resources Commission, Raleigh.

Kendall, P. (1999). Wounded coyote captured by rush of loop traffic, *Chicago Tribune,* online edition, March 25, 1999.

Kerby, G. and Macdonald, D. W. (1988). Cat society and the consequences of colony size, in *The Domestic Cat: The Biology of Its Behavior,* Turner, D. C., and Bateson, P., Eds., Cambridge University Press, New York, pp. 67–81.

Kessler, C. C. (2002). Eradication of feral goats and pigs and consequences for other biota on Sarigan Island, Commonwealth of the Northern Mariana Islands, in *Turning the Tide: The Eradication of Invasive Species,* Veitch, C. R. and Clout, M. N., Eds., IUCN SSC Invasive Species Specialist Group, IUCN, Gland, Switzerland and Cambridge, UK, pp. 132–140.

Killian, G., Miller, L., Rhyan, J., Dees, T., Perry, D., and Doten, H. (2003). Evaluation of GNRH contraceptive vaccine in captive feral swine in Florda, in *Proceedings of the 10th Wildlife Damage Management Conference,* Fagerstone, K. A. and Witmer, G. W., Eds., pp. 128–133.

Klimstra, W. D., Hardin, J. W., Silvy, N. J., Jacobson, B. N., and Terpening, V. A. (1974). *Key Deer Investigations Final Report: Dec 1967–Jun 1973,* U.S. Fish and Wildlife Service, Big Pine Key, FL.

Lampa, K. (2001). Coyotes in the city, *Discovery Magazine,* Vancouver Natural History Society.

Lepczyk, C. A., Mertig, A. G., and Liu, J. (2004). Landowners and cat predation across rural-to-urban landscapes, *Biological Conservation,* 115:191–201.

Liberg, O., and Sandell, M. (1988). Spatial organization and reproductive tactics in the domestic cat and other felids. Pages 83–98 in D.C. Turner and P. Bateson, editors. *The Domestic Cat: The Biology of Its Behavior.* Cambridge University Press, New York.

Littauer, G. A. (1993). Control techniques for feral hogs, in *Feral Swine: A Compendium for Resource Managers,* Hanselka, C. W. and Cadenhead, J. F., Eds., Texas Agricultural Extension Service, Kerrville, pp. 139–148.

Long, E. L., Sweitzer, R. A., and Ben-David, M. (2001). Historic changes of mountain lion diets in relation to introduced feral pigs, *Abstracts of the Wildlife Society 8th Annual Conference,* Reno/Tahoe, NV, abstract.

Lucas, E. G. (1977). Feral hogs — problems and control on national forest lands, in *Research and Management of Wild Hog Populations,* Wood, G. W., Ed., Belle Baruch Forest Science Institute of Clemson University, Georgetown, SC, pp. 17–21.

Lopez, R. R., Peterson, T. R., and Silvy, N. J. (2001). Emphasizing numbers in the recovery of an endangered species: a lesson in underestimating public knowledge, in *Proceedings of the Second International Wildlife Management Congress,* Field, R., Warren, R. J., and Okarma, H., Eds., Gödöllõ, Hungary, pp. 353–356.

Lopez, R. R. (2001). Population ecology of the Florida Key Deer, Dissertation, Texas A&M University, College Station.

Lopez, R. R., Silvy, N. J., Pierce, B. L., Frank, P. A., Wilson, M. T., and Burke, K. M. (2003a). Population density of the endangered Florida Key deer, *Journal of Wildlife Management,* 68(3):570–575.

Lopez, R. R., Viera, M. E. P., Silvy, N. J., Frank, P. A., Whisenant, S. W., and Jones, D. A. (2003b). Survival, mortality, and life expectancy of Florida Key deer, *Journal of Wildlife Management,* 67:34–45.

Lopez, R. R., Silvy, N. J., Wilkins, R. N., Frank, P. A., Peterson, M. J., and Peterson, N. M. (2004). Habitat use patterns of Florida Key deer: Implications of urban development, *Journal of Wildlife Management,* 68:900–909.

Lovell, C. D., Leopold, B. D., and Shropshire, C. S. (1998). Trends in Mississippi predator populations, 1980–1995, *Wildlife Society Bulletin,* 26(3):552–556.

Maffei, M. D., Klimstra, W. D., and Wilmers, T. J. (1988). Cranial and mandibular characteristics of the Key deer (*Odocoileus virginianus clavium*), *Journal of Mammalogy,* 69:403–407.

Mahlow, J. C. and Slater, M. R. (1996). Current issues in the control of stray and feral cats, *Journal of American Veterinary Medical Association,* 209(12):2016–2020.

Martin, D. (1999). Wild coyote in captured in Central Park, *New York Times,* April 2, 1999.

Marzluff, J. M., McGowan, K. J., Donnelly, R., and Knight, R. L. (2001). Causes and consequences of expanding American Crow populations, in *Avian Ecology and Conservation an Urbanizing World,* Marzluff, J. M., Bowman, R., and Donnelly, R., Eds., Kluwer Academic Press, Norwell, MA, pp. 331–363.

Mayer, J. J., and Brisbin, I. L., Jr. (1991). *Wild pigs of the United States: Their History, Morphology, and Current Status,* The University of Georgia Press, Athens.

McClennen, N., Wigglesworth, R. R., Anderson, S. H., and Wachob, D. G. (2001). The effect of suburban and agricultural development on the activity patterns of coyotes (*Canis latrans*), *American Midland Naturalist,* 146(1):27–36.

McCracken, G. F. (1999). Brazilian free-tailed bat, in *The Smithsonian Book of North American Mammals,* Wilson, D. E. and Ruff, S., Eds., Smithsonian Institution Press, Washington, DC, pp. 127–129.

McGowan, K. J. (2001). Demographic and behavioral comparisons of suburban and rural American Crows, in *Avian Ecology and Conservation an Urbanizing World,* Marzluff, J. M., Bowman, R., and Donnelly, R., Eds., Kluwer Academic Press, Norwell, MA, pp. 365–381.

McMurray, F. B. and Sperry, C. C. (1941). Food of feral house cats in Oklahoma, a progress report, *Journal of Mammalogy,* 22(2)185–190.

McShea, W. J., Underwood, H. B., and Rappole, J. H. (1997). Deer management and the concept of deer overabundance, in *The Science of Overabundance: Deer Ecology and Population Management,* McShea, W. J., Underwood, H. B., and Rappole, J. H., Eds., Smithsonian Institution Press, Washington, D.C., pp. 1–11.

Meffe, G. K. and Carroll, C. R. (1997). *Principles of Conservation Biology,* Sinauer Associates, Inc, Sunderland, MA.

Miller, J. E. (1993). A national perspective on feral swine, in *Feral Swine: a Compendium for Resource Managers,* Hanselka, C. W. and Cadenhead, J. F., Eds., Texas Agricultural Extension Service, Kerrville, pp. 9–16.

Myers, P. (1993). The Wild Boar Conservation Association, in *Feral Swine: A Compendium for Resource Managers,* Hanselka, C. W. and Cadenhead, J. F., Eds., Texas Agricultural Extension Service, Kerrville, pp. 130–131.

Murphy, M. (1990). The bats at the bridge, *BATS,* 8(2):5–7.

Natoli, E. (1985). Spacing patterns in a colony of urban stray cats (*Felis catus,* L.) in the historic centre of Rome, *Applied Animal Ethology,* 14:289–304.

Nelson, M. (1999). Habitat conservation planning, *Endangered Species Bulletin,* 24(6):12–13.

Nettles, V. F., Quist, C. F., Lopez, R. R., Wilmers, T. J., Frank, P., Roberts, W., Chitwood, S., and Davidson, W. R. (2002). Morbidity and mortality factors in Key deer, *Odocoileus virginianus clavium, Journal of Wildlife Diseases,* 38:685–692.

Nowak, R. M. (1991). *Walker's Mammals of the World,* 5th ed., vol. 2, Johns Hopkins Univ, Baltimore, MD.

Nunley, G. L. (1999). The Cooperative Texas Wildlife Damage Management Program and feral swine damage management, in *Proceedings of the Feral Swine Symposium,* June 2–3, Texas Animal Health Commission, Fort Worth, pp. 27–30.

Oppenheimer, E. C. (1978). *Felis catus*: Population densities in an urban area. *Carnivore Genetic Newsletter,* 4:72–80.

Parmalee, P. W. (1953). Food habits of the feral house cat in east-central Texas, *Journal Wildlife Management,* 17(3):375–376.

Passanisi, W. C. and Macdonald, D. W. (1990). The fate of controlled feral cat colonies, UFAW Animal Welfare Research Report No. 4.

Patronek, G. J. (1998). Free-roaming and feral cats — their impacts on wildlife and human beings, *Journal of American Veterinary Medical Association,* 212(2):218–226.

Peine, J. D. (2001). Nuisance bears in communities: strategies to reduce conflict. *Human Dimensions of Wildlife,* 6(3):223–237.

Peterson, M. N., Peterson, T. R., Peterson, M. J., Lopez, R. R., and Silvy, N. J. (2002). Cultural conflict and the endangered Florida Key deer, *Journal of Wildlife Management,* 66:947–968.

Remfry, J. (1996). Feral cats in the United Kingdom, *Journal of American Veterinary Medical Association,* 208(4):520–523.

Ricklefs, R. E. and Miller, G. L. (1999). *Ecology,* W. H. Freeman, New York.

Riley, S. P.D., Sauvajot, R. M., Fuller, T. K., York, E. C., Kamradt, D. A., Bromley, C., and Wayne, R. K. (2003). Effects of urbanization and habitat fragmentation on bobcats and coyotes in southern California, *Conservation Biology,* 17(2):566–576.

Roemer, G. W., Donlan, C. J., and Courchamp, F. (2002). Golden eagles, feral pigs, and insular carnivores: How exotic species turn native predators into prey, *Proceedings of the National Academy of Sciences,* 99:791–796.

Rogers, L. L. (1989). Black bears, people, and garbage dumps in Minnesota, Bear-People Conflicts, in *Proceedings of a Symposium on Management Strategies,* Northwest Territories Department of Natural Resources.

Rollins, D. (1993). Statewide attitude survey on feral hogs in Texas, in *Feral Swine: A Compendium for Resource Managers,* Hanselka, C. W. and Cadenhead, J. F., Eds., Texas Agricultural Extension Service, Kerrville, pp. 1–8.

Rollins, D. and Carroll, J. P. (2001). Impacts of predation on quail, in *The Role of Predator Control as a Tool in Game Management,* Texas Agricultural Extension Service Publication SP-113.

Rosatte, R. C., Power, M. J., and MacInnes, C. D. (1991). Ecology of urban skunks, raccoons and foxes in Metropolitan Toronto, in *Proceedings Symposium: Wildlife Conservation in Metropolitan Environments,* Adams, L. W. and Leedy, D. L., Eds., National Institute for Urban Wildlife Publishing, Columbia, MD, pp. 31–38.

Rubin, E. S, Boyce, W. M., Stermer, C. J., and Torres, S. G. (2002). Bighorn sheep habitat use and selection near an urban environment, *Biological Conservation,* 104(2):251–263.

Ryser, G. R. and Popovici, R. (2000). The fiscal impact of the Congress Ave, bridge bat colony on the city of Austin, Unpublished report to Bat Conservation Internations, Inc., Austin, Texas, http://www.batcon.org/home/congressreport.html.

Schuyler, P. T., Garcelon, D. K., and Escover, S. (2002). Eradication of feral pigs (*Sus scrofa*) on Santa Catalina Island, California, USA, in *Turning the Tide: The Eradication of Invasive Species,* Veitch, C. R. and Clout, M. N., Eds., IUCN SSC Invasive Species Specialist Group, IUCN, Gland, Switzerland and Cambridge, UK, pp. 274–286.

Sillero-Zubiri, C. and Laurenson, M. K. (2001). Interactions between carnivores and local communities: conflict or co-existence? in *Carnivore Conservation,* Gittleman, J. L.,

Funk, S. M., Macdonald, D., and Wayne, R. K., Eds., Cambridge University Press, Cambridge, pp. 282–312.

Singer, F. J., Swank, W. T., and Clebsch, E. E. C. (1984). Effects of wild pig rooting in a deciduous forest, *Journal of Wildlife Management,* 48:464–473.

Soulé, M. E., Bolger, D. T., Alberts, A. C., Wright, J., Sorice, M., and Hill, S. (1988). Reconstructed dynamics of rapid extinctions of chapparal-requiring birds in urban habitat islands, *Conservation Biology,* 2(1):75–92.

Springer, M. D. (1975). Food Habits of Wild hogs on the Texas Gulf Coast, Thesis, Texas A&M University, College Station.

Springer, M. D. (1977). Ecological and economic aspects of wild hogs in Texas, in *Research and Management of Wild Hog Populations,* Wood, G. W., Ed., Belle Baruch Forest Science Institute of Clemson University, Georgetown, SC, pp. 37–46.

Stevens, R. L. (1996). *The Feral Hog in Oklahoma,* Samuel Roberts Noble Foundation, Ardmore, OK.

Stokes, D. and Stokes, L. (1996). *Field Guide to Birds: Eastern Region,* Little, Brown, Boston.

Sweeney, J. M. and Sweeney, J. R. (1982). Feral hog, in *Wild Mammals of North America: Biology, Management, and Economics,* Chapman, J. A. and Feldhammer, G. A., Eds., The Johns Hopkins University Press, Baltimore, MD, pp. 1099–1113.

Sweeney, J. M., Sweeney, J. R., and Provost, E. E. (1979). Reproductive biology of a feral hog population, *Journal of Wildlife Management,* 43:555–559.

Sweitzer, R. A. and Van Vuren, D. H. (2002). Rooting and foraging effects of wild pigs on tree regeneration and acorn survival in California's oak woodland ecosystems, USDA Forest Service General Technical Report PSW-GTR-184.

Tate, J., Ed. (1984). Techniques for controlling wild hogs in the Great Smoky Mountains National Park, Proceedings of a workshop, November 29–30, Research/Resources Mgmt, Rpt, SRE-72. United States Department of the Interior, National Park Service, Southeast Regional Office, Atlanta, GA.

Taylor, R. B., Hellgren, E. C., Gabor, T. M., and Ilse, L. M. (1998). Reproduction of feral pigs in southern Texas, *Journal of Mammalogy,* 79:1325–1331.

The Urban Coyote: Exploring Attitudes, Perceptions, and Behaviors toward Urban Coyotes, produced by Webber, K. and Delta Cable in 1999.

Thomas, M. and Young, N. (1999). Preliminary trial of a water-resistant bait for feral pig control, *Science for Conservation,* 127:49–55.

Tigas, L. A., Van Vuren, D. H., and Sauvajot, R. M. (2002). Behavioral responses of bobcats and coyotes to habitat fragmentation and corridors in an urban environment, *Biological Conservation,* 108:299–306.

Tinsley, A. M. (2002). City will trap, kill wild hogs, Fort Worth *Star-Telegram,* 16 September 2002; section B:1.

Toner, G. C. (1956). House cat predation on small animals, *Journal of Mammalogy,* 37(1):119.

Towne, C. W. and Wentworth, E. N. (1950). *Pigs from Cave to Cornbelt,* University of Oklahoma Press, Norman.

Tuttle, M. D. (1979). Status, causes of decline, and management of endangered gray bats. *Journal of Wildlife Management,* 43(1):1–17.

USFWS. (1998). Recovery plan for upland species of the San Joaquin Valley, California, Region 1. U.S. Fish and Wildlife Service, Portland, OR.

USFWS. (2000). Recovery plan for bighorn sheep in the Penisular Ranges, California, U.S. Fish and Wildlife Service, Portland, OR.

Utah Division of Wildlife Resources, (2000). Utah Black Bear Management Plan, Publication No, 00–23. Salt Lake City, UT.

Wahlenberg, W. G. (1946). *Longleaf Pine,* Charles Lathrop Pack Foundation, Washington, DC.

Wakely, P. C. (1954). Planting the southern pines, United States Department of Agriculture Forest Service Agricultural Monograph 18. Washington, DC.

Weatherhead, P. J. (1984). Two principal strategies in avian communal roosts, *American Naturalist,* 121(2):237–243.

Webber, K. (1997). Urban Coyotes in the Lower Mainland, British Columbia: Public Perceptions and Education, Thesis, University of British Columbia.

Whitehouse, D. B. (1999). Impacts of feral hogs on corporate timberlands in the south eastern U.S., in *Proceedings of the Feral Swine Symposium,* June 2–3. Texas Animal Health Commission, Fort Worth, pp. 108–110.

Wilson, D. E. and Ruff, S. (1999). *The Smithsonian Book of North American Mammals,* Smithsonian Institution Press, Washington.

Wood, G. W. and Barrett, R. H. (1979). Status of wild pigs in the United States, *Wildlife Society Bulletin,* 7:237–246.

Wood, G. W. and Brenneman, R. E. (1977). Research and management of feral hogs on Hobcaw Barony, in *Research and Management of Wild Hog Populations,* Wood, G. W., Ed.,Belle Baruch Forest Science Institute of Clemson University, Georgetown, SC, pp. 23–35.

Wood, G. W. and Roark, D. N. (1980). Food habits of feral hogs in coastal South Carolina, *Journal of Wildlife Management,* 44:506–511.

Woods, M., Macdonald, R. A., and Harris, S. (2003). Predation of wildlife by domestic cats, *Felis catus,* in Great Britain, *Mammal Review,* 33(2):174–188.

Varley, R. (2003). From across the state, protestors flock to Auburn, *The Citizen* online, http://www.auburbpub.com/.

Yumane, A., Emoto, J., and Ota, N. (1997). Factors affecting feeding order and social tolerance to kittens in the group-living feral cat (*Felis catus*). *Applied Animal Behavior Science,* 52:119–127.

Zack, C. S., Milne, B. T., and Dunn, W. C. (2003). Southern oscillation index as an indicator of encounters between humans and black bears in New Mexico, *Wildlife Society Bulletin,* 31(2):517–520.

Zaunbrecher, K. I. and Smith, R. E. (1993). Neutering of feral cats as an alternative to eradication programs, *Journal of American Veterinary Medical Association,* 203(3):449–452.

CHAPTER **12**

Distribution, Abundance, and Management Considerations of Resident Canada Geese and Urban White-Tailed Deer

I think people see the need to control geese, but they don't want to see it happening.

John Moriarty, Natural Resource Manager, Ramsey County, Minnesota

CONTENTS

Figure 12.1 Urban deer invading a golf course. (John M. Davis/Texas Parks and Wildlife Department)

Two animals of special significance in urban and suburban America are resident giant Canada goose (*Branta* sp.) and white-tailed deer (*Odocoileus virginianus*). These two animals (Figure 12.1 and Figure 12.2) share a common legacy in terms of: (1) those factors that contributed to their prominence in urban America; (2) distribution in the continental U.S.; (3) the human response to their presence; (4) ecological impacts on their habitats; (5) human health and safety issues; and (6) feasible and acceptable management strategies.

Thousands of professional and popular articles have been written on geese and deer problems in urban America. It has been our practice throughout this book to find and identify synthesis documents that compile the body of literature related to

Figure 12.2 A large flock of urban Canada geese. (USDA APHIS Wildlife Services)

a specific topic or to go through the task of assimilating all of the most recent relevant literature on a topic under a single chapter. For this case study, we found an entire issue of the *Wildlife Society Bulletin* (1997, 25:2) that was devoted to urban deer management. The magnitude of the issues related to Canada goose management was compiled by the U.S. Fish and Wildlife Service in a draft environmental impact statement cited below. These two sources contain citations of the research conducted by the leading researchers on resident Canada geese and urban white-tailed deer management.

12.1 FACTORS THAT CONTRIBUTED TO GEESE AND DEER ABUNDANCE IN URBAN AMERICA

Geese were nearly eliminated in most parts of the U.S. by unrestricted hunting, harvesting of eggs, and draining of wetland habitat (Smith et al. 1999). In 1999, the U.S. Fish and Wildlife Service (USFWS) estimated the population of resident Canada geese in North America at 3.5 million (USFWS 2002). There used to be only 100,000 deer in the U.S. in the early 1900s, reduced by commercial hunting, but now they number over 30 million. The restoration of geese and deer populations in North America has led to one of the most challenging problems facing wildlife managers today and in the future—geese and deer overabundance (Warren 1997; USFWS 2002). Both geese and deer are superbly adapted to exploit the resources in urban areas and appear to be in exponential growth patterns caused by basically the same factors, which include:

1. Low abundance of natural predators. Large predators are the first species eliminated during urban sprawl.
2. Healthy breeding habitat conditions that include alternative sheltered nesting sites for geese and patchy and defined edge habitat for deer.
3. Tolerance of urban disturbances including human presence and their activities, e.g., golfing or picnicking.
4. Living longer in the city than they do in the country.
5. Lack of hunting in urban areas, and regulatory protection at the state and/or Federal levels.
6. High production and survival rates in offspring.
7. Abundant alternative food resources in the form of ornamental shrubs, garden plants, succulent grasses, small plants, and supplemental feed.

12.2 DISTRIBUTION OF RESIDENT CANADA GEESE AND WHITE-TAILED DEER IN THE CONTINENTAL U.S.

There is a remarkable similarity in the distribution of resident Canada geese and white-tailed deer in the continental U.S. (Figures 12.3 and 12.4). This similarity is recognized in the states where they occur, and in some cases, the numbers of geese and deer that occur within specific states. The highest numbers (rank from 1 to 5) of resident Canada geese were in Michigan, Virginia, Pennsylvania, Minnesota, and

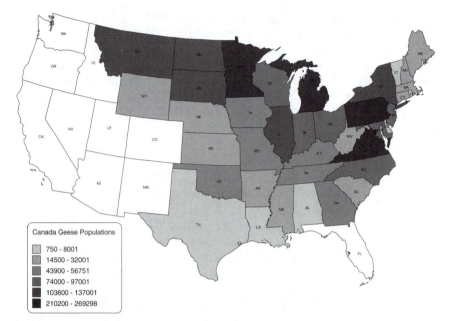

Figure 12.3 Distribution and number of resident Canada geese in the continental U.S. There are no Canadian geese in nonshaded areas, e.g., Florida. (From USFWS 1999)

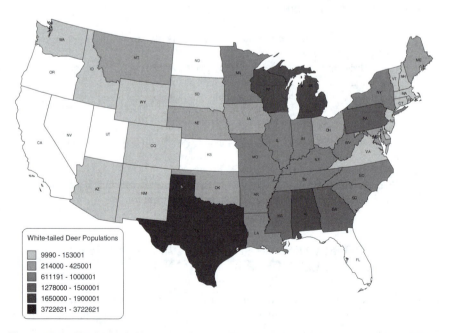

Figure 12.4 Distribution and number of white-tailed deer in the continental U.S. in 1999. White-tailed deer are not present in nonshaded states or census data are not available. (From the Quality Deer Management Association)

New York. The major factor that influenced the distribution of geese was the migratory patterns that include the Atlantic, Mississippi, Central, and Pacific flyways. However, resident Canada geese do not migrate and, in fact, probably recruit additional nonmigrants among migrating flocks by acting as living decoys. Furthermore, it is unlikely that offspring of resident Canada geese heed the environmental signals to head north or south.

The highest numbers (rank from 1 to 5) of white-tailed deer were in Texas, Michigan, Alabama, Wisconsin, and Mississippi. In general, the highest numbers of white-tailed deer were found in Texas and in states adjacent to and east of the Mississippi River (Figure 12.4). Even though white-tailed deer are present in Oregon, North Dakota, and Kansas, there is no information available on the size of the deer herd in these states.

The methods of determining the number of geese and deer within each state vary in terms of agency census techniques and methods of generalizing actual count data to the entire state. Resident Canada geese and white-tailed deer are not run through turnstiles that give actual counts. The published data on geese and deer numbers within each state are, at best, estimates. In addition, the estimates may be based on a census of the geese and deer that live outside city borders. However, the process of urban sprawl has connected the rural populations of geese and deer with people.

This led us to determine whether some of the measurable variables that denote the process of urbanization (e.g., number of golf courses, human population size, percent of population living in urban areas) could be better predictors of geese and deer numbers than farmland acreage, and square miles of land and water area.

We ranked each state in terms of the number of resident Canada geese ($N = 38$), white-tailed deer ($N = 41$), and golf courses, farmland acres, people per sqare mile, and the amount of land and water per square mile ($N = 43$). A Spearman rank correlation revealed that the number of geese, deer, and people were correlated ($r = 0.34$ to 0.59, $P = 0.05$) with the number of golf courses and square miles of surface water. The number of geese and deer were correlated ($r = 0.42$ and 0.50, respectively, $P = 0.05$) with the number of people. Apparently, people, geese, and deer have a lot in common when it comes to the type of landscape they prefer.

12.3 THE HUMAN RESPONSE TO RESIDENT CANADA GEESE AND URBAN WHITE-TAILED DEER

The human response to the presence of resident Canada geese or white-tailed deer is measured in terms of a range of values they attribute to each species in terms of esthetic, recreational, ecological, educational, utilitarian, and potential danger. Both species have aesthetic value when viewed as part of the inherent natural beauty of the landscape. The recreational value is based on the enjoyment derived from viewing a pair of geese gliding effortlessly across a city lake followed by eight to ten goslings or a doe attending twin or triplet fawns in the front yard. The ecological value is realized by determining the relationship of geese and deer to other species and the urban environment, community, and ecosystem. Geese and deer have educational value in what can be learned about their life history, methods of resource utilization in urban environments, and how to manage their numbers within the carrying capacity of their ecosystem.

If geese and deer can be considered useful to humans in some tangible way, e.g., food, then they have utilitarian value. White-tailed deer are revered as the state mammal in 11 states and appear on the Vermont state flag and seal. Both geese and deer represent a potential for danger in terms of negative impacts on human health or economics, particularly when their population size increases beyond a level considered tolerable by urban residents. In general, human values are fickle, given a host of variables that influence human attitudes, activities, knowledge, and expectations concerning Canada geese and white-tailed deer in urban communities.

The majority of residents in urban communities are three to four generations removed from rural life and all direct functional ties with the natural world. As such, the value systems of urban residents are based largely on a sanitized, synthetic, unrealistic portrayal of wildlife in the media, literature, and commercial aspects of their urban communities. As such, urban residents have a profound ignorance of the natural world and how it works. Marchinton (1997) coined the term "urbanism" to define this condition as a way of looking at life by urban residents. This way of looking at life has led to almost insurmountable problems in the effective management of urban white-tailed deer populations.

12.4 ECOLOGICAL IMPACTS OF RESIDENT CANADA GEESE AND URBAN WHITE-TAILED DEER

The key to understanding the ecological impacts of resident Canada geese and urban white-tailed deer is to consider effects on habitat when a species pushes or exceeds the carrying capacity. The lesson on population dynamics in Chapter 4 demonstrated why no population of organisms can be allowed to increase exponentially indefinitely without causing serious damage to other species within the biotic community and itself. There is ample evidence suggesting that resident Canada geese and urban white-tailed deer are pushing the carrying capacity of the urban ecosystems they occupy.

For example, Canada geese are "eating-excreting" machines. A large flock of geese can compact soil and/or denude an area of vegetation, which leads to erosion. An adult Canada goose excretes up to 1 pound of feces per day. Imagine the fecal mess that can be produced by a flock of 200 Canada geese in one park every day! The ecological consequence of the fertilizer load in aquatic ecosystems is nutrient enrichment (eutrophication) of city lakes and ponds. The enormous amount of goose feces deposited in city lakes and ponds turns them into a thick green soup (caused by green feces and excessive algae growth), destroys much of the plant and animal diversity that once existed, and alters the water chemistry (e.g., O_2 concentration), so that only those organisms that have the widest ranges of tolerance to the altered abiotic conditions can survive. Agricultural and natural resource impacts include losses to grain crops, overgrazing of pastures, and degrading water quality. On land, the high nitrogen content of the feces leads to overfertilization of plants and, if concentrated in one area, leaves dead spots on grassy areas (Smith et al. 1999).

The ecological impact of overabundant white-tailed deer in urban areas either: (1) affects the distribution or abundance of many other species; (2) can affect community structure by strongly modifying patterns of relative abundance among competing species; or (3) affects community structure by affecting the abundance of species at multiple trophic levels (Waller and Alverson 1997:218). The heavy and constant browsing of urban white-tailed deer can destroy entire vegetative communities (e.g., forest understory) upon which other species depend for their survival. In addition, browsing deer promote the selection of vegetative communities with less plant and animal biodiversity. These profound impacts on the ecology of urban ecosystems led Waller and Alverson (1997) to designate urban white-tailed deer as a "keystone" species.

12.5 HEALTH AND SAFETY ISSUES RELATED TO RESIDENT CANADA GEESE AND URBAN WHITE-TAILED DEER

There is a large body of literature that documents the human health and safety problems caused by resident Canada geese and urban white-tailed deer. Both species transmit diseases to other animals. For example, geese can transmit coccidiosis, avian influenza, schistosomes, chlamydiosis, salmonella, and avian cholera to other birds and cholera to cattle (Smith et al. 1999).

Conflicts between geese and people affect or damage several types of resources, including property, human health and safety, agriculture, and natural resources. Common problem areas include public parks, airports, public beaches and swimming facilities, water-treatment reservoirs, corporate business areas, golf courses, schools, college campuses, private lawns, athletic fields, amusement parks, cemeteries, hospitals, residential subdivisions, and along or between highways (USFWS 2002).

Property damage usually involves landscaping and walkways, most commonly on golf courses, parks, and waterfront property. In parks and other open areas near water, large goose flocks create local problems with their droppings and feather litter. Surveys have found that, while most landowners like seeing some geese on their property, eventually, increasing numbers of geese and the associated accumulation of goose droppings on lawns cause many landowners to view geese as a nuisance, which results in a reduction of both the aesthetic value and recreational use of these areas (USFWS 2002).

Negative impacts on human health and safety occur in several ways. At airports, large numbers of geese can create a serious threat to aviation. Resident Canada geese have been involved in a large number of aircraft strikes resulting in dangerous landing\take-off conditions, costly repairs, and loss of human life. As a result, many airports have active goose control programs (USFWS 2002).

Excessive goose droppings are a disease concern for many people. Public beaches in several states have been closed by local health departments due to excessive fecal coliform levels that in some cases have been traced back to geese and other waterfowl. Additionally, during nesting and brood-rearing, aggressive geese have bitten and chased people, and injuries have occurred due to people falling or being struck by wings (USFWS 2002).

12.6 WHITE-TAILED DEER AND LYME DISEASE

Urban white-tailed deer carry one of the stages in the Lyme disease life cycle, which can be transmitted to humans (Figure 12.5). Lyme disease exists because of a complex set of interactions between oak trees, white-footed mice, white-tailed deer, gypsy moths, ticks, and a bacterium. Lyme disease is transmitted to humans by the bite of deer ticks (*Ixodes* ticks) carrying the bacterium *Borrelia burgdorferi*. For this disease to exist in an area, at least three closely interrelated elements must be present: the Lyme disease bacteria, ticks that can transmit them, and mammals (such as mice and deer) to act as a host for the ticks in their various life stages.

The life cycle of an *Ixodes* tick requires 2 years to complete. Adult ticks feed and mate on large animals, especially deer, in the fall and early spring. Female ticks then drop off the host animals to lay their eggs on the ground. By summer, eggs hatch into larvae. The larvae feed on mice and other small mammals and birds in the summer and early fall, then go dormant until the next spring when they molt into nymphs. Nymphs feed on small rodents and other small mammals and birds in the late spring and summer and molt into adults in the fall, completing the 2-year life cycle.

Larvae and nymphs typically become infected with Lyme disease bacteria when they feed on infected small animals, particularly the white-footed mouse. The bacteria remain in the tick as it changes from larva to nymph or from nymph to adult. Infected nymphs and adult ticks then bite and transmit Lyme disease bacteria to other animals and humans in the course of their normal feeding behavior (http://www.cdc.gov/).

So how do gypsy moths, mice, deer, and oak trees come into the picture? Ostfeld et al. (1999) documented a web of biotic interrelationships involving episodic acorn production by oak trees, how white-footed mice and white-tailed deer respond to acorn abundance, and how the former two events affected the population dynamics of tick parasites and defoliating insects (Figure 12.6). Gypsy moths are well known

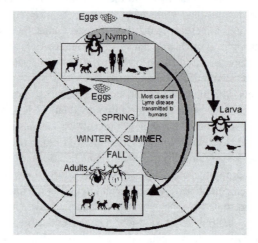

Figure 12.5. Life cycle of Lyme disease ticks. (L. Causey and Center for Disease Control)

1) many acorns are produced. 2) mice in oak forest eat and store acorns and survive winter well. 3) deer immigrate into oak forest and 4) transport their burdens of adult ticks into oak forest. In spring/ summer following a mast year: 5) mouse density in oak forest is high. 6) many mice emigrate from oak to non-oak forest carrying both larval and nymphal ticks; 7) density of larval ticks in oak forest is high, mice are infested. 8) gypsy moth survive poorly due to high mouse populations, resulting in low egg mass densities by the end of summer.

9) Mouse density in oak forest us low due to high predation pressure; 10) deer avoid oak forest and concentrate in non-oak forest; 11) adult ticks infest deer in non-oak forest, ticks are infected with Lyme bacterium due to emigration of mice carrying infected nymphs from oak forest the prior spring/ summer. In the spring/ summer following a mast year: 12) mouse density in oak forest is low; 13) density of nymphal ticks in oak forest is high and infection prevalence with Lyme bacterium is high; 14) gypsy moth pupae survive well, resulting in a large increase in egg mass densities by the end of the summer; 15) density of larval ticks is high in non-oak forest.

Figure 12.6 Dynamic interrelationships between oak trees, white-footed mice, and white-tailed deer that promote the spread of Lyme disease during a mast year. (From Ostfeld et al. (1999). *Integrative Biology,* 1:178–186.)

Table 12.1 Top 10 U.S. States Affected by Lyme Disease in 2002

State	Number of Cases Reported	Annual Incidence[a]
New York	5,535	28.9
Connecticut	4,631	133.8
Pennsylvania	3,989	32.3
New Jersey	2,349	27.3
Massachusetts	1,807	28.1
Wisconsin	1,090	20.0
Minnesota	867	17.3
Rhode Island	852	79.7
Maryland	738	13.5
New Hampshire	261	20.8

[a] Per 100,000 population.
Source: http://www.cdc.gov.

for the destructive effect they can have by defoliating entire forests. Bumper crops of acorns increase mouse populations, which are voracious consumers of the pupal stage of the gypsy moth life cycle. Predation by mice provides control of gypsy moth populations. When mice and deer are drawn to the acorn crop, it concentrates these two prime tick vectors in a relatively small area, increasing the chance for transmission of Lyme disease.

In contrast, an unhealthy forest ravaged by gypsy moths does not produce bumper acorn crops. Mice and deer must go elsewhere for food or starve, which increases the chance of transmission of the disease. To some extent, the choice is between a healthy forest (which people find aesthetically and economically pleasing) or healthy people (Wright 2004:407).

Lyme disease is the leading cause of vector-borne infectious illness in the U.S. with about 15,000 cases reported annually, though the disease is greatly underreported. Based on reported cases, during the past ten years 90% of cases of Lyme disease occurred in ten states (Table 12.1). There is a much higher incidence of Lyme disease in the northeastern Atlantic seaboard states and in Minnesota, Wisconsin, and Michigan than in the rest of the U.S. (Figure 12.7).

It seemed reasonable to assume that the number of cases of Lyme disease would correlate with the number of white-tailed deer (vectors of the disease) in each state. However, the factor that correlated ($r = 0.74$, $P = 0.05$) the highest with the rank of the number of cases of Lyme disease in each state was the rank of human population per square mile followed by the rank of deer per square mile ($r = 0.33$, $P = 0.05$). As mentioned earlier in this chapter, the methods of estimating the number of deer in each state are less than a perfect census, unlike the methods used to determine the actual number of people in each state. The deer census methods may be inaccurate enough to offset the expectation of a higher correlation between deer densities and incidence of Lyme disease in each state. There may be a problem with underreporting the actual cases of Lyme disease in each state. However, the reader can take note of the occurrence of Lyme disease in their state of residence and be aware how it is transmitted and understand the role that white-tailed deer play in the disease life cycle.

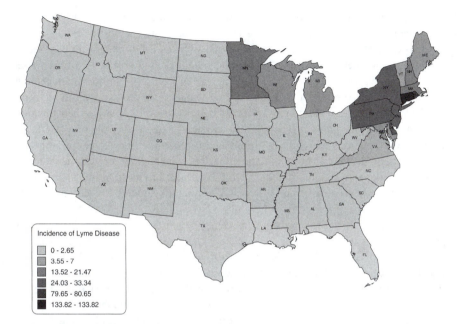

Figure 12.7   Incidence (per 100,000 population) of Lyme disease in the continental U.S. in 2002. (http://www.cdc.gov)

White-tailed deer are involved in 1.5 million car collisions per year, killing between 100 and 200 motorists annually and causing in excess of $2 billion in damage (Rondeau and Conrad 2003). McShea et al. (1997) and Warren (1997) present comprehensive overviews of the problems associated with deer overabundance and various programs that have been employed to manage these overabundant deer herds (as cited by Warren 2000).

12.7 FEASIBLE AND ACCEPTABLE MANAGEMENT STRATEGIES FOR OVERABUNDANT RESIDENT CANADA GEESE AND URBAN WHITE-TAILED DEER POPULATIONS

There is a point when the number of resident Canada geese and urban white-tailed deer has negative impacts on some aspect of urban residents' quality of life. In recent years, numbers of both species have undergone dramatic population growth and have increased to levels that are increasingly coming into conflict with people and causing personal and public property damage. In many communities, increasing numbers of locally breeding Canada geese and white-tailed deer have resulted in an example of the conflict and disagreement that can occur among various publics when wildlife becomes locally overabundant and exceeds the tolerance level of some people and communities. Overall, complaints related to personal and public property damage, agricultural damage, public safety concerns, and other public conflicts have increased as resident Canada goose and white-tailed deer populations increased (Warren 1997; USFWS 2002).

Interestingly, the first considerations in resident Canada geese and urban white-tailed deer management are the urban residents' attitudes, activities, expectations, and knowledge concerning the species in question, i.e., the human dimensions of wildlife management. Any urban residential neighborhood will have a diverse set of stakeholder groups that embrace compatible or contrary wildlife management agendas. The key to effective urban wildlife management is to bring all the various stakeholder groups together, so everyone has an opportunity to voice their position on the management problem. The wildlife manager takes on the role of a facilitator to identify the various stakeholder groups and to open the lines of communication between them during a town hall meeting. The stakeholder groups need to be organized under some banner of recognition (e.g., Deer or Goose Action Committee) that identifies them as a group ready and willing to address the urban wildlife management problem that affects them all.

The town meeting is a relatively consistent phenomenon in terms of participants and the points of view expressed by them (Kirkpatrick and Turner 1997). Its participants will include: (1) those who want to save the geese or deer; (2) those who object to hunting in general; (3) those who object to management of any kind; (4) those who hate the geese or deer because of the damage they have done or the potential that they can spread diseases; (5) city and county officials who want to be reelected; (6) at least one representative of the state game agency; (7) some shotgun hunters; (8) some bow hunters; (9) a representative from either an animal rights or animal-welfare organization; and (10) the media. Expect discussions between the various stakeholder groups to be spirited, contentious, and sometimes combative. Also expect the town hall meeting events to be extended in the form of a media feeding frenzy for several weeks, and that dialogue will supersede sound science in the end.

The second consideration in resident Canada geese and urban white-tailed deer management is the availability of background data that defines the nature of the goose or deer overabundance problem. The types of information that are needed include:

1. The actual numbers in the neighborhood;
2. Each species' rate of annual increase and the projected growth in population size with and without management intervention;
3. Problems they are causing with property, human health and safety, agriculture, and natural resources;
4. Legal ramifications including deed restrictions, city ordinances, state hunting laws, and Federal migratory bird acts;
5. Factors (listed at the beginning of this chapter) that contribute to each species' overabundance;
6. Identified or suspected ecological, economic, sociological, and political consequences (Coffey and Johnston 1997).

The integrated pest management (IPM) strategy proposed by Coffey and Johnston (1997) was designed to address white-tailed deer overabundance, but applies equally to resident Canada geese overabundance. IPM is a problem-solving approach that begins by acquiring the information listed above. This information allows the development of a management plan based on a specific problem. Problem definition facilitates the statement of clear, precise IPM goals and objectives. Goals identify the desired end points in the management process, and objectives provide the blueprint for action in achieving the goals. The final component in IPM is the development and implementation of a monitoring program. The monitoring program provides the baseline data needed by managers to measure one action against another; progress, success, or failure; and to evaluate future management actions.

IPM addresses the criteria that define overabundance. Coffey and Johnston (1997) identified a set of consequences that managers can use to demonstrate that populations have or will exceed the environmental or cultural carrying capacity of the neighborhood. These consequences include (1) unacceptable damage; (2) changes in ecological processes; (3) destruction of native plants, agricultural crops, and landscapes; (4) unacceptable health or safety conditions; (5) displacement or loss of native species; or (6) prevention of the attainment of established goals and objectives.

The identification of the specific problem associated with resident Canada geese and urban white-tailed deer management guided by the goals and objectives leads to the selection of a specific management strategy. However, urban residents will opt for strategies that do not harm humans or animal (e.g., goose or white-tailed deer), do not cost a lot of money, do not impact negatively on their quality of life, and solves the problem permanently. Needless to say, there is no management strategy that can satisfy these criteria. So a management compromise needs to be designed that represents the best management practice.

The management strategies that can be applied to resident Canada geese and urban white-tailed deer are identical in terms of lethal and nonlethal alternatives (Coffey and Johnston 1997; Conover 2002). In general, management strategies should be focused on the reduction, management, and control or geese and deer populations and to reduce related damage. The difficult part of the selection of a best management strategy is deciding whether to avoid the problem altogether, get at the root cause of the problem, attack the symptoms, or do nothing.

12.7.1 Avoiding the Problem

This is the image of the anthropocentric man: he seeks not unity with nature but conquest.

Ian L. McHarg 1969

A thing is right when it tends to preserve the integrity, stability, and beauty of the biotic community. It is wrong when it tends to do otherwise.

Aldo Leopold 1976

There are ways to avoid the problems of wildlife overabundance in urban communities altogether, i.e., a proactive management strategy. This involves urban development that integrates rather than excludes nature. It involves community designs that preserve the natural habitat, are sustainable in the use of natural resources, and reconnect human society with the natural world they live in. There are several rubrics that identify the alternative forms of urban community development that include sustainable communities, smart growth, design with nature, and conservation design for subdivisions.

Sustainable communities improve the quality of human life by living within the carrying capacity of supporting ecosystems. Sustainability can be achieved by not exceeding the carrying capacity of natural resources and ecosystems; reducing the impact that human activities have on the environment (e.g., rates at which renewable and nonrenewable resources are used); integrating long-term economic, social, and environmental goals; and preserving biological, cultural, and economic diversity (Farrell and Hart 1998). When serious consideration is given to sustainability issues in community development decisions, the conditions that promote wildlife overabundance are considered immediately and factored out of the decision-making process. For example, the issues of sustainability and overabundance present a paradox of terms. In sustainable communities, living within the carrying capacity of the ecosystem precludes allowing any species to exceed the carrying capacity.

In communities across the nation, there is a growing concern that current development patterns—dominated by what some call "sprawl"—are no longer in the long-term interest of our cities, existing suburbs, small towns, rural communities, or wilderness areas. Communities that are supportive of smart growth are questioning the economic costs of abandoning infrastructure in the city only to rebuild it further out. Factors promoting the smart growth movement are demographic shifts, a strong environmental ethic, increased fiscal concerns, and more nuanced views of growth. The result is both a new demand and a new opportunity for smart growth (http://www.smartgrowth.org/).

Design with nature and conservation design are similar concepts. Each promotes a form of urban development around an open space framework that includes meadows, fields, and woodlands that would otherwise be cleared, graded, and converted into households and streets. The open space that is conserved in this way can be laid out so that it will ultimately coalesce to create an interconnected network of protected lands (McHarg 1969; Arendt 1996). The "design with nature" and "conservation design" approaches to urban development integrate the existing habitat into the urban development process. The "business as usual approach" produces habitats that are fragmented, unnatural, and conducive to invasions of suburban adapter species such as resident Canada geese and white-tailed deer.

12.7.2 Getting at the Root Cause

Getting at the root causes of animal overabundance in urban areas requires an analysis of factors that promote and prevent the presence of geese and deer in a "typical" urban community. Many aspects of urban sprawl provide a landscape and resources that invite the presence of resident Canada geese and white-tailed deer. The most effective management strategy to get at the root causes of the overabundance of geese and deer is to "clean up the neighborhood," which, in general means to alter the existing habitat so it is less inviting to geese and deer (Humane Society of the United States 1997). Examples of "cleaning up the neighborhood" would include:

1. Removing food sources altogether or changing the landscape vegetation to plants that are less palatable to geese and deer,
2. Surrounding lakeshores with tall grass and/or dense hedges to restrict goose access to residential lawns,
3. Removing accumulated nesting materials or nests that do not contain eggs,
4. Planting agricultural crops that are not preferred deer food.

12.7.3 Attack the Symptoms

The most prevalent management strategies to reduce or control geese and white-tailed deer overabundance in urban communities do little more than attack the symptoms, i.e., a reactive management strategy. The cliché "winning the battle, but losing the war" is used appropriately in this case. Management strategies that attack the symptoms include those that (1) clean up the mess—feces and automobile or plane wrecks, (2) cull the flock or herd of some members, (3) move the problem elsewhere—translocate, (4) high or electric fences, (5) behavioral modification—aversion techniques, (6) introduce pseudo natural predators—dogs, and (7) fertility control—egg addling or contraceptives (Smith 1999; Conover 2002). These strategies are usually applied using the triage approach, i.e., they are applied only in the most problematic areas. Further, the most humane strategies are the most expensive on a per/animal basis.

When resident Canada geese or urban white-tailed deer become so overabundant in an urban community that the prior listed management strategies are unsuccessful in reducing animal numbers or the damage they cause, more aggressive techniques that significantly reduce the geese or deer populations may have to be used. For example, large numbers of geese can be herded into a pen when they are in molt and taken to a local slaughtering house, where the meat can be processed for the food for the needy program (Smith et al. 1999). Likewise, urban white-tailed deer can be harvested by sharpshooters (sometimes off-duty police), and the meat used similarly. During a three-year-long deer control program in a coastal Georgia residential community, Butfiloski et al. (1997) donated nearly 20 metric tons of edible venison to needy families and organizations.

12.7.4 Do Nothing

There are groups of stakeholders that advocate a "live and let live (or die)" or "let nature take its course" philosophies of management regardless of population size. These philosophies cannot or do not consider the environmental impacts (habitat degradation) when a species exceeds the carrying capacity of its environment. The height of the browse line is an indicator of the degree to which overabundant deer populations are capable of deriving sufficient food from the urban habitat. Starvation is a slow and uncomfortable way to die. Emaciated animals (particularly young ones) are a pitiful site. A do-nothing approach to resident Canada geese and urban white-tailed deer management will probably bring little satisfaction to anyone as a credible approach to the problem.

12.8 IN SUMMARY

There is an abundant body of information on the issues pertaining to overabundant geese and deer in urban communities. This chapter attempted to reduce the complexities of the issues by highlighting the principal components involved in the management of these two species. Both geese and deer can be cast in the same light regarding their capabilities of exponential growth, environmental damage, and danger to human health and safety in urban environments. It seems that the only difference between them is feather or fur. Nevertheless, it is important for the urban resident to be exposed to the fundamental ecological, economic, cultural, and political issues that need to be considered with any species that exceeds the carrying capacity of its environment. Resident Canada geese and urban white-tailed deer are charismatic personifications of this problem, which is why we chose to focus on these two species in this chapter.

REFERENCES

Arendt, R. (1996). *Conservation Design for Subdivisions: A Practical Guide to Creating Open Space Networks*, Island Press, Washington, DC.

Butfiloski, J. W., Hall, D. I., Hoffman, D. M., and Forster, D. L. (1997). White-tailed deer management in a coastal Georgia residential community, *Wildlife Society Bulletin*, 25:491–495.

Coffey, M. A. and Johnston, G. H. (1997). A planning process for managing white-tailed deer in protected areas: integrated pest management, *Wildlife Society Bulletin*, 25:433–439.

Conover, M. R. (2002). *Resolving Human-Wildlife Conflicts: The Science of Wildlife Damage Management*, Lewis Publishers, Boca Raton, FL.

Farrell, A. and Hart, M. (1998). What does sustainability really mean? The search for useful indicators, *Environment*, 40:4–9;26–31.

Humane Society of the United States. (1997). *Wild Neighbors: The Humane Approach to Living with Wildlife*, Fulcrum Publishing, Golden, CO.

Kirkpatrick, J. F. and Turner, J. W., Jr. (1997). Urban deer contraception: the seven stages of grief, *Wildlife Society Bulletin*, 25:515–519.

Leopold, A. (1976). *A Sand County Almanac*, Ballantine Books, New York.

Marchinton, R. L. (1997). Obstacles to future deer management, *Quality Whitetails*, 4(1):21–23.

McHarg, I. L. (1969). *Design with Nature*, J. Wiley, New York.

McShea, W. J., Underwood, H. B., and Rappole, J. H. (1997). *The Science of Overabundance: Deer Ecology and Population Management*, Smithsonian Institution Press, Washington, DC.

Ostfeld, R. S., Keesing, F., Jones, C. G., Canham, C. D., and Lovett, G. M. (1999). Integrative ecology and dynamics of species in oak forests, *Integrative Biology*, 1:178–186.

Rondeau, D. and Conrad, J. M. (2003). Managing urban deer, *American Journal of Agricultural Economics*, 85:266–281.

Smith, A. E., Craven, S. R., and Curtis, P. D. (1999). *Managing Canada Geese in Urban Environments*, Jack Berryman Institute Publication 16, and Cornell University Cooperative Extension, Ithaca, NY.

USFWS. (2002). Draft environmental impact statement: resident Canada goose management, U.S. Fish and Wildlife Service, FWS/AMBS-DMBM/006380.

Waller, D. M. and Alverson, W. S. (1997). The white-tailed deer: a keystone herbivore, *Wildlife Society Bulletin*, 25(2):217–226.

Warren, R. J. (1997). The challenge of deer overabundance in the 21st century, *Wildlife Society Bulletin*, 25(2):213–214.

Warren, R. J. (2000). Overview of fertility control in urban deer management, in *Proceedings of the (2000) Annual Conference of the Society for Theriogenology*, San Antonio, Texas, Society of Theriogenology, Nashville, TN, pp. 237–246.

Wright, R. T. (2004). *Environmental Science: Toward a Sustainable Future*, 9th ed., Pearson Education, Upper Saddle River, NJ.

Index